BIOTECHNOLOGY IN THE PULP AND PAPER INDUSTRY

8th ICBPPI Meeting

Progress in Biotechnology

Volume 1	New Approaches to Rescarch on Cereal Carbohydrates (Hill and Munck, Editors)
Volume 2	Biology of Anaerobic Bacteria (Dubourguier et al., Editors)
Volume 3	Modifications and Applications of Industrial Polysaccherides (Yalpani, Editor)
Volume 4	Interbiotech '87. Enzyme Technologies (Blazej and Zemek, Editors)
Volume 5	In Vitro Immunization in Hybridoma Technology (Borrebaeck, Editor)
Volume 6	Interbiotech '89. Mathemathical Modelling in Biotechnology (Blazej and Ottová Editors)
Volume 7	Xylans and Xylanases (Visser et al. Editors)
Volume 8	Biocatalysis in Non-Conventional Media (Tramper et al. Editors)
Volume 9	ECB6: Proceedings of the 6th European Congress on Biotechnology (Alberghina et al., Editors)
Volume 10	Carbohydrate Bioengineering (Petersen et al., Editors)
Volume 11	Immobilized Cells: Basics and Applications (Wijffels et al., Editors)
Volume 12	Enzymes for Carbohydrate Engineering (Kwan-Hwa Park et al., Editors)
Volume 13	High Pressure Bioscience and Biotechnology (Hayashi and Balny, Editor
Volume 14	Pectins and Pectinases (Visser and Voragen, Editors)
Volume 15	Stability and Stabilization of Biocatalysts (Ballesteros et al., Editors)
Volume 16	Bioseparation Engineering (Endo et al., Editors)
Volume 17	Food Biotechnology (Bielecki et al., Editors)
Volume 18	Molecular Breeding of Woody Plants (Morohoshi and Komamine, Editors)
Volume 19	Trends in High Pressure Bioscience and Biotechnology (Hayashi, Editor)
Volume 20	Biotechnology Organizations in Action: Turing Knowledge into Business (Norus)
Volume 21	Biotechnology in the Pulp and Paper Industry: 8th ICBPPI Meeting (Viikari and Lantto, Editors)

Progress in Biotechnology 21

BIOTECHNOLOGY IN THE PULP AND PAPER INDUSTRY: 8th ICBPPI

Edited by

L. Viikari
VTT Biotechnology, P.O. Box 1500, FIN-02044 VTT, Finland

R. Lantto
VTT Biotechnology, P.O. Box 1500, FIN-02044 VTT, Finland

2002

ELSEVIER
Amsterdam - London - New York - Oxford - Paris -Tokyo - Boston -
San Diego - San Francisco - Singapore - Sydney

ELSEVIER SCIENCE B.V.
Sara Burgerhartstraat 25
P.O. Box 211, 1000 AE Amsterdam, The Netherlands

First edition 2002

Library of Congress Cataloging-in-Publication Data

International Conference on Biotechnology in the Pulp and Paper Industry (8th : 2001 : Helsinki, Finland)
Biotechnology in the pulp and paper industry / edited by L. Viikari.
p. cm. -- (Progress in biotechnology ; 21)
Includes bibliographical references and index.
ISBN 0-444-51078-8 (alk. paper)
1. Wood-pulp--Biotechnology--Congresses. I. Viikari, Liisa, 1949- II. Title. III. Series.

TS1176.6.B56 I57 2002
676--dc21

2002068369

British Library Cataloguing in Publication Data
A catalogue record from the British Library has been applied for.

ISBN: 0-444-51078-8
ISSN: 0921-0423

∞ The paper used in this publication meets the requirements of ANSI/NISO Z39.48-1992 (Permanence of Paper).
Printed in The Netherlands.

Preface

Biocatalysts have significant potential for improving traditional pulp and paper manufacturing processes, achieving environmental benefits and introducing unique properties to the fibre raw materials. The general goals of pulp and paper industry today are to increase the cost efficiency, to develop environmentally benign processes and to improve the product quality. Fibre raw materials can be modified by enzymes or micro-organisms to improve chemical or mechanical process steps in pulp and paper manufacture. The challenge for commercial success is to identify superior enzymes and microorganisms and develop their applications. Advances in molecular genetics have already enabled the production and application of several new enzymes in industrial scale. In some applications, however, the identified enzymes represent products of the first generation, which may not act optimally under the harsh industrial conditions. The catalytic activity and stability of potential enzymes can further be improved by new powerful methods, such as directed evolution. During the last years the number of new applications of enzymes in pulp and paper manufacture has grown steadily, and several have reached or are approaching commercial use. These include enzyme-aided bleaching with xylanases, direct delignification with oxidative enzymes, energy saving refining with cellulases, pitch reduction with lipases, freeness enhancement with cellulases and hemicellulases as well as enzymatic slime control of the paper machine. In addition to enzymes, microbial treatments are potential for increasing pulping efficiency, reduction of pitch problems and enhancing process water reuse. This book is based on the written versions of selected papers presented at the 8th International Conference on Biotechnology in the Pulp and Paper Industry, held in Helsinki in June 2001. The conference gathered 300 participants from 31 countries. It fulfilled well its commission as a forum for scientists and technologists to meet, share their views and discuss the future lines of biotechnical research. About 200 participants represented universities and other public research organisations, the rest coming from enzyme companies and pulp and paper industry. European Union was powerfully presented. Experts from 12 member countries participated in the conference.

We wish to acknowledge the European Union, 5th Framework Programme, Quality of Life and Management of Living Resources, the European COST Organisation, the Academy of Finland, the TAPPI, as well as the Finnish Forest Industries Association for their financial support. In addition, we wish to thank AB Enzymes Finland, Genencor, Hercules, Iogen and Novozymes for sponsoring the conference. We also want to thank all the writers for their excellent contribution and members of the International Programme Committee for planning and acting as reviewers of publications of this book.

We hope that you enjoy this book.

Espoo, 15.12.2001

Liisa Viikari　　　　Raija Lantto

COST

Academy of Finland

VTT

Biotechnology in the Pulp and Paper Industry
L. Viikari and R. Lantto (Editors)

Trends in pulp and paper biotechnology

L. Viikari

VTT Biotechnology, P.O.Box 1500, FIN-02044 VTT, Finland
E-mail: liisa.viikari@vtt.fi

During the last decade new techniques and innovations in biotechnology have been overwhelming. Not only have the new developments in molecular biology made the production of new bulk enzymes economically feasible and enabled the design of improved enzymes, but also allowed us to understand the genetic background of biosynthetic and metabolic routes more thoroughly. The new methods include genomics, proteomics, metabolomics and bioinformatics. In the light of these developments, pulp and paper area should represent a huge market potential for engineering raw materials and biocatalysts. The key questions are, how the new techniques could be applied in the pulp and paper sector, and what the new breakthroughs will be.

This book is based on selected papers presented in the 8th International Conference on Biotechnology in the Pulp and Paper Industry. Within the last two decades, this triannual conference has become the major forum for reviewing latest achievements in the field. The need for an international forum was originally recognized by the European Liaison Committee for Pulp and Paper (EUCEPA) in 1980, and the first conference was held in 1981 in Canada. During the last decade, each of these meetings has attracted about 250-300 participants from all over the world. This series of meetings covers both basic and applied sciences in a rather specified area. The basic science of lignocellulose enzymology and plant genetics is covered also in many other meetings, whereas the application of biotechnology in process and product development is uniquely reviewed in the ICBPPI meetings.

The needs of the industry to find biotechnical solutions have obviously changed. Today, the general goals of pulp and paper industry are to increase the cost efficiency, to develop environmentally benign processes and to improve the product quality. Biotechnical methods can help to reach these goals, but seldom solve problems alone. The chlorine issue was hot in the late 1980s, raising the search for environmentally benign bleaching technologies. This was obviously the first time in the history of the pulp and paper sector that the consumers' attitudes started deeply to affect the design of industrial processes. Since then, also other technologies, offering environmental advantages, have been developed. These methods compete with biotechnology. To some extent, the competitive advantage of intrinsically cleaner biotechnical processes has diminished. The uniqueness and specificity of biocatalysts offering solutions also for new product design has, however, become even more relevant.

Due to the fast development of new methodologies within biosciences, it is often difficult to judge the competitiveness and economical viability of biotechnical approaches at an early phase. The unforeseeable potential for improvement, generally typical for biotechnical processes, is a relevant argument often presented. It has, however, been observed that the time needed for commercial breakthroughs is unpredictable. Thus, any interesting observation or result achieved may be a starting point for a future success story.

Raw materials

Forest trees are one of the world's most important natural resources. Fast growing trees with low lignin content could provide significant practical benefits. Removal of lignin from the wood cell walls is the most capital intensive and environmentally problematic step in wood processing for pulp and paper. Significant advances in understanding the genetic background of lignin biosynthesis have already been achieved. The genome of major tree species will be sequenced within the coming years. This will slowly lead to improved understanding of the genetic background of various traits, and allows exploration of this knowledge for designing future raw materials for various industries. Any increase of efficiency that allows production of more wood and wood products would help to conserve the natural forests and reduce the environmental impact of processing wood into pulp and paper products. There is still, however, a long way to practical exploitation of these improvements. The future exploitation of genetically modified forest trees will depend on the answers given to the scientific and ethical questions, as well as on the dialogue and consensus reached between industry and the public, especially in Europe. However, basic understanding on the biosynthetic mechanisms of wood components and fibres will help to improve fibre quality even by traditional breeding technologies already used in forestry. In connection with the 8th ICBPPI meeting, a special presymposium was arranged to cover more comprehensively the basic sciences of lignin biosynthesis and biodegradation. Thus, this area, including the latests advances in plant genetics, is not included in this special issue.

Before tackling the raw material quality genetically, the intermediate methodology involves utilization of microbial and enzymatic pretreatment technologies. The most recent results on biopulping technology have shown that the process appears to be economically feasible for mechanical pulp production. Results on benefits achievable by microbial pretreatments for chemical pulping are more contradictory.

Enzymes

Most of the biotechnical applications proposed for pulp and paper industry are based on the use of enzymes. The modifications in fibre material, which can be achieved by enzymes or micro-organisms, have so far been combined with chemical or mechanical treatments. The real challenge for new commercial successes is to identify superior enzymes and their applications. The specificity of enzymes makes them unique tools for targeted modification of specific components of fibres and their catalytic nature makes them efficient even in small dosages. The limitations of the use of enzymes in pulping and papermaking processes are related to the size and properties of the enzymes. The enzymatic action is limited to accessible surfaces of the fibre matrix due to the macromolecular size of enzymes, a fact that has been successfully exploited in the surface analysis of pulp fibres. The inactivation and destruction of protein structure of enzymes in conditions used in many process stages in pulping and papermaking has led to the development of separate enzymatic pretreatments rather than process stages. This concept allows to adjust the conditions more suitable for the enzymes, but does not exclude the constant need for morestable enzymes. The laccase-mediator bleaching concept which aims at direct delignification of pulp would be the first truly enzymatic process phase, enabeling replacement of current bleaching chemical stages such as oxygen or ozone stage.

The enzymatic degradation of lignocellulosic material includes a set of different enzymes. Cellulases have been studied intensively already for decades and the reaction mechanisms of key enzymes have been revealed on molecular level. In spite of the long history, new interesting enzymes are still being found, such as the swollenins and expansins. The various potential uses and effects of cellulose binding domains attract constant interest. The prices of commercial cellulases have decreased several fold due to improved production technologies, and further improvements can be expected. The enzymology of hemicellulases is also well established with the same general trends. Except for xylanases, few applications based on other hemicellulolytic enzymes have been found. Further improvements are expected especially in functionality at extreme conditions. Following the commercial success of hydrolytic enzymes, oxidoreductases are now emerging as the next enzyme generation. Oxidative enzymes have great potential in many applications and first commercial products have been launched. Efficient secretion of oxidoreductases has turned out to be problematic. Detailed mechanistical studies of oxidoreductases have been performed on model substrates, but the reactions on insoluble substrates have not yet been elucidated. There are still many aspects that should be understood in order to improve their performance. The concept of using oxidative enzyme with electron transfering mediator molecules opens up new possibilities for enzyme based oxidation and delignification technologies.

Applications

It is interesting to observe the changing interests in the applied field during the two last decades. A simple statistical analysis based on the oral and poster presentations of the applied research in ICBPPI meetings held in 1983-2001, reveals that a number of approaches seem to attract rather constant interest, such as the biopulping and delignification by enzymes. This study does, however, not include the basic scientific papers which have rather constantly covered cellulolytic, hemicellulolytic and ligninolytic enzymes. The continuing intrerest on some areas reflects the complexicity of the problem and to some extent the failure of commercial breakthrough of the applications. It is noteworth that at present, no single area seems to be dominating. The trends, however can be clearly seen (Fig 1.)

Obviously, the most challenging applications are related to delignification, the basic process step in the production and bleaching of chemical pulps. The interest in lignin degrading enzymes is still remarkable, and continuing scientific efforts are needed for success also in larger scale. On the contrary, applications which have been commercialized naturally attract less interest of the scientific community. Sometimes commercialization slows down further improvement which could be reached by pertinent scientific exploration. The exact mechanisms and scientific basis for some applications still remain unsolved, like in the case of the hemicellulase-aided bleaching.

It will be interesting to see, how the commercialization of biomechanical pulping will proceed. The biopulping concept is one of the longest-term efforts for applying biotechnology in the pulp and paper industry. The development of microbial treatments of wood chips for biopulping purposes or for management of pitch problems is still an interesting issue worldwide. If the problems related to specificity and performance of micro-organisms and technical implementation of the treatment into mill scale processes will be solved, the microbial pre-treatment of chips will be a very tempting method to modify fibres prior to pulping process. Some areas, such as waste water purification have found their own specialized forums, and have not been central interest areas since 1980s.

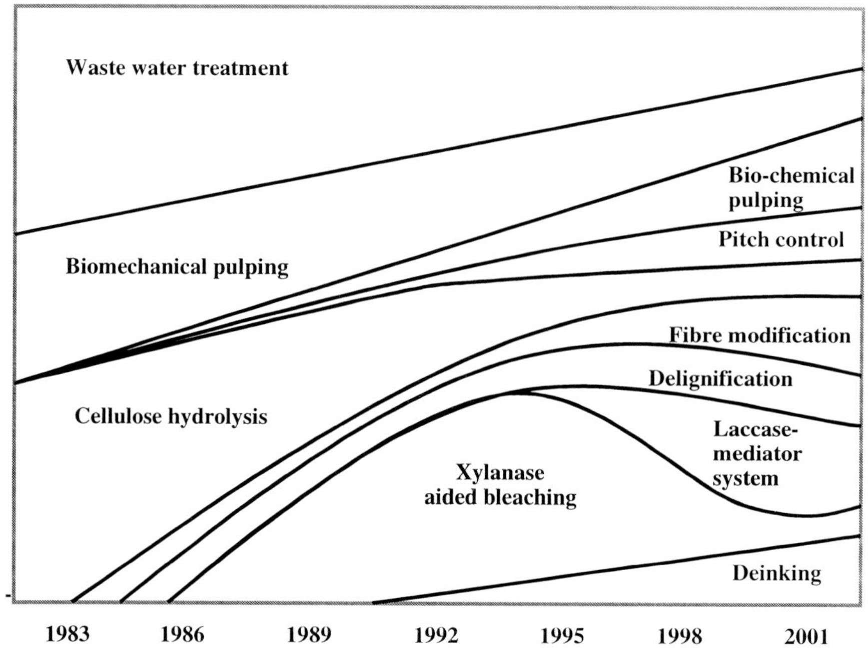

Figure 1. Trends of applied biotechnology in the pulp and paper area. Relative interest based on the number of oral and poster presentations on applied areas of the ICBPPI meetings in 1983-2001.

The use of lignocellulosic waste raw materials as source of sugars for the production of energy and chemicals has been an important goal for research on cellulases and hemicellulases since the start of the ICBPPI meetings in early 1980s. The diminishing activity of this area does not, however, reflect the decrease of importance, neither interest in the area. Rather, it reflects increased specialization and focusing of various conferences. During the last decade, conversion of renewable biomass into fuel energy and chemicals has again received growing attention as a means of replacing energy, chemicals and materials presently derived from fossil fuels. The issues of limited oil supply and the prospects of improvements in the relative economics of biomass feedstock have motivated the research carried out especially in USA. An important driving force is the Kyoto protocol. The price of sugar based on grain or other agricultural products as raw material has up to this date been prohibitive for economical production of several chemicals, but the expected decreasing costs of new biomass based sugar resources have raised interest in developing new production technologies for commodity chemicals, solvents and polymers. The importance of utilizing side streams and waste materials will obviously grow in the future.

The use of different enzymes in fibre modification or "fibre engineering" instead of their use as process aids is an interesting and potential field of application in the case of both chemical and mechanical pulps. By the targeted modification of fibre surface by enzymatic or combined enzymatic and chemical treatments improved fibre properties or completely new fibre characteristics for various applications can be created. Fibre engineering could be used both in improving paper and board manufacturing properties of pulp fibres, as well as for modification of fibres suitable for non-paper applications. Expectedly, this is one of the most fast growing areas in fibre based industries. Fibre modification also reflects the trend of using biotechnical means for product design rather than for process improvement. Successful engineering of different fibres for various purposes will, however, require new concepts and broad cooperation between fibre, polymer and surface chemists as well as papermakers and biotechnologists. New packaging materials and sensors indicating the status of the contents of the package offer challenges for biotechnologists. The safety of processes and products are among the important questions in future. The importance of hygiene and control of microbial growth in paper manufacture has not been reflected in the number of papers presented in the ICBPPI meetings.

Future

Biotechnical applications for pulp and paper industry have been developed for the past twenty years. The first introduction of enzymes at mill scale took place at 1980s, rapidly after the discovery and validation of the xylanase-aided bleaching concept. Since then, the development of industrial enzyme preparations having pH and temperature ranges suitable for target processes has been fast. In addition, novel enzymatic and microbial applications have been investigated for improving the processing of wood fibres, as well as for improving the runnability of the pulping, bleaching and papermaking processes.

There have been high expectations for the rising of new, edge-cutting technologies based on biotechnology in pulp and paper industry. The implementation of economically and technically viable biotechnical stages or treatments to mill scale operation has, however, been found to be difficult. The biotechnical applications competing with the chemical applications must overdue the performance of traditional chemistry and result in economical benefits without compromising the product quality. It seems that the most potential future applications of biotechnical methods will be found in the fields of speciality products, targeted modification of the fibres and controlling the safety of products.

Biotechnology in the Pulp and Paper Industry
L. Viikari and R. Lantto (Editors)

Biotechnology in the pulp and paper industry. A challenge for change.

Arthur J. Ragauskas

Institute of Paper Science and Technology, 500 10th Street, NW, Atlanta, GA, USA

The globalisation of the pulp and paper industry is a relatively new phenomenon that has been a harbinger of change and opportunities. Today's paper industry employs advanced chemical- and mechanical-based technologies to provide high-quality consumer products that are in worldwide demand and support the lifestyles of our new global economy. And yet, from these successes arise many of the current and future difficulties of the industry. Various paper industry leaders have stated that the capital requirements of manufacturing paper products are too high, are limiting creativity, and the entrepreneurial spirit of the industry. Coupled with these challenges, the emergence of low-cost fiber resources outside the Northern Hemisphere has contributed to further pressures on the paper industry to significantly reduce its manufacturing costs through a major redesign of its core manufacturing technologies. Within these difficulties, are disguised but unparalleled opportunities for researchers to efficaciously develop new biotechnology-based processes for our industry. These new technologies must reduce the capital costs of pulp production, be readily implemented in today's mill, and provide exceptional return for the resources invested if they are to be commercially feasible. The pathway to the future will entail the development of new global alliances between industry, governments, and research organizations. These new research partnerships will be linked by the Web, coordinated by industry and government, and will utilize the best research expertise and facilities available in the world. These collaborative efforts will generate higher value fiber resources, lower total manufacturing costs, and develop new materials from which new products can be designed and manufactured. In summary, the business of pulp and paper has provided R&D a challenge for change, and biotechnology is destined to address this opportunity.

INTRODUCTION

The production and usage of paper products is a cornerstone of modern societies and is interwoven into most major societal activities including education, government, business, and leisure. Indeed, it is well established that the consumption of paper on a national basis can be correlated to a nation's GDP [1]. Although many pulping technologies have been developed over the past century, the dominant pulping process used globally is the kraft process, producing more than 65% of 1997 virgin pulp production [2]. The German chemist C. F. Dahl invented this process in 1879 and the first kraft mill was built in 1890 in Sweden [3]. The discovery and subsequent implementation of the kraft process was undoubtedly broadcast on

Table 1. The role of major process variables in the treatment of natural fibers with cellulase enzymes

Application	Desirable Outcome	Treatment Parameter		
		Time	Cellulase Composition	Cellulase Loading
Fiber Modification	Reduced cell wall thickness, more flexible & collapsible fibers	Short	Monocomponents	Low
Cellulose hydrolysis to glucose	Full conversion of cellulose to glucose	Longer (12-24 h)	Complete (Endo, exo, and ß-G)	Relatively High
Textile biopolishing	Depilling and ageing of fabric	Medium (20 min-8 h)	Complete, EG or EG-rich	Low

For example, complete cellulase mixtures are used in depilling/cleaning of cotton fabrics, whereas pure endoglucanase (EG) or EG-rich mixtures are used to produce aged and soft fabrics demanded by the fashion market [1,7]. It is postulated that during depilling, enzymes attack and hydrolyze the microfibrils that hold the pills to the fiber surface, whereas in fabric ageing, the attack occurs on the fiber surface and results in fiber defibrillation [7]. The accompanying mechanical action removes the dye bound to the surface and imparts an aged appearance [7]. In both cases, the accessibility of cellulose surface to the enzymes plays a key role. Commercial cellulases have also been shown to enhance the whiteness, brightness and color characteristics of cotton fabrics [8].

In comparison with crude cellulase preparations, the cellulase mono-components were shown to be more effective in enhancing fiber collapsibility while circumventing the yield and strength losses, although they decreased individual fiber integrity [9]. The partial hydrolysis of some cell wall components weaken the fibers' natural integrity and "peel off" the cell wall layers, thereby enhancing the swelling and flexibility of the fibers. It has also been shown that cellulase treatment can increase the handsheet density and tensile strength of long, strong subalpine fir fibers, however, improvements in the tensile strength was dependent on the degree of fiber coarseness in the original pulp [9].

Cellulases have also been used to remove ink from papers and to enhance papermaking properties of recycled fibers. Enzymatic deinking can lower the need for deinking chemicals and reduce the adverse environmental impacts of the paper industry [10]. While in general, enzymatic deinking results in little or no loss in fiber strength [10-14], the overall effectiveness of the treatment depends on variables, such as toner quality and type, the type and amount of sizing, and the presence of other contaminants [15,16]. Although strength properties have not been compromised substantially, the excessive use of enzymes must be avoided [14], as it has been shown that significant hydrolysis of the fines [14,17-21] could reduce the bondability of the fibers [22-29].

Mechanistically, it has been postulated that improvements in dewatering and deinking of various pulps results in the peeling of the individual fibrils and bundles, which have a high affinity for the surrounding water and ink particles [30]. It appears that cellulase treatments can release ink particles bound to the fines and to the fiber, and enhance the removal of ink by flotation [31]. While cellulases clearly enhance the deinking process, the mechanical agitation still plays a critical role in the efficiency of ink removal [31-33]. These claims are consistent with similar findings concerning enzymatic stone washing of cotton fabrics, which indicated that enzymatic treatments in combination with mechanical agitation improve the efficacy of the process [34-35]. During textile bioprocessing, the small fiber ends protruding from the yarn are weakened by the action of the enzymes [36,37], while the simultaneous mechanical action completes the process by releasing the short fibers from the surface of the fabric [35] similar to the phenomenon occurring during deinking.

Refining, a mechanical action necessary for improving the physical properties of primary or secondary fibers, can generate small particles (fines) that can reduce the drainage rate of pulps during papermaking operations. Cellulases seem to preferentially attack and hydrolyze the fines produced during the refining operation, and therefore, improve the pulp's drainage property. For example, one mill trial revealed that the freeness of the refined stock could be increased to allow greater incorporation of the recycled fibers into a corrugating medium [38]. Other mill trials on recycled kraft fibers and old corrugated container pulp successfully demonstrated savings in refining energy requirements [39].

The retention of water by fibers during refining reduces the softening temperature of hemicellulose and lignin present between adjacent fibers and weakens inter-fiber bonding, hence improving the separation of fibers from one another and reducing the energy consumption during refining operation [40]. It ahs been shown that cellobiohydrolase I, a cellulase monocomponent, could selectively reduce the crystallinity of cellulose and subsequently produce more amorphous material with a higher affinity for water. Treatment with CBH I was able to reduce the refining energy demands by 40% [40].

While applications of cellulases in the textile and pulp and paper industry revolve around low dosage, partial cellulose hydrolysis by full or individual cellulase components, the biomass utilization operations require the use of complete cellulase systems at relatively high loading to achieve complete cellulose hydrolysis. The objective is to maximize the yield of glucose recovery and its fermentation to ethanol. Biomass-derived ethanol, either in pure form or in blend with gasoline, can be used as a renewable fuel by the transportation sector.

1.2. Cellulases in Bioconversion of Lignocellulosic Biomass

Cellulose constitutes the majority of polysaccharides present in the cell wall of herbaceous and woody plants. There is a growing interest in recovering and converting the polysaccharide fraction of plant materials to fermentable sugars to produce bioethanol as a sustainable form of energy. Unlike cellulose, which is a linear polymer of only glucose and has a relatively tight structure, hemicelluloses are made up of various pentoses and hexoses, have a branched structure, and are easily degraded under mild acidic conditions at higher temperatures. Therefore, in bioethanol production applications, lignocellulosic feedstocks, such as softwoods and hardwoods, various types of perennial grasses (e.g., switchgrass), and agricultural residues (e.g., corn stover, rice husk or wheat straw) are first pretreated. Pretreatment solubilizes the hemicelluloses and renders the cellulose fraction more amenable to further processing for glucose production.

Cellulose depolymerization (hydrolysis) can be achieved by both chemical (acids and bases) and enzymatic methods. Cellulose acid hydrolysis occurs at a much faster rate, but may involve the risk of sugar degradation under non-optimum reaction conditions, such as excessively long residence times or fluctuations in temperature or acid concentration. Degradation of sugars, which in turn results in formation of byproducts with inhibitory effects on fermentative microorganisms, does not occur in enzymatic hydrolysis operations. Moreover, enzymes, unlike acids, do not pose any deleterious corrosive effects, and therefore, do not require the use of costly, corrosion-resistant equipment.

The major drawback of the enzymatic process is the high cost of enzymes [41]. Research is currently underway to address this problem by: 1) improving the productivity of cellulase producing microorganisms and increasing the specificity of the enzyme components through genetic engineering; 2) enhancing the susceptibility of cellulosic substrates by developing more effective pretreatment and post-treatment methods which enhances the accessibility of cellulose structure to cellulase enzymes and reduces the amount of enzyme required; and 3) developing novel process schemes to recover and recycle the enzymes after each round of hydrolysis. It is likely that the development of an economically feasible technology for the enzymatic conversion of cellulose will require advancements in all three dimensions concurrently.

An important question relevant to any industrial application of cellulases is the mechanism of cellulose hydrolysis. Research conducted over the past five decades has advanced our understanding of the interactions between cellulosic substrates and cellulolytic proteins. The effect of many enzyme and substrate related factors on the rate and extent of hydrolysis has been elucidated [42, 43], however, the results have varied depending on the substrate used, and it is still unclear what determines the initial rate of cellulose hydrolysis and what governs the extent to which it can be hydrolyzed. The rate of cellulose hydrolysis, in both model (bacterial microcrystalline cellulose [BMCC] or Avicel) and fibrous substrates (wood, herbaceous materials or pulp), diminishes the time and does not allow complete cellulose conversion in short periods of time. The following section discusses how the structural barriers of cellulosic substrates can limit the accessibility of cellulose by enzymes and hinder the overall reaction.

1.3. Cellulose Hydrolysis

Cellulolytic microorganisms produce an array of ß-1,4-glucanases during growth on cellulosic substrates in nature. These include endoglucanases that attack ß-1,4-glucan chains randomly and exoglucanases, usually cellobiohydrolases, with a strong preference for acting at chain ends, as well as ß-glucosidase which mediates the cleavage of cellobiose to glucose monomers (Table 2). Despite intensive research in recent years, a comprehensive model that can fully describe the mechanisms by which cellulases disintegrate and depolymerize cellulose has not been developed. The structural intricacies of the lignocellulosic substrates also add to the complexity of the cellulose hydrolysis phenomenon. The completion of the reaction usually requires a relatively long residence time (12 h to a few days), during which the properties of the enzyme system as well as the structural features of the substrate can undergo significant changes [44-49]. These changes are manifested in the dwindling rate of hydrolysis and the incomplete hydrolysis at shorter reaction times.

Table 2. The role of various cellulase components during the enzymatic hydrolysis of cellulose

Cellulase Component	Postulated Role during Cellulose Hydrolysis
Endoglucanases	Cleavage of internal ß-glycosidic bonds to produce free chain ends that will be acted upon by cellobiohydrolases
Exoglucanases (Cellobiohydrolases, CBH)	Cleavage of cellobiose units from the cellulose chain ends
ß-glucosidase	Cleavage of cellobiose units to glucose monomers

Numerous attempts have been made to identify a single property of the substrate or enzyme as the primary determinant of the hydrolysis rate and yield. However, a close review of the literature reveals that the hydrolysis appears to be controlled by the interactions among various factors including a) processing variables, b) enzyme related factors, c) substrate characteristics. The processing variables include temperature [50], solids concentration [51], mixing pattern and speed [51-53], and the presence of air interface in the reaction vessel [54]. The shear forces generated by the mixing mechanism and/or prolonged exposure to high temperature can inactivate enzymes and reduce the reaction rate. These parameters can usually be adjusted to an optimum level to avoid, or at least limit, their adverse impacts.

The importance of the enzyme related factors stems from the multiplicity of enzyme components. Although these individual enzymes are of different sizes and have different levels of activity and affinity for crystalline and amorphous regions of cellulose, they are all necessary to generate the synergistic effects required for complete hydrolysis. It has been suggested that the hydrolysis reactions are most efficient when the pores within a substrate are large enough to accommodate both large and small enzyme components to maintain the synergistic action of the enzyme system [55, 56]. However, when the pores are too small to accommodate the larger molecules, the enzyme components are segregated and the synergistic action of the enzyme system is reduced [44, 45]. Inhibition of enzymes due to the accumulation of cellobiose, and to a lesser extent glucose, is another example of an important, but not the primary, rate-limiting factor of cellulose enzymatic hydrolysis. The non-specific binding of enzymes to lignin in lignaceous substrates can also reduce the activity of the enzyme system, and the rate at which they hydrolyze cellulose.

Numerous studies [43, 55, 57, 58] have shown that the rate and extent of hydrolysis is influenced by the physiochemical properties of the substrate at three different levels; microfibril (e.g., crystallinity and degree of polymerization), fibril (e.g., lignin content and distribution), and fiber (pore size and distribution, available surface area and degree of swelling). None of these factors alone has been able to provide a universal explanation for the drop-off in the rate of cellulose hydrolysis after the initial rapid reaction.

In an attempt to shed some light on the impact of cellulose accessible surface area on the hydrolyzability of cellulosic substrates, steam-exploded Douglas-fir wood with and without bleaching as well as Avicel were hydrolyzed using a complete cellulase system. We altered the accessibility of this lignocellulosic substrate, steam-exploded Douglas fir wood chips, by post-treatment with an alkaline, hydrogen peroxide solution at a high temperature (80 ^{o}C), and examined the effect of the enzyme loading on the rate and extent of hydrolysis. Avicel was used in parallel as a model substrate.

2. MATERIALS AND METHODS

2.1. Substrate

The three substrates used in this study were Avicel (Sigma), steam-exploded Douglas fir (*Pseudotsuga menziesii)* wood chips, and bleached steam-exploded Douglas fir wood chips. The Douglas fir chips were subjected to SO_2-catalyzed steam-explosion at medium severity conditions (4.5 min, 195 °C, 4.5% (w/w) SO_2/original dry wood) as described previously [59]. The steam-exploded material was then washed with 50-fold (v/w) water. Some of the steam-exploded substrate was bleached using a two-stage alkaline-peroxide protocol described below.

2.2. Bleaching

The bleaching procedure consisted of chelation and hydrogen peroxide treatment stages. Residual metal ions which could interfere with subsequent peroxide treatment were removed by incubating a 5% (dry w/v) substrate slurry with 0.2% EDTA at 45°C for 1 hour. Following chelation, 10 g (dry weight) of substrate was used to make a 10% (dry w/v) slurry containing 0.2% (w/w) EDTA, 0.5% (w/w) $MgSO_4$, 1.5% (w/w) NaOH, and 10% (w/w) hydrogen peroxide solution. The pH of this solution was adjusted to 11.5 by adding sodium hydroxide, and was incubated at 80°C for 45 minutes. The bleached substrate was then washed with 20-fold (v/w) water.

2.3. Enzymes

All experiments were conducted using Celluclast, a complete *Trichoderma reesei* cellulase system, and a commercial ß-glucosidase preparation, Novozym 188 (Novo Nordisk, Denmark. The activities of Celluclast and Novozym 188 were 67 FPU/ml and 80 CBU/ml respectively, as measured by standard procedures [60].

2.4. Protein Adsorption

Adsorption experiments on the lignaceous and bleached steam-exploded Douglas fir substrates were carried out in 50-ml centrifuge tubes containing 0.3 g (dry weight) of the substrate in 15 ml of 0.05 M sodium acetate buffer (pH 4.8). Different volumes of Celluclast (20, 50, 100, 150, 200, 400, 600, 1000, 1200, 1400, 1600, 2000 μl) were added, in duplicate, to the tubes, which were then incubated at 4°C for 90 minutes. All tubes were shaken vigorously every 15 minutes to ensure proper mixing of the substrate. The tubes were then spun down and 1.5-ml samples of the supernatant were taken and analyzed for protein content using the Biorad Protein Assay (Biorad Laboratories, Hercules, California).

Due to the fine nature of Avicel powder, adsorption experiments on Avicel were conducted with a 20 mg/ml substrate slurry in 0.05 M sodium acetate buffer (pH 4.8) to ensure consistent substrate loading. In a series of 15 ml centrifuge tubes, 2.5 ml of the Avicel slurry was added to 2.5 ml of a series of enzyme solutions containing protein contents equivalent to those used with the woody substrates. The tubes were incubated at 4°C for 90 minutes, and were shaken vigorously every 15 minutes. The tubes were then centrifuged and the supernatant liquid was analyzed for protein content as described previously.

2.5. Enzymatic Hydrolysis

All hydrolysis experiments were conducted in 125-ml Erlenmeyer flasks incubated in a gyratory water bath at 45°C shaking at 150 rpm. Substrate was loaded at a 2% (w/v) solid concentration in a total reaction volume of 75 ml of 0.05 M sodium acetate buffer (pH 4.8). The cellulase loading was varied as described below and supplemented with β-glucosidase (CBU:FPU = 2:1) to prevent product inhibition by cellobiose. Microbial contamination was prevented by adding sodium azide to the reaction medium at a concentration of 0.5% (w/v). Aliquots of 500 μL taken from the supernatant liquid at different time points were immediately boiled for 5 minutes to inactivate the enzymes before being chilled for storage. The samples were micro-centrifuged at 11000 rpm for 5 minutes prior to sugar analysis.

3. RESULTS AND DISCUSSION

In order to study the effect of substrate accessibility on the hydrolyzability of cellulosic materials, we hydrolysed Avicel as well as steam-exploded Douglas fir wood chips, with and without bleaching at varying enzyme loadings. The hydrogen-peroxide bleaching step significantly reduced the amount of lignin in the steam-exploded wood (Table 3), producing a substrate with a presumably more accessible cellulose fraction.

Table 3. Chemical composition of Avicel and steam-exploded (SE) Douglas fir wood chips before and after bleaching.

Component (%)	Avicel	SE wood	Delignified SE wood
Glucose	94.9	58.4	92.7
Mannose	1.5	0.9	0.6
Galactose	0.0	0.8	0.0
Xylose	1.3	0.4	0.2
Arabinose	0.0	0.0	0.0
Lignin	0.2	42.7	2.6
Ash	0.1	0.1	0.0

Adsorption of sufficient amounts of enzyme onto the substrate is generally regarded as the prerequisite step for effective cellulose hydrolysis by comparing the enzyme adsorption isotherms and the enzymatic hydrolysis profiles of untreated, sulfonated, and delignified mechanical pulps, we [61] have shown that a substrate's enzyme adsorption capacity is not necessarily a good predictor of its hydrolyzability. We showed that while sulfonation significantly increased the enzyme adsorption capacity of the pulp, it did not improve the enzymatic digestibility of the pulp to the same extent. On the other hand, bleaching of the pulp caused marginal improvements in the pulp's enzyme adsorption capacity, but significantly improved the hydrolysis efficiency. It was postulated that in the sulfonated pulp, a large proportion of enzymes were adsorbed onto the lignin, and therefore, were not actively catalyzing the ß-1,4 linkages, whereas in the delignified pulp, cellulose was made more accessible to the enzymes by the removal of lignin shielding the cellulose, and thus was hydrolysed more efficiently [61].

In the current investigation, we observed a similar pattern, the steam exploded wood, which contained about 40% lignin, adsorbed more enzymes (g/g available cellulose) than did the bleached substrate in which the lignin content had been reduced to less than 2% (Fig. 1). It is probable that the non-specific binding of enzymes onto the lignin fraction caused the higher adsorption capacity. This assumption is in agreement with our hydrolysis results, which will be described later, where the high-lignin substrate exhibited a lower hydrolyzability (Fig. 2).

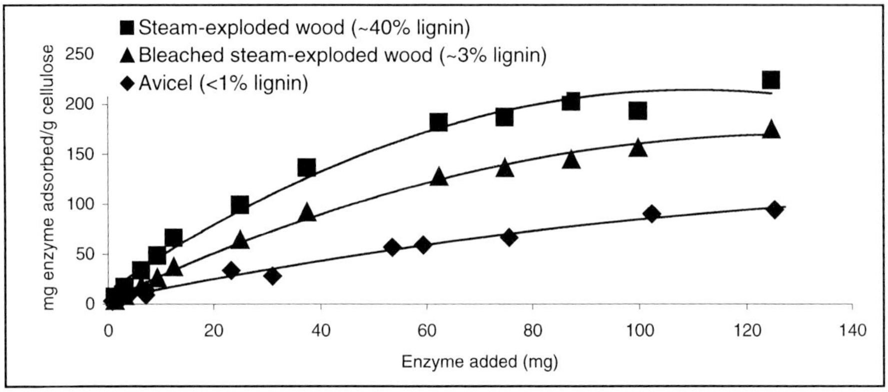

Figure 1. Adsorption of cellulase enzymes onto Avicel and steam-exploded Douglas fir wood (with and without bleaching).

Avicel, which had the lowest lignin content and the smallest particle size among the three substrates, exhibited the lowest capacity for enzyme adsorption, and had the lowest hydrolyzability (Fig. 1 & 2). It is known that smaller particles have a higher surface area-to-mass ratio, and as a result, exhibit better protein adsorption capacity and hydrolyzability [62]. However, the surface area and availability of reaction sites on the cellulose structure can also be influenced by the type and intensity of physical or chemical treatments applied to the substrate.

For instance, Avicel is derived from wood using a process similar to pulping, the objective of which is to solubilize lignin and separate cellulose fibers while maintaining the integrity of individual fibers. The objective of steam explosion, however, is to solublize the hemicellulose fraction and disrupt the structure of fibers as much as possible to make them more amenable to downstream processing. In the present study, the difference observed between the enzyme adsorption capacity and hydrolyzability of Avicel and steam-exploded wood can be partly explained by the difference in the way these substrates were produced.

The structure of steam-exploded wood is extensively disrupted, which makes it a highly accessible substrate, whereas Avicel, which has a fibrous form, contains less damaged areas, and is therefore less accessible. Moreover, during the production process, Avicel is dried, whereas the steam-exploded materials had never been dried. Drying is known to negatively influence substrate surface accessibility by removing the interfibrillar water, and consequently, increasing the extent of cross-linking between adjacent microfibrils due to formation of internal hydrogen bonds [63-66]. The collapse of fiber pores stiffens the fibers, and renders the internal structure less accessible to enzymes, a phenomenon known as hornification. It has been shown that the wetting of hornified fibers cannot restore fibers' original accessbility even after prolonged wetting periods [5, 17, 63-65, 67, 68]. Therefore, the lower enzyme adsorption capacity and reduced hydrolyzability of Avicel, in comparison with the never-dried, steam-exploded wood, is in part due to the hornifying effect of the drying step.

It was apparent that the higher enzyme adsorption onto the substrate did not translate into better hydrolysis, as the lignified and bleached substrates were both hydrolysed at similar rates and to the same extent (Fig. 2). The rate and yield of hydrolysis was even slightly higher for the bleached sample, which displayed a lower enzyme adsorption capacity. It has been suggested that enzymes can bind non-specifically to the lignin fraction of the substrate [69, 70]. The lignin-bound enzymes, however, may not be hydrolytically productive, either due to their inability to access the cellulose structure or due to deactivation upon adsorption on lignin [69, 70]. Therefore, it appeared that the accessibility, and subsequently the hydrolyzability, of a substrate is not necessarily dictated and cannot be accurately predicted by the amount of enzyme adsorbed. Likewise, it appears that the lignin content alone could not be used to predict the accessibility and hydrolyzability of a substrate. Bleaching can generally enhance the accessible surface area of a substrate, however, exposure to other treatments, such as drying in the case of Avicel, can also have a significant influence on the enzyme adsorption capacity and hydrolyzability of the substrate.

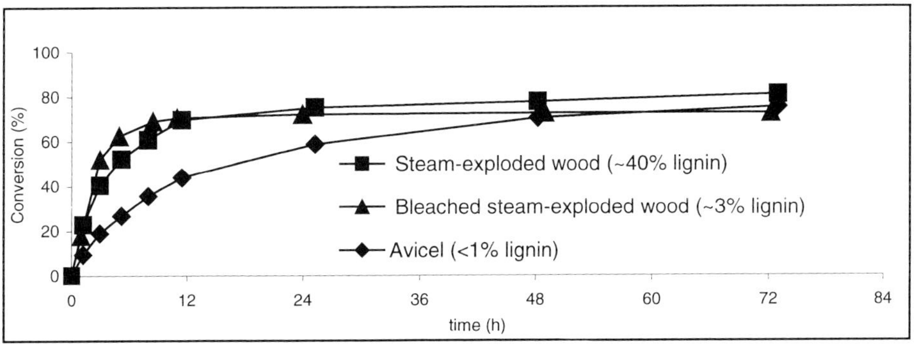

Figure 2. Enzymatic hydrolysis of Avicel and steam-exploded Douglas fir wood (with and without bleaching) at 50 FPU/g cellulose enzyme loading.

It is generally known that increasing the concentration of enzyme increases the rate and extent of enzymatic hydrolysis. In order to determine the correlation between the substrate accessibility and the degree to which higher enzyme loadings can improve hydrolysis efficiency, we hydrolyzed the three substrates under multiple levels of enzyme loading (10, 30, 50, 100 and 200 FPU/g cellulose). The initial reaction rate in each experiment was determined using the empirical equation (1) developed by Ooshima et al. (1983), which describes glucose production as a function of time:

$$\mathbf{G = k * \ln (1 + m*t)} \qquad \mathbf{(1)}$$

G: glucose produced (mg.ml^{-1})
k (mg.ml^{-1}) & m (t^{-1}): constants
t: time elapsed (min)

The hydrolysis reaction rate can be determined by equation (2):

$$\textbf{Reaction Rate (R)} = dG / dt = (m * k) / (1 + m * t) \qquad \mathbf{(2)}$$

The initial reaction rate (eq. 3) is calculated by setting the time in equation (2) to zero:

$$\textbf{Initial Reaction Rate } (R_o) = k*m \qquad \mathbf{(3)}$$

ACKNOWLEDGMENT

This work was supported by grants from Natural Resources Canada and Natural Science and Engineering Research Council of Canada (NSERC). The technical assistance of Dexter Lam and Emily Lai (UBC Co-op students) in conducting the experiments is acknowledged.

REFERENCES

1 J. S. Tolan and B. Foody. "Cellulase from Submerged Fermentation" *Advances in Biochemical Engineering, vol.* 65, pp. 41-67, 1999.

2 G. Smook. "Secondary Fiber" in *Handbook for Pulp and Paper Technologists.* Vancouver: Angus Wilde, 1992, pp. 209-218.

3 A. L. Moekerbak and W. Zimmermann. "Applications of Enzymes in Paper Deinking Processes" in *Enzyme Applications in Fiber Processing, vol.* 687, K. Eriksson and A. Cavaco-Paulo, Eds. Washington, D.C.: American Chemical Society, 1998, pp. 133-141.

4 G. Bhat, J. Heitmann, and T. Joyce. "Novel Techniques for Enhancing the Strength of Secondary Fiber" *Tappi, vol.* 74, pp. 151-156, 1991.

5 J. Pommier, J. Fuentes, and G. Goma. "Using Enzymes to Improve the Process and the Product Quality in the Recycled Paper Industry, Part 1: The Basic Laboratory Work" *Tappi,* vol. 72, pp. 187-191, 1989.

6 S. D. Mansfield, D. J. Swanson, N. Roberts, J. A. Olson, and J. N. Saddler. "Enhancing Douglas-fir Pulp Properties with a Combination of Enzyme Treatments and Fiber Fractionation" *Tappi,* vol. 82, pp. 152-158, 1999.

7 A. Cavaco-Paulo. "Processing Textile Fibers with Enzymes: An Overview" in *Enzyme Applications* in *Fiber Processing,* ACS Symposium Series 687, K. E. Eriksson and A. Cavaco-Paulo, Eds. Washington, DC: American Chemical Society, 1998, pp. 180-189.

8 E. Csiszar, G. Szakacs, and I. Rusznak. "Cotton Wall Structure and Enzymatic Treatments" in *Enzyme Applications in Fiber Processing, American Chemical Society Symposium Series 687,* K. E. Eriksson and A. Cavaco-Paulo, Eds. Washington, DC: American Chemical Society, 1998, pp. 204-211.

9 C. D. Edgar, S. D. Mansfield, G. M. Gubitz, and J. N. Saddler. "The Synergistic Effects of Endoglucanase and Xylanase in Modifying Douglas Fir Kraft Pulp," in *Enzyme Application in Fiber Processing, vol.* 687, K. E. Eriksson and A. Cavaco Paulo, Eds. Washington, DC: American Chemical Society, 1998, pp. 75-87.

10 G. Stork and J. Puls. "Changes in Properties of Different Recycled Pulps by Endoglucanase Treatment," in *Biotechnology in the Pulp and Paper Industry: Recent Advances in Applied and Fundamental Research, vol. 1,* E. Srebotnik and K. Mesner, Eds., 1996, pp. 145-150.

11 J. Sarkar, D. Cosper, and E. Hartig. "Applying Enzymes and Polymers to Enhance the Freeness of Recycled Fiber" *TAPPI,* vol. 78, pp. 89-95, 1995.

12 J. M. Sarkar. "Recycle Paper Mill Trials Using Enzyme and Polymer for Upgrading Recycled Fiber" presented at 50th APPITA Annual General Conference, 1996.

13 J. M. Sarkar. "Recycle Paper Mill Trial Using Enzyme and Polymer for Upgrading Recycled Fiber" *APPITA,* vol. 50, pp. 57-60, 1997.

14 G. Stork, H. Pereira, T. Wood, E. Dusterhoft, A. Toft, and J. Puls. "Upgrading Recycled Pulps Using Enzymatic Treatment" *TAPPI,* vol. 78, pp. 79-88, 1995.

15 T. W. Jeffries, M. S. Sykes, K. Rutledge-Cropsey, J. H. Klungness, and S. Abubakr. "Enhanced Removal of Toners form Office Waste Papers by Microbial Cellulases" in *Biotechnology in the Pulp and Paper Industry: Recent Advances in Applied and Fundamental Research,* vol. 1, E. Srebotnik and K. Messner, Eds. Vienna, Austria, 1996, pp. 141-144.

16 S. K. Ow, J. M. Park, and S. H. Han. "Effects of Enzyme on Ink Size and Distribution during the Enzymatic Deinking Process of Old Newsprint" in *Biotechnology in the Pulp and Paper Industry: Recent Advances in Applied and Fundamental Research,* E. Srebotnik and K. Messner, Eds. Vienna, Austria: Facultas-Universitatsverlag, 1996, pp. 163-168.

17 L. S. Jackson, J. A. Heitmann, and T. W. Joyce. "Enzymatic Modifications of Secondary Fiber" *Tappi,* vol. 76, pp. 147-154, 1993.

18 D. Y. Prasad, J. A. Heitmann, and T. Joyce. "Enzyme Deinking of Black and White Letterpress Printed Newsprint Waste" *Progress* in *Paper Recycling,* vol. 1, pp. 21-30, 1992.

19 D. Y. Prasad. "Enzymatic Deinking of Laser and Xerographic Office Wastes" *APPITA,* vol. 46, pp. 289-292,1993.

20 D. Y. Prasad, J. Heitman, and T. Joyce. "Enzymatic Deinking of Colored Offset Newsprint" *Nordic Pulp and Paper Research Journal,* vol. 8, pp. 284-286, 1993.

21 G. Stork and J. Puls. "Untersuchungen zur Verbesserung der Altpapierqualitat durch Enzymatische Behandlung" *Das Papier,* vol. 6, pp. 311-323, 1994.

22 R. P. Beatson, W. Bichard, and N. Muradali. "The Internal Bond Strength of Sulphite Treated TMP" *Paperi* ja *Puu-Paper and Timber,* vol. 70, pp. 66-72, 1988.

23 J. Gorres, R. Amiri, J. R. Wood, and A. Karnis. "Mechanical Pulp Fines and Sheet Structure" Presented at International Paper Physics Conference, Niagara-on-the-Lake, 1995.

24 A. Karnis. "The Role of Latent and Delatent Mechanical Pulp Fines in Sheet Structure and Pulp Properties" *Paperi* ja *Puu-Paper and Timber,* vol. 77, pp. 491-497, 1995.

25 R. P. Kibblewhite. "Interrelations between Pulp Refining Treatments, Fiber and Fines Quality and Pulp Freeness" *Paperi ja Puu-Paper and Timber,* vol. 57, pp. 519-526, 1975.

26 R. P. Kibblewhite, A. D. Bawden, and C. L. Brindley. "TMP Fiber and Fines Qualities of 13 Radiata Pine Wood Types" *APPITA,* vol. 48, pp. 367-377, 1995.

27 P. A. Moss and E. Retulainen. "The Effect of Fines on Fiber Bonding: Cross-sectional Dimension of TMP Fibers at Potential Bonding Sites" presented at 1995 International Paper Physics Conference, 1995.

28 L. Paavilainen. "Importance of Particle Size - Fiber Length and Fines - for the Characterization of Softwood Kraft Pulp" *Paperi ja Puu-Paper and Timber,* vol. 72, pp. 516-526, 1990.

29 E. Retulainen and K. Nieminen. "Fiber Properties as Control Variables in Papermaking? Part 2: Strengthening Interfibre Bonds and Reducing Grammage" *Paperi ja Puu-Paper and Timber,* vol. 78, pp. 305-312, 1996.

30 S. B. Lee and I. H. Kim. "Structural Properties of Cellulose and Cellulase Reaction Mechanisms" *Biotechnology & Bioengineering*, vol. 25, pp. 33-51, 1983.

31 T. Welt and R. J. Dinus. "Enzymatic Deinking: A Review" *Progress in Paper Recycling,* vol. 4, pp. 36-47, 1995.

32 C. Zeyer, T. Joyce, J. W. Rucker, and J. A. Heitman. "Enzymatic Deinking of Cellulose Farbic: A Model Study of Enzymatic Paper Deinking" *Progress in paper Recycling,* vol. 2, pp. 36-44, 1993.

33 C. Zeyer, T. Joyce, J. A. Heitmann, and J. W. Rucker. "Factors Influencing Enzyme Deinking of Recycled Fiber" *TAPPI,* vol. 77, pp. 169-177, 1994.

34 R. M. Tyndall. "Upgrading Garment Washing Techniques" *American Dyestuff Reporter,* vol. 79, pp. 22-30, 1990.

35 R. M. Tyndall. "Improving the Softness and Surface Appearance of Cotton Fabrics and Garments by Treatment with Cellulase Enzymes" *Textile Chemist* and *Colorist, vol.* 24, pp. 23-26, 1992.

36 C. Byrne. "Biotechnology in Textiles" *Technology in the Textile Market*, vol. 25, pp. 30-49, 1996.

37 S. Chikkodi. "Enzymatic Treatments on Cotton and Cotton and Wool Mixes" in *Industrial Textiles,* vol. 1265, pp. 29-31, 1995.

38 B. R. Moran. "Enzyme Treatment Improves Refining Efficiency, Recycled Fiber Freeness" *Pulp* and *Paper,* vol. 70, pp. 119-121, 1996.

39 F. C. Caram, J. M. Sarkar, H. P. Didwania, E. Espinoza, and J. C. Benavides. "Paper Mill Evaluation of a Cellulolytic Enzyme and Polymers for Improving the Properties of Waste Paper Pulp" Presented at Papermakers Conference, 1996.

40 J. Pere, S. Liukkonen, M. Siika-aho, J. Gullichsen, and L. Viikari. "Use of Purified Enzymes in Mechanical Pulping" presented at TAPPI Pulping Conference, Nashville, TN, 1996.

41 D. J. Gregg, A. Boussaid, and J. N. Saddler. "Techno-economic Evaluations of a Generic Wood-to-Ethanol Process: Effect of Increased Cellulose Yields and Enzyme Recycle" *Bioresource Technology,* vol. 63, pp. 7-12, 1998.

42 A. Esteghlalian, V. Srivastava, N. Gilkes, D. J. Gregg, and J. N. Saddler. "An Overview of Factors Influencing the Enzymatic Hydrolysis of Lignocellulosic Feedstocks," in *Glycosyl Hydrolases for Biomass Conversion, ACS Symposium Series 769,* M. E. Himmel, J. O. Baker, and J. N. Saddler, Eds. Washington, DC: American Chemical Society, 1999, pp. 100-111.

43 S. D. Mansfield, C. Mooney, and J. N. Saddler. "Substrate and Enzyme Characteristics that Limit Cellulose Hydrolysis" *Biotechnology Progress,* vol. 15, pp. 804-816, 1999.

44 J. Woodward, M. K. Hayes, and N. E. Lee. "Hydrolysis of Cellulose by Saturating and Non-saturating Concentrations of Cellulase: Implications for Synergism" *Bio/Technology,* vol. 6, pp. 301-304, 1988.

45 J. Woodward, M. Lima, and N. E. Lee. "The Role of Cellulase Concentration in Determining the Degree of Synergism in the Hydrolysis of Microcrystalline Cellulose" *Biochemistry Journal,* vol. 255, pp. 895-899, 1988.

46 N. Din, H. G. Damude, N. R. Gilkes, R. C. Miller, R. A. J. Warren, and D G Kilburn. "C1-Cx Revisited: Intramolecular Synergism in a Cellulase" proceedings *of National Academy of Science,* vol. 91, pp. 11383-11387, 1994.

47 B. Henrissat, H. Driguez, C. Viet, and M. Schulein. "Synergism of Cellulases from *Trichoderma Reesei* in the Degradation of Cellulose" *Bio/Technology,* vol. 3, pp. 722-726, 1985.

48 D. C. Irwin, M. Spezio, L. P. Walker, and D. B. Wilson. "Activity Studies of Eight Purified Cellulases: Specificity, Synergism, and Binding Domain Effects" *Biotechnology and Bioengineering,* vol. 42, pp. 1002-1013, 1993.

49 K. Riedel and K. Bronnenmeier. "Intramolecular Synergism in an Engineered

Exo-Endo-1,4 □-glucanase Fusion Protein" *Molecular Biology, vol.* 28, pp. 767-775, 1998.

50 C. Tengborg, M. Galbe, and G. Zacchi. "Influence of Enzyme Loading and Physical Parameters on the Enzymatic Hydrolysis of Steam-Pretreated Softwood" *Biotech. Prog.* vol. 17, pp. 110-117, 2001.

51 H. Ingesson, G. Zacchi, B. Yang, A. Esteghlalian, and J. N. Saddler. "The Effect of Shaking Regime on the Rate and Extent of Enzymatic Hydrolysis of Cellulose" *Journal of Biotechnology*, vol. 88, pp. 177-182, 2001.

52 E. T. Reese and D. Y. Ryu. "Shear Inactivation of Cellulase of *Trichoderma reesei*" *Enzyme and Microbial Technology,* vol.: 2, pp. 239-240, 1980.

53 Q. Nguyen. "Tower Reactors for Bioconversion of Lignocellulosic Material" US Patent 5,733,758, USA, 1998.

54 S. Mukataka, M. Tada, and J. Takahashi. "Effects of Agitation on Enzymatic Hydrolysis of Cellulose in a Stirred-Tank Reactor" *Journal of Fermentation Technology,* vol. 61, pp. 615-621, 1983.

55 A. O. Converse, R. Matsuno, M. Tanaka, and M. Taniguchi. "A Model of Enzyme Adsorption and Hydrolysis of Microcrystalline Cellulose with Slow Deactivation of the Adsorbed Enzyme" *Biotechnology and Bioengineering*, vol. 32, pp. 38-45, 1988.

56 M. Tanaka, M. Ikesaka, and R. Matsuno. "Effect of Pore Size in Substrate and Diffusion of Enzyme on Hydrolysis of Cellulosic Materials with Cellulases" *Biotechnology and Bioengineering,* vol. 32, pp. 698-706, 1988.

57 L. P. Walker and D. B. Wilson. "Enzymatic Hydrolysis of Cellulose: An Overview" *Bioresource Technology,* vol. 36, pp. 3-14, 1991.

58 L. T. Fan, Y. Lee, and D. H. Beardmore. "The Influence of Major Structural Features of Cellulose on Rate of Enzymatic Hydrolysis" *Biotechnology and Bioengineering,* vol. 23, pp. 419-424, 1981.

59 A. Boussaid, A. R. Esteghlalian, D. J. Gregg, K. H. Lee, and J. N. Saddler. "Steam Pretreatment of Douglas-Fir Wood Chips" *Applied Biochemistry and Biotechnology,* vol. 84-86, pp. 693-705, 2000.

60 T. M. Wood and K. M. Bhat. "Methods for Measuring Cellulase Activities" in *Methods in Enzymology*, vol. 160, W. Wood and S. Kellogg, Eds.: Academic Press, 1988, pp. 87-112.

61 C. Mooney, S. D. Mansfield, M. G. Touhy, and J. N. Saddler. "The Effect of Initial Pore Volume and Lignin Content on the Enzymatic Hydrolysis of Softwoods" *Bioresource Technology,* vol. 00, pp. 1-7, 1997.

62 L. T. Fan, Y. H. Lee, and D. H. Beardmore. "Effects of Pretreatments on Major Structural Features of Cellulose and on the Rate of Enzymatic Hydrolysis" presented at Bioconversion and Biochemical Engineering, 1980.

63 G. V. Laivins and A. M. Scallan. "The Influence of Drying and Beating on the Swelling of Fines" *Journal of Pulp and Paper Science*, vol. 22, pp. 178-183, 1996.

64 T. Oksanen, J. Buchert, and L. Viikari. "The Role of Hemicelluloses in the Hornification of Bleached Kraft Pulps" *Holzforschung,* vol. 51, pp. 355-360, 1997.

65 U. Weise. "Hornification: Mechanisms and Terminology" *Paperi Ja Puu*, vol. 80, pp. 110-115, 1998.

66 A. M. Scallan. "The Accommodation of Water within Pulp Fibres" in *Fiber Water Interactions in Papermaking,* vol. 1, H. Corte, Ed., 1977, pp. 9-27.

67 M. M. Nazhad and L. Paszner. "Fundamentals of Strength Loss in Recycled Paper" *Tappi* vol. 77, pp. 171-179, 1994.

68 M. M. Nazhad, L. P. Ramos, L. Paszner, and J. N. Saddler. "Structural Constraints Affecting the Initial Enzymatic Hydrolysis of Recycled Paper" *Enzyme and Microbial Technology*, vol. 17, pp. 68-74, 1995.

69 D. Lee, A. H. C. Yu, K. K. Y. Wong, and J. N. Saddler. "Evaluation of the Enzymatic Susceptibility of Cellulosic Substrates using Specific Hydrolysis Rates and Enzyme Adsorption" *Applied Biochemistry and Biotechnology*, vol. 45/46, pp. 407-415, 1994.

70 K. W. Lin, M. R. Ladisch, M. Voloch, J. A. Patterson, and C. H. Noller. "Effect of Pretreatment and Fermentation on Pore Size in Cellulosic Materials" *Biotechnology and Bioengineering*, vol. 27, pp. 1427-1433, 1985.

71 T. Sawada, Y. Nakamura, and F. Kobayashi. "Effects of Fungal Pretreatment and Steam Explosion Pretreatment on Enzymatic Saccharification of Plant Biomass" *Biotechnology* and *Bioengineering,* vol. 48, pp. 719-724, 1995.

72 H. H. Brownell and J. N. Saddler. "Steam-Explosion Pretreatment for Enzymatic Hydrolysis" *Biotechnology* and *Bioengineering Symposium,* vol. 14, pp. 55-68, 1984.

Biotechnology in the Pulp and Paper Industry
L. Viikari and R. Lantto (Editors)

Lignocellulose oxidation by low molecular weight metal-binding compounds isolated from wood degrading fungi: A comparison of brown rot and white rot systems and the potential application of chelator-mediated Fenton reactions*

Barry Goodell[a], Yuhui Qian[a], Jody Jellison[b], Michael Richard[c], and Weihong Qi[b]

[a]Professor and Graduate Research Assistant, respectively, Wood Science and Technology
[b]Professor and Graduate Research Assistant, respectively , Biology, University of Maine, Orono, Maine 04469-5755 USA.
[c]Managing Member, Advanced Technology Development LLC, Charlotte, North Carolina, USA.

Email: Goodell@UMIT.Maine.edu

This review focuses on the non-enzymatic mechanisms employed by brown rot fungi in the biodegradation of wood. Specifically we review a hypothesis for a system employing fungal produced catecholates and their function in chelator-mediated Fenton chemistry. This system would allow the production of hydroxyl radicals within the wood cell wall to in-part mimic the action of brown rot fungi. Similarities and differences between brown rot fungi, white rot fungi, and molds are discussed with regard to pH modification of wood as well as the types of low molecular weight chelators produced. Potential application of the chelator-mediated Fenton system chemistry in remediation and paper recycling are discussed.

1. BROWN ROT DEGRADATION MECHANISMS

Researchers have made significant advances in understanding brown rot degradation of wood over the last 30 years, in particular with regard to understanding the components involved in nonenzymatic degradation which initiate brown rot attack. Wood degradation processes are of importance because of the tremendous losses in value to property caused by fungal attack on wood, but also because of potential application of fungal mechanisms in bioremediation systems and bioindustrial processing (1, 2, 3, 4). Early work (5, 6) explored the role of catecholate or phenolate siderophores - iron binding compounds that were known to be produced by bacteria - showing that some basidiomycetous fungi also produced a type of low molecular weight catecholate chelator. In the early and mid-1990's a hypothesis was advanced proposing that these compounds functioned in a manner different than what had

* This is paper 2519 of the Maine Agricultural and Forest Experiment Station. We thank the Wood Utilization Research program at the University of Maine for support of this work. We also appreciate the assistance of Mr. Duan Hui and Ms. Jing Bian in the laboratory.

previously been described in bacteria (5, 7, 8, 9). This research indicated that the fungal biochelators served to reduce metals, particularly iron, outside of the fungal hyphae and that this reduction permitted the Fenton-like chemical reactions that had been shown by Koenigs (1974), Highley (10), Schmidt and Nicholas (11), and others to occur in brown rot processes.

In 1997 details were published on a mechanism mediated by the catecholate chelators produced by the brown rot fungus *Gloeophyllum trabeum* (7) that permitted a single mole of chelator to reduce multiple moles of iron. This phenomenon has not been well described in the free radical literature, nor in the medical literature, where much of the published work on metal sequestration and oxygen radical production is found. However, previous work (12) suggested a greater than 1:1, non-stochiometric, reduction of iron by caffeic acid. More recent work has also shown that catechols reduce iron via oxidation and complete or partial mineralization of the phenolic ring to CO_2 (13). This suggests that orthodihydroxy phenolic groups produced by wood degrading fungi may function in a similar manner. This can be demonstrated (Figure 1˙) using the model chelator 2, 3 dihydroxy benzoic acid (DHBA) which is produced as a siderophore by bacteria and which we have previously employed (14) to mimic the *G. trabeum* (Gt) chelator system in iron reduction and hydroxyl radical production studies. Later work (15) has shown that both 2,5– dimethoxyhydroquinone as well as 4.5- dimethoxycatechol are produced by *G. trabeum*. Kerem et al. (16) also found that this hydroquinone as well as its benzoquinone form were produced by *G. trabeum*. Other low molecular weight compounds are also produced by this fungus (7).

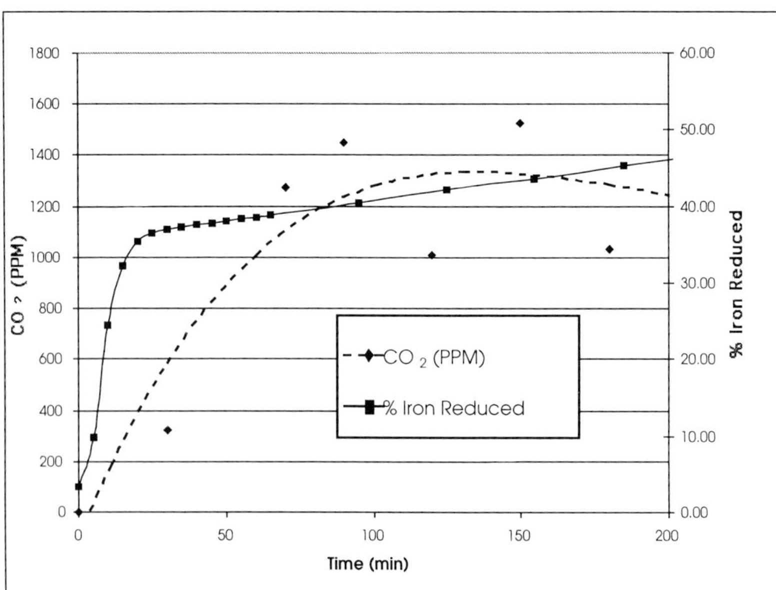

Figure 1. The production of CO_2 concurrent with iron reduction by 2,3 DHBA. $FeCl_3$ (50μM) was reacted with 2, 3 DHBA (10μM) and CO_2 production was monitored using gas chromatography. CO_2 is produced through oxidation and mineralization of the DHBA to reduce iron in a non-stochiometric (greater than 1:1 ratio) manner.

Our preliminary work has shown that superoxide may be generated from molecular oxygen in the process typically described as 'redox cycling' of the orthodihydroxy chelators (17). This term is somewhat of an anachronism however, as the process is now known to occur through the reduction of iron with subsequent autoxidation of ferrous iron to produce superoxide (18). The superoxide may then dismutate to produce limited amounts of hydrogen peroxide. Although this process to produce superoxide has been described with catechols, and the sequestration and reduction of metals by phenolates with orthodihydroxy groups is well known, this chemistry has not been reviewed previously as a mechanism for hydrogen peroxide production in the fungi.

It is clear however, that simple addition of catechol compounds to wood does not promote lignocellulose degradation. This may be because iron autoxidation is a pH mediated reaction (19, 20) and production of oxalate with attendant pH control of the fungal-wood cell wall environment is important. Hydrogen peroxide production through this mechanism appears to be limited at pH levels preferred by the fungi (21), however within the wood cell wall where pH may be as high as 5.5 to 6.0 the hydrogen peroxide produced may help to initiate non-enzymatic degradative processes. In combination with hydrogen peroxide production through a number of enzymatic fungal mechanisms, the pseudo-catalytic mechanism for redox cycling of iron by catechols, as described above, permits a near steady-state production of free radicals to be generated from relatively small quantities of free iron and catecholate chelator. This enhanced production of hydroxyl radical, termed the chelator-mediated Fenton system (CMFS), permits many-fold greater production of free radicals than simple Fenton chemistry alone. Because hydroxyl radicals survive only a matter of micro-seconds in the environment (22, 23) they are capable of diffusing only 4 to 7 molecular diameters. The increase in hydroxyl radical production with CMFS is therefore important as it permits enhanced degradation of various organic substrates (4). Because Fenton generated hydroxyl radicals are not site directed to the structural components of wood, many will react in the microvoid interstitial spaces of the wood cell wall away from cellulosic and lignin structural components. We hypothesize that the continued production of free radicals mediated by fungal chelators is one of the mechanisms that allows brown rot fungi to be so effective in degrading and oxidizing wood cell wall components.

Oxalate production and regulation is also important in the brown rot fungi. Oxalate functions in the control of pH of the fungal environment but also may serve in other ways. It would be undesirable to have free iron in the immediate environment of the fungal cell membrane, and oxalate in insoluble crystalline as well as soluble forms is concentrated in this region. Free transition metals would be readily sequestered by oxalate in this environment. Fungi may also use oxalate to create a pH gradient which would allow iron to be sequestered in the low pH environment surrounding the fungal hyphae but allow phase transfer of the iron to catecholate chelators in the higher pH environment of the wood cell wall (7, 21). [Under high pH alkaline conditions where wood is infiltrated with alkaline materials, hyperproduction of oxalate crystals occurs in *G. trabeum* (24), presumably in an effort to reduce the pH of the fungal environment for metabolic function as well as to regulate pH for decay mechanisms.] Oxalate degradation (hydrolysis) of hemicellulose has been reviewed by Green et al. (25) and Shimada et al. (26, 27). Depolymerization of cellulose by oxalate to mimic the action of brown rot fungi is less certain however (28, 29), as very high concentrations of oxalate and/or high temperatures would need to be involved for this, and the resulting degradation products are not similar to those produced in brown rot degradation.

Site-directed hydroxyl radical production is another factor that must be considered in the understanding of decay processes. Xu and Goodell (21) showed that ferric iron is bound to the wood cell wall with a significantly greater binding capacity (k_b= 0.11 – $5.3x10^{-2}$) compared to ferrous iron (k_b= $1.5x10^{-3}$ – $2.1x10^{-3}$). The binding of iron to cellulose in the presence of the catecholate chelator fraction isolated from *G. trabeum* (Gt chelator) however was similar to that of the ferrous iron. This indicates that Gt chelator was able to sequester iron from cellulose and reduce it to the ferrous oxidation state. Lignin in wood is composed primarily of guaiacyl or syringyl phenylpropane units (30). Because of the lack of adjacent hydroxyl groups on the rings and because most units are linked in polymeric form, few bonding sites for iron are present in the native lignin matrix. (Some extracted lignin compounds would be expected to bind iron, but because these are chemically modified to produce iron binding sites, these are not considered in this discussion.) The binding of ferric but not ferrous iron to cellulose is important as is the lack of iron binding to lignin.

Previous researchers (31, 32) have discussed the importance of reactant binding to DNA prior to effective oxidative damage by oxygen radicals. Because oxygen is ubiquitous in biological systems, oxidative damage to nuclear material would be expected to be quite high in these systems. However DNA damage does not usually occur unless promoted by metals, drugs, or other chemicals which can bind to DNA to promote oxidative reactions within a molecular distance that can be traversed by the active hydroxyl radical. In wood we would expect the same principles to hold. Where iron, or perhaps other reactive transition metals can bind, oxidative reactions will be promoted. Within the wood cell wall this would occur in the cellulose and hemicellulose rich regions. Because the Gt chelator catechols can reduce iron bound at these sites, if hydrogen peroxide is also present, then Fenton reactions will occur to produce hydroxyl radical with consequent oxidative damage to cellulose. This helps to explain why cellulose is selectively depolymerized in brown rot decay; however it does not explain why lignin is demethylated, demethoxylated, or the propyl side chain is cleaved. Hydroxyl radicals have been reported to depolymerize as well as repolymerize lignin (33). The lignin matrix which surrounds the cellulosic portions of the wood cell wall is likely to be within the diffusion limits of hydroxyl radicals produced near sites where iron is bound to cellulose and the lignin may be depolymerized and repolymerized initially in this manner. It has been speculated by Filley et al. (34) that after initiation of cell wall depolymerization, lignin degradation products could potentially participate in chelator-mediated Fenton chemistry as described above.

Our current understanding of the chelator-mediated Fenton system is summarized in Figure 2. This schematic is supported by the data presented above; however, it should be considered a working hypothesis. In previous work (15, 16), a role for quinone reductase was proposed which would allow redox cycling of orthodihydroxy catechols, to allow repeated reduction of iron and thus promote chelator-mediated Fenton reactions. The quinone reductase system would permit the reduction of the quinones produced *de noveau* by the fungus as well as those produced extracellularly in the reduction of iron. Quinone reductase enzymes occur in membrane bound, intracellular, and extracellular forms in the wood degrading fungi (20, 35, 36, 37, 38). It is undesirable to have reactive quinones in near proximity to the fungus and it is likely that these enzymes function near the fungal membrane to limit exposure of the fungus to these moieties. However, for hydroxyl radical production to occur in the wood cell wall, the chelator-mediated Fenton system must be active within the lignocellulose matrix, and quinones would therefore be produced in that matrix. Quinones from catechol reduction of iron would not be produced in the vicinity of the fungal membrane because of iron sequestration by oxalate and because of the low pH of that environment. Quinone reductase enzymes, like all known enzymes, are too large to penetrate into the wood cell wall matrix (39, 40, 41, 42, 43) and they would not be able to directly reduce quinones produced or diffusing into the wood cell wall. It is possible but unlikely that the quinones formed would repeatedly diffuse from the wood cell wall to the lumen void and then back. This suggests that other mechanisms for the iron reduction, as discussed above, likely fulfill this role. Quinone reductases could potentially be indirectly involved in the production of reducing potential.

It is important when examining the roles played by any metabolites produced by fungi in cell wall degradation schemes, to assess proposed mechanisms from a number of viewpoints. Understanding the biochemical nature of enzymatic systems, for example, was the basis for the hypothesis developed on enzymatic degradation of the wood cell wall in white rots and brown rots. The original understanding however did not include the perspective that known enzymes were too large to penetrate the wood cell wall matrix and therefore could not be responsible for some of the degradation patterns observed (44). This is one of the reasons why degradative mechanisms employing enzyme-mediator systems for the white rot fungi and non-enzymatic systems in brown rot fungi have been proposed. It is important therefore, to link chemical and biochemical data to the biological and spatial relationships between fungi and the wood cell wall in the development of hypotheses explaining fungal wood cell wall degradation. A schematic representation (Figure 3) attempts to incorporate the known chemical and spatial relationships for the CMFS in brown rot fungi as discussed in the sections above.

2. ARE THERE DIFFERENCES BETWEEN LOW MOLECULAR WEIGHT METABOLITES AND DEGRADATIVE SYSTEMS IN THE BROWN ROT AND WHITE ROT FUNGI?

Although brown and white rot fungi both are capable of lowering the pH of their immediate environment, in general it is the brown rot fungi that are associated with the more drastic pH reductions (20, 25, 45, 46). Brown rot fungi, such as *Postia placenta* and *G. trabeum*, can be shown to decrease the overall pH values of degrading wood from initial values of 6 or above to overall values in the 3-4 range (47). These are average values taken from extracted or ground wood samples and localized pH values in the immediate environment of the fungal hyphae would be expected to be much lower. These changes occur fairly rapidly as the wood degrades from zero to 25% weight loss and continue more gradually as the wood continues to degrade to higher weight losses. (47)

Oxalate has been implicated in many facets of the degradation process including, as noted earlier, the solubilization and subsequent cyclic reduction of iron. Oxalate has also been implicated indirectly in the regulation of lignin degradation by the white rot fungi (48) In addition, oxalate may play a role in the development of pH gradients by the fungi. Oxalate can be shown to be produced by both white rot and brown rot fungi and calcium oxalate crystal formation has been observed for both types of fungus (49). Detailed electron microscopy studies from multiple isolates of two brown rot fungi (*G. trabeum* and *Fomitopsis pinicola*) and one white rot (*Trichaptum abietinum*) showed extensive isolate and environmentally induced variability in calcium oxalate crystal formation, suggesting that greater crystal formation/accumulation occurs in the early decay stages by the brown rot (49). This implies that the brown rot fungi may accumulate insoluble forms of oxalate prior to as well as during the regulation of their environment for decay initiation. Dissolution of the crystals, as occurs with *G. trabeum* with decay progression, may allow the fungus to control both pH as well as metal sequestration. The pH gradient set up by the fungus may change spatially within the wood cell wall as decay progresses to optimize the chemistry involved in degradative processes.

Both brown and white rot fungi have also been shown to produce biochelators, although the pattern of production and properties of the biochelators produced appears to differ (37). Studies examining liquid cultures of three representative genera each of white rot, brown rot and non-degrading wood inhabiting fungi showed production of iron-binding compounds by all the genera tested. This is to be expected because of the universal need for microorganisms to have a mechanism to sequester iron from the environment for metabolic functions. The nature and chemical composition of these iron-binding compounds however, was variable. Among the fungi evaluated the brown rot fungi displayed higher iron-reducing activity than filtrates from the white rot fungi (37). All species tested produced hydroxamic acid type iron chelators (50) . Phenolate (or catecholate) chelator production however, was very low in the white rot *P. chrysosporium* which produced more hydroxamic acid type chelator than phenolate type chelator. The reverse was seen for the brown rot *G. trabeum* which produced more of the phenolate type compounds associated with iron reduction (37).

Just as some non-wood degrading deuteromycete fungi (molds) such as *Trichoderma* spp. will produce cellulolytic enzymes, mold fungi that have been examined can also produce low molecular weight metal binding chelators (37, 51). The chemical nature of these biochelators has not been thoroughly explored, however it is important to recognize that environmental control of pH, as well as the production of oxalate, hydrogen peroxide, and other metabolites all may play a role in non-enzymatic wood degradation mechanisms. The absence or repression of any one key component may inhibit site directed oxidation of lignocellulose in the wood cell wall.

In the presence of cellulose, cellobiose dehydrogenase (CDH) and quinone reducing activity was found in *P. chrysosporium*, *G. trabeum* and the non-decay fungus *Trichoderma viride* (37). CDH is wide spread in the fungi and has been previously reported in numerous white rot fungi including the brown rot fungus *Coniophora puteana* (20) and a thermophillic soft rot. CDH may function in the generation of hydroxyl radicals (52) which could initiate Fenton chemistry and inhibit lignin repolymerization in white rot degradation (53). The overall role CDH and reductase enzymes may play in non-enzymatic biodegradative systems has still not been clearly elucidated.

3. POTENTIAL APPLICATIONS OF A CHELATOR-MEDIATED FENTON SYSTEM?

Although some of the mechanisms discussed above are still hypothetical, the chemistry discussed to support the CMFS hypothesis may have potential application, particularly in the waste remediation field. Previous work by Krishnamurthy (54) and Goodell et al. (7) has shown that pentachlorophenol can be mineralized by the CMFS using the low molecular weight phenolate fraction from *G. trabeum* (Gt chelator). Other workers (55, 56) had previously shown that Fenton chemistry could be employed in the breakdown of xenobiotics in soils and waters. Oxidation of pentachlorophenol using the CMFS however was enhanced over that of Fenton chemistry alone.

Free radical reactions using the CMFS have also been employed in the treatment of pulp mill effluents (57). Under pH 4.0 conditions, using 2,3 DHBA in Fenton reactions these researchers were able to decrease AOX and reduce toxicity of wastewater samples beyond that possible with the Fenton system alone. Other work in the water quality field using CMFS has centered on the decolorization of dyes. Goodell and Qian (58) showed that both stillbene and azo-type dyes could be rapidly decolorized by CMFS using 2,3 DHBA. Other dye systems and mixed dyes were more recalcitrant to breakdown however. In all cases the CMFS performed better, reducing color more rapidly and more efficiently, with up to 10 times greater reduction in absorbance, than Fenton chemistry reactions alone.

More limited work has been attempted using brown rot derived degradative systems for the treatment of lignocellulose pulp. Because brown rot fungi are known primarily for their ability to preferentially depolymerize cellulose while adding color to the wood, and both characteristics are undesirable in pulp and paper production, little interest has been shown in brown rot application in this field. However if hydroxyl radical production can be controlled or site-directed to limit the amount of cellulose degradation which occurs, then limited application in the pulp and paper field may be possible. Qian and Goodell (59) showed that removal of ink particles in paper recycling processes can be enhanced using the CMFS. In this work, repulped laser printed copy paper was treated with a standard sodium hydroxide treatment as well as with the CMFS and NaOH plus CMFS. Deinking was significantly improved after floatation when either treatment using the CMFS process was used, with TAPPI dirt count reduced by greater than 50% over the NaOH treatment. At the same time freeness of the pulp increased, and tear strength, tensile index, and fiber length were unchanged with the CMFS treatments. Brightness of the pulp however was reduced from 80.2 % ISO to 77.5% ISO. As the brightness reduction was attributed to the production of oxidized phenolics and the addition of iron in the CMFS process, brightness recovery with an additional oxalate bleaching step was proposed. Deinking improvement and freeness increase was attributed to cleavage of fibrils attached to ink particles, allowing these particles to be more easily removed in the floatation step.

Additional testing of the CMFS system on hardwood kraft pulp has not shown favorable results to date (60). In this work the test pulp was first beaten to produce pulp with three different freeness levels and fines content. The density of all three pulps increased after CMFS treatment but other properties including viscosity, tear strength, tensile index, and zero span breaking length all declined or remained unchanged. Although the tests were not promising for hardwood pulp, using this system in tests on coarse softwood pulp was suggested as the CMFS may improve properties of this type of pulp in some applications.

4. CONCLUSIONS

The non-enzymatic mechanisms employed by wood degrading fungi have not been fully elucidated. Clearly however, low molecular weight metabolites are produced by both brown and white rot fungi and may participate in radical generating reactions within the wood cell matrix. The chemical nature of these metabolites and their regulation will vary between brown and white rots. We have focused here primarily upon the brown rot fungi and have presented our working hypothesis for a chelator driven, iron-dependent mechanism for hydroxyl and other radical generation. Such a mechanism would have implications for developing an understanding of the chemical processes involved in brown rot degradation. In addition, this mechanism is of potential interest in the context of waste remediation, including the remediation of wood preservatives, dyes and pulp mill effluents. Possible application of a modified CMFS treatment for deinking or freeness improvement has also been investigated. The ultimate goal of this research is to enhance our understanding of the basic processes involved in biodegradation and the ultimate application of this knowledge to the improved utilization and processing of wood and wood fiber.

REFERENCES

1. W. J. Bao, S. N. Usha and V. Renganathan. 1993. Purification and characterization of cellobiose dehydrogenase, a novel extracellular hemoflavo-enzyme from the white-rot fungus *Phanerochaete chrysosporium*. Arch. Biochem. Biophys. 300:705-713.
2. D. P. Barr, and S. D. Aust, S. D. 1994. Mechanisms white rot fungi use to degrade pollutants. Environ. Sci. Technol. 28:78A-87A.
3. J. Brock, S. Rieble and M. H. Gold. 1995. Purification and characterization of a 1,4-benzo quinone reductase from the basidiomycete *Phanerochaete chrysosporium*. Appl. Envior. Micro. 61:3076-3081.
4. E. Cadenas, E. 1995. Mechanisms of oxygen activation and reactive oxygen species de-toxification. In: Ahmad, S. (Ed.), Oxidative Stress and Antioxidant Defences in Biology. Chapman and Hall, International Thompson Publishing Co. N.,Y. pp 1-61.
5. M. D. Cameron and S. D. Aust. 1999. Degradation of chemicals by reactive radicals produced by cellobiose dehydrogenase from *Phanerochaete chrysosporium* Archives of Biochemistry and Biophysics 367:115-121.
6. J. M. Carberry D. K. Sang. 1994. Emerging Technologies in Hazardous Waste Management IV, Tedder and Pohland (Eds.), ACS Symp. Ser. 554. pp. 197-223
7. V. Chandhoke, B. Goodell, J. Jellison and F. Fekete. 1992. Oxidation of 2-keto-4-thio-methylbutyric acid (KTBA) by iron-binding compounds produced by the wood-decaying fungus *Gloeophyllum trabeum*. FEMS Microbiology Letters 90:263-266.
8. J. Connolly and J. Jellison. 1996. Scanning electron microscopic examination of calcium oxalate crystal formation in six degradative fungi. Scanning Microscopy 10:385-400.
9. E. B. Cowling. 1961. Comparative biochemistry of the decay of sweet-gum by white- and brown-rot fungi. USDA Tech. Bull. 1258.
10. G. Daniel, T. Nilsson and B. Petterson. 1989. Intra- and extracellular localization of lignin peroxidase during the degradation of solid wood and wood fragments by *Phanerochaete chrysosporium* by using transmission electron microscopy and immuno-gold labeling. Appl. Environ. Microbiol. 55:871-881.
11. S. Deiana, C. Gessa, B. Manunza, M. Marchetti, and M. Usai. 1992. Mechanism and stoichiometry of the redox reaction between iron (III) and caffeic acid. Plant and Soil 145:287-294.
12. E. Dreosti. 1991. Free radical pathology and the genome. In Dreosti, I. E. (Ed.), Trace Elemants, Micronutrients, and Free Radicals. Humana Press. Clifton, NJ, pp. 149-170.
13. F. A. Fekete, V. Chandhoke and J. Jellison, J. 1989. Iron-binding compounds produced by wood-decaying basidiomycetes. Appl. Environ. Microbiol. 55:2720-2722.
14. T. R. Filley, P. G. Hatcher, W. C. Shortle, and R. T. Praseuth. 2001 The application of ^{13}C-labeled tetramethylammonium hydroxide (^{13}C-TMAH) thermochemolysis to the study of the fungal degradation of wood. Organic Geochemistry (in press).
15. T.R. Filley, G.D. Cody, B. Goodell, J. Jellison, C. Noser and A. Ostrofsky. 2001. Microbial production of phenolic-rich lignin residues in coarse woody debris: A laboratory

degradation of red spruce wood by two common brown rot fungi (*Gloeophyllum trabeum* and *Postia placenta*). Organic Geochemistry. *In Press.*

16. D. S. Flournoy, T. K. Kirk and T. Highley. 1991. Wood decay by brown-rot fungi: changes in pore structure and cell wall volume. Holzforschung 45: 383-388.
17. B. Goodell. 1989. The potential of biotechnology applications in the forest products industry. In: Schniewind, A. (Ed.), Advances in Materials Science and Engineering, Encyclopedia of Wood and Wood-Based Materials. Pergamon Press, Oxford, 354 pp.
18. B. Goodell and J. Jellison. 1998. The role of biological metal chelators in wood degradation and in xenobiotic degradation. In: Bruce, A. and Palfreyman, J. [Eds.]. Forest Products Biotechnology . Taylor and Francis, London. pp. 235- 250.
19. B. Goodell, J. Jellison, J. Liu and S. Krishnamurthy. Degradation and protection of organic compounds mediated by low molecular weight chelators. April 4, 2000. US Patent #6,046,375.
20. B. Goodell, J. Jellison, J., Liu, G. Daniel, A. Paszczynski, F. Fekete , S. Krishnamurthy, L. Jun and G. Xu. 1997a. Low molecular weight chelators and phenolic compounds isolated from wood decay fungi and their role in the fungal biodegradation of wood. Journal of Biotechnology. 53: 133-162.
21. B. Goodell, J. Jellison, Q. Yuhui and G. Daniel. 1997b. Redox cycling chelators isolated from *Gloeophyllum trabeum* and their effect on wood fibers. TAPPI proceedings 1997 Biological Sciences Smposium pp 463-471.
22. B. Goodell, G. Daniel, J. Liu, L. Mott and R. Frank. 1997c Decay resistance and microscopic analysis of wood-cement composites. Forest Products Journal 47:75-80.
23. B. Goodell and Y. Qian. 2001. Decolorization and degradation of dyes using a mediated Fenton system. In review.
24. F. Green, M. J. Larsen, J. E. Winandy and T. L. Highley. 1991. Role of oxalic acid in incipient brown-rot decay. Mater. Org. 26, 191-213.
25. Halliwell and J. M. C. Gutteridge. 1990. Role of free radicals and catalytic metal ions in human disease: An overview. Methods Enzymol. 186, 1-85.
26. D. v. d. Helm and G. Winkelmann. 1994. Hydroxamates and polycarboxylates as iron transport agents (siderophores) in fungi. In: Winkelmann, G., and Winge, D. R. (Eds.), Metal Ions in Fungi. Marcel Dekker, Inc. NY, pp. 39-98.
27. T. Highley. 1980. Degradation of cellulose by *Poria placenta* in the presence of compounds that affect hydrogen peroxide. Mater. Org. 15: 81-90.
28. S. M. Hyde P. M. and Wood. 1997. A mechanism for production of hydroxyl radicals by the brown rot fungus *Coniophora puteana*: Fe(III) reduction by cellobiose dehydrogenase and Fe(II) oxidation at a distance from the hyphae. Microbiology. 143:259-266.
29. J. Jellison, V. Chandhoke, B. Goodell, and F. Fekete. 1991. The isolation and immunology of iron-binding compounds produced by *Gloeophyllum trabeum.* Appl. Microbiol. Biotechnol. 35: 805-809.
30. Jellison, J. Connolly, B. Goodell, B. Doyle, B. Illman, F. Fekete, and A. Ostrofsky, A. 1997. The role of cations in the biodegradation of wood by the brown rot fungi. Int. Biodegrad. Biodet. 39:165-179.
31. J. Jellison, K.C. Smith and W.T. Shortle. 1992. Cation analysis of wood degraded by white rot and brown rot fungi. International Research Group on Wood Preservation Box 5607, S-11486. Stockholm, Sweden. Series Document IRG/WP/1552-92
32. Z. Kerem, K. A. Jensen and K. E. Hammel. 1999. Biodegradative mechanism of the brown-rot basidiomycete *Gloeophyllum trabeum*: evidence for an extracellular hydroquinone-driven fenton reaction. FEBS 446:49-54.

33. Y. S. Kim, B. Goodell and J. Jellison. 1991. Immuno-electron microscopic localization of extracellular metabolites in spruce wood decayed by brown-rot fungus *Postia placenta*. Holzforsch. 45, 389-393.
34. Y. S. Kim, B. Goodell and J. Jellison. 1993. Immunogold labelling of extracellular metabolites from the white-rot fungus *Trametes versicolor*. Holzforsch. 47, 25-28.
35. T. K. Kirk, T. Highley, R. Ibach and M. D. Mozuch. 1989. Identification of terminal structures in cellulose degraded by the brown-rot fungus *Postia placenta*. International Research Group on Wood Preservation Series. Box 5607, S-114 86. Stockholm, Sweden. Document IRG/WP 1389.
36. T. K. Kirk, R. Ibach, M. D. Mozuch, A. H. Conner and T. Highley. 1991. Characteristics of cotton cellulose depolymerized by a brown-rot fungus, by acid, or by chemical oxidants. Holzforsch. 45, 239-244J.
37. W. Koenigs. 1974. Hydrogen peroxide and iron: A proposed system for decomposition of wood by brown-rot basidiomycetes. Wood Fiber 6: 66-80.
38. S. Krishnamurthy. 1997. Degradation of chlorinated xenobiotics, mediated by iron binding biochelators secreted by the wood rot fungus *Gloeophyllum trabeum*. MS Thesis. Graduate School. University of Maine. USA 78 pp.
39. P. Leanderson and C. Tagesson.1993. Mineral fibers, cigarette smoke, and oxidative DNA damage. In: Halliwell, B. and Aruoma, O. I. (Eds.), DNA and Free Radicals. Ellis Horwood Ltd, Simon and Schuster, West Sussex, England, pp. 293 - 314.
40. J. Lu, B. Goodell, J. Liu, A. Enoki, J. Jellison, H. Tanaka and F. Fekete. 1994. The role of oxygen and oxygen radicals in the one-electron oxidative reactions mediated by low-molecular weight chelators isolated from *Gloeophyllum trabeum*. International Working Group on Wood Preservation Series. Box 5607, S-11486. Stockholm, Sweden. Document IRG/WP/1086.
41. D. B. Ma, T. Hattori, Y. Akamatsu, M. Adachi and M. Shimada. 1992. A novel equation for the noncompetitive inhibition of the Li-P catalyzed reactions by oxalic acid. In: Fifth International Conference on Biotechnology in the Pulp and Paper Industry, May 27-30, 1992. Kuwahara M and Shimada M (eds.) Uni Publishers, Tokyo, Japan pp. 321-326.
42. D. M. Miller, G. R. Buettner, and S. D. Aust. 1990. Transition metals as catalysts of "Autoxidation" reactions. Free Racical Biology & Medicine. 8: 95-108.
43. J. W. Palfreyman, E. M. Phillips and H. J. Staines. 1996. The effect of calcium ion concentration on the growth and decay capacity of *Serpula lacrymans* and *Coniophora puteana*. Holzforschung 50:3-8.
44. A. Paszczynski and R. Crawford. 1995. Potential for bioiremediation of xenobiotic compounds by the white-rot fungus *Phanerochaete chrysosporium*. Biotechnol. Prog. 11:368-379.
45. A. Paszczynski, R. Crawford, D. Funk and B. Goodell. 1999. De Novo synthesis of 1,5-dimethoxy catechol by the brown-rot fungus *Gloeophyllum trabeum*. Applied and Environmental Microbiology. 65:674-679.
46. J. Pignatello. 1993. Degradation of pesticides by ferric reagents and peroxide in the presence of light. U. S. Patent 5,232,484. US Patent and Trademark Office.
47. J. Pracht. J. Boenigk, M. Isen-Schröter, F. Keppler, and H. F. Schöler. 2001. Abiotic Fe(III) induced minerizalization of phenolic substances. Chemosphere. In press.
48. C. Pueyo and R. R. Arizo. 1993. Role of reactive oxygen species in the mutagenicity of complex mixtures of plant origin. In : Halliwell, B. and Aruoma, O.I. (Eds.) DNA and Free Radicals. Ellis Horwood, Chichester, UK. Pp. 275-291.

49. Y. Qian and B. Goodell. 2001. Deinking of Laser Printed Copy Paper with Mediated Fenton System. In Review.
50. Y. Qian, B. Goodell, J. Genco. 2001. The Effect of a ChelatorMediated Fenton System on the Fiber and Paper Properties of Hardwood Kraft Pulp. In review.
51. W. Qi and J. Jellison. 2000. Chelator production and cellobiose dehydrogenase activity of wood inhabiting fungi. International Research Group on Wood Preservation Box 5607, S-11486. Stockholm, Sweden. Series Document 00-10364.
52. J. Rodriguez, D. Contreras, C. Parra, J. Freer, J. Baeza, and N. Duran. 1999. Pulp mill effluent treatment by fenton-type reactions catalyzed by iron complexes. Elsevier Science Ltd. International Association of Water Quality (IAWQ). pp 351-355.
53. C. J. Schmidt, B. K. Whitten and D. D. Nicholas. 1981. A proposed role for oxalic acid in nonenzymatic wood decay by brown-rot fungi. Proc. Am. Wood-Preserv. Assoc. 77: 157-164.
54. M. Shimada, D. B. Ma and Y. Akamatsu. 1994. A proposed role of oxalic acid in wood decay systems of wood-rotting basidiomycetes. FEMS Microbiol. Rev. 13l, 285.
55. M. Shimada, Y. Akamatsu, D. B. Ma and M. Takahashi. 1992. New biochemical aspects of oxalic acid production and decomposition by wood destroying fungi. In: Kuwahara. M., and Shimada, M. (Eds.), Biotechnology in the Pulp and Paper Industry, 5th International Conference. Kyoto, Japan. May 27-30. UNI Publishers Co. Ltd. Tokyo Japan, pp. 273-278
56. E. Sjöström.1993. In: Wood chemistry: Fundamental and applications. Academic press. 2nd edition, pp. 81-83.
57. E. Srebotnik and K. Messner. 1991. Immunoelectron microscoical study of the porosity of brown-rot degraded wood. Holzforsch. 45:95-101.
58. U. Srinivasan, T. L. Highley and A. Bruce. 1995. The role of siderophore production in the biological control of wood decay fungi by *Trichoderma spp.* In Biodeterioration and Biodegradation 9:226-230.
59. D. M. Stearns , L. J. Kennedy, K. D. Courtney, P. H. Giangrande , L. S. Phieffer and K. E. Wetterhahn . 1995. Reduction of chromium(VI) by ascorbate leads to chromium-DNA binding and DNA strand breaks in vitro. Biochemistry 34(3):910-9
60. J. D. Stahl, S.D. Rasmussen and S. D. Aust. 1995. Reduction of quinines and radicals by a plasma membrane redox system of *Phanerochaete chrysosporium.* Arch. Biochem. Biophys. 322:221-227.
61. U. Temp and C. Eggert. 1999. Novel interaction between laccase and cellobiose dehydrogenase during pigment synthesis in the white rot fungus *Pycnoporus cinnabarinus.* Applied and Envir. Micro. 65:389-395.
62. G. Xu and B. Goodell. 2001.Mechanisms of wood degradation by brown-rot fungi: Chelator-mediated cellulose degradation and binding of iron by cellulose. J. Biotechnol. 87:43-57.

Biotechnology in the Pulp and Paper Industry
L. Viikari and R. Lantto (Editors)

Dislocations and balloon swelling in spruce kraft pulp fibres – Effect of cellulases, xylanase and laccase/HBT

Paul Ander

WURC, Wood Ultrastructure Research Centre, Dept of Wood Science,
Swedish University of Agricultural Sciences, PO Box 7008, SE-75007 Uppsala, Sweden.
E-mail: paul.ander@trv.slu.se

The effect of stirring, cellulase, xylanase and laccase/HBT on the formation of dislocations and fibre ballooning in spruce kraft pulp fibres was studied using polarised light microscopy. Stirring in small flasks under non-shearing conditions did not increase the number of dislocations, while stirring under shear conditions disintegrated the fibres and consequently lead to the disappearance of inherent dislocations. Stirring with HCl at pH 1.8 at 80ºC did not cleave fibres, while a mixture of endo- and exoglucanase cleaved fibres at dislocations. Xylanase treatment resulted in delamination of dislocations. Laccase/HBT did not affect dislocations or fibre appearance. Phosphoric acid was found to be an excellent agent for inducing fibre ballooning. Fibres made from chips cut with blunt steel, bleached pulps or cellulase treated pulps gave rapid ballooning and then further swelling and dissolution, dependending on the severity of fibre degradation. Pulp fibres with 3-4% lignin often gave stable balloons while pulp fibres with 10% lignin did not give balloons at all.

1. INTRODUCTION

Deformations in pulp fibres have been called nodes, dislocations, slip planes, micro-compressions, minute compression failure, misaligned areas or zones. Although they make fibres more flexible they are also points for chemical, mechanical and enzymatic attack which can decrease paper strength (Allison et al. 1998; Gurnagul et al. 1992; Hartler 1995). It is therefore of importance to study the properties and appearance of fibre deformations (Nyholm et al. 2001). The objective of this paper is not to clarify whether dislocations are good or bad for pulp quality, only to show some features of dislocations and some related phenomena such as fibre balloon swelling.

Deformations, here called dislocations, appear when there is a change in microfibril alignment as a result of microfibril compression. Dislocations are most easily studied using polarised light microscopy, while scanning (SEM) and transmission electron microscopy (TEM) may be used for more detailed investigations on the involved different cell wall layers in dislocations.

Due to the less ordered or exposed cellulose in dislocations they are more readily attacked by cellulases. During kraft cooking hemicellulose can be reprecipitated on pulp fibre surfaces. Remaining and reprecipitated hemicellulose in and on fibres may also be affected by xylanases giving a change in fibre dislocation structure. The surface of spruce kraft pulp fibres contain high concentrations of lignin, and should therefore be possible to investigate using the laccase/hydroxybenzotriazole system reported to degrade lignin in unbleached pulp fibres. In this paper the connection between dislocations and balloon swelling in phosporic

with HBT gave no absorbance at 408 nm. The laccase/HBT treated fibres were more brownish than control fibres. Before sampling for dislocation and balloon studies the fibres were washed with water.

3. RESULTS

A dislocation is a structure easily seen in polarised light microscopy, and is the result of a localised change or distorsion of the microfibrils in the fibre cell wall. Their formation is often due to a compressive stress induced parallel to the fibre direction. In *Figure 1* a slip plane and a minute compression failure is shown as discussed by Wardrop and Dadswell (1947) and Keith and Côté (1968). This slip plane is very similar to a small dislocation and traverses at least half the fibre width through both S1 and S2 at an angle of about 70° to the fibre axis. The minute compression failure forms a larger dislocation. In reality there is a gliding scale of the dislocation size as is seen in *Figure 2,* which shows dislocations in a typical spruce fibre. Several dislocations may also form a dislocated area which can be as long as the fibre width or even longer.

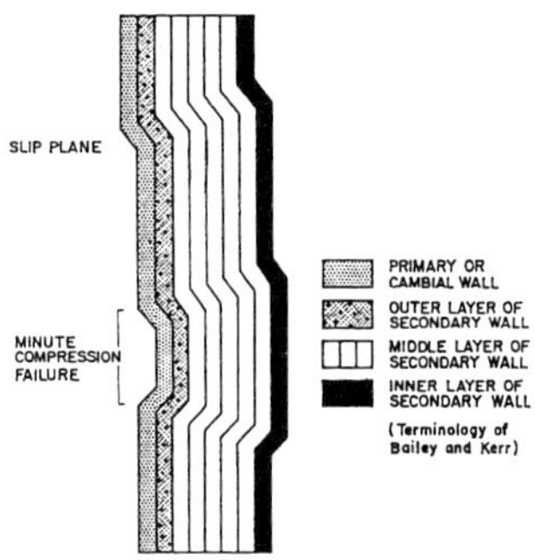

Figure 1. Slip plane and minute compression failure (Wardrop and Dadswell 1947; Keith and Côté 1968).

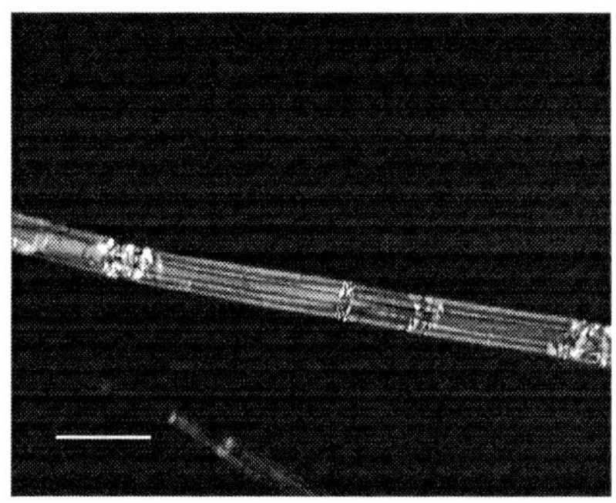

Figure 2. Dislocations of different sizes in a spruce pulp fibre (WU5) visualized under polarised light. Scale bar 50 µm.

3.1. Effect of cellulases

In order to study the effects of cellulases on fibre dislocations, spruce fibres were initially (Nyholm and Ander 1999), incubated with a mixture of endoglucanase (EG1) and exoglucanase (CBH1). CBH alone did not release sugar from the fibres, while a mixture of EG1+CBH1 resulted in a clear synergistic effect and sugar release. Thus kinetic studies on different spruce pulp fibres using less than 5 mg pulp is possible with the BCA method. Spruce pulp fibres made from chips cut with sharp or blunt steel were also tested for sugar release using a mixture of EG1+CBH1 (Ander and Nyholm 2000). Although it was observed that pulp fibres made from chips cut with blunt steel had slightly more dislocations, this did not result in greater sugar release from "blunt steel pulp". Probably the major cellulase action was on the fibre surfaces without visible dislocations (see also 3.3). It was found that the lignin content strongly regulated sugar release. Thus a high lignin content on the fibre surface restricts the action of the cellulases and less sugar release is obtained.

3.2. Stirring effect on dislocations

In the investigations on the structure and importance of dislocations it was of great interest to study the mechanism of their induction. Therefore, some experiments on the effect of different stirring conditions were performed. The pulps WU1 and WU6 were chosen since they had relatively few natural dislocations. The pulp fibres were incubated under non-shearing conditions in Erlenmeyer flasks with 40 ml water and in test tubes with a small inner diameter to achieve shearing conditions with 4 ml water.

Non-shearing conditions: The number of dislocations seen under polarised light already after 5 min did not increase after longer incubation times. For WU1, the fibres were unaffected after 1h (*Figure 3A*). After 4h swellings appeared, and after 23h the fibres were disintegrated with the dislocations almost disappearing. During this change in fibre structure, the red/blue latewood fibres (as seen under polarised light without staining) transferred into more grey fibres indicating a change in the polarised pattern. The fibres of WU6 changed more slowly, and almost no effect of the stirring was observed. Many dislocations remained even after 23h.

Shearing conditions: For WU1, swellings and disintegrations were seen already after 1h (*Figure 3B*). Simultaneously, dislocations began to disappear, and after 23h the fibres were totally disintegrated. Again, WU6 was more resistant to the shear forces. During stirring an inner cord of unaffected twisted fibres were formed in WU6. The chemical composition of WU1 (conventional kraft) and WU6 (isothermal cooking) were similar (Table 1). The above mentioned swellings are not the same as balloon swelling (see Ch. 4).

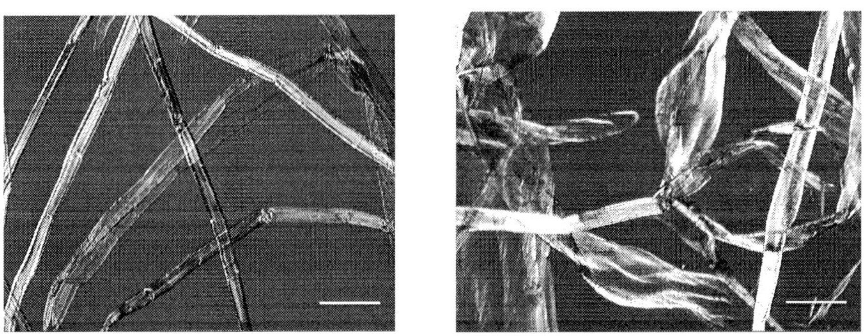

A. *B.*

Figures 3A-B. Stirring of WU1 after 1h under non-shearing (A, non-affected fibres with dislocations) and shearing conditions (B, many partly disintegrated fibres without dislocations). Polarised light microscopy. Scale bars 50 μm.

3.3. Stirring and influence of HCl and Cellulase

These experiments were performed in order to compare pulps made from chips cut with sharp or blunt steel regarding the influence of HCl and cellulase treatments. Fully bleached pulps (pulps E & F) were chosen since it was thought that bleached pulps would show a stronger effect. Pulps WU1 and WU4 were also used.

Treatment of pulps E and F with HCl at 80°C and stirring for 4h, did not give more dislocations, shorter fibres or more sugar release when compared to controls in water slightly stirred at room temperature. The low consistency 0.25% may be one reason for this (during the conditions used more than 1% consistency could not be tested). In the industry or in special mixers with 9% consistency a significant increase in dislocation number/mm fibre length from about 40 to 49 have been obtained (Allison et al. 1998; Ellis et al. 1997).

As a result of dislocation cleavage, treatment of the fibres with EG1+CBH1 gave a decrease in fibre length from 2-3 mm to 0.5-1 mm as measured in the FibreMaster. Dislocations became more apparent and increased slightly in number. Pulp F "blunt steel fibres" had more cut fibres than pulp E "sharp steel fibres". The kink value did not increase probably since cellulase cleavage ocurred just in the kinks.

Cellulase treatment (EG2+CBH1, see 2.7) of WU1 and WU4 is shown in *Figure 4*. This treatment gave many shorter fibre fragments which rapidly sedimented in the test tube and such severed fibres are seen in Fig. 4A. In Fig. 4B, a grey earlywood fibre probably just about to be cleaved by the cellulases is shown, whereas in Fig. 4C, an unevenly cleaved fibre is apparent. At close inspection of the earlywood fibres in Figures 4B&C, a mosaic pattern is clearly visible indicating a surface erosion by the cellulases. Thus they are acting not only on dislocations but over the whole fibre surface. Figure 4D shows a nicely cut WU4 fibre.

A.

B.

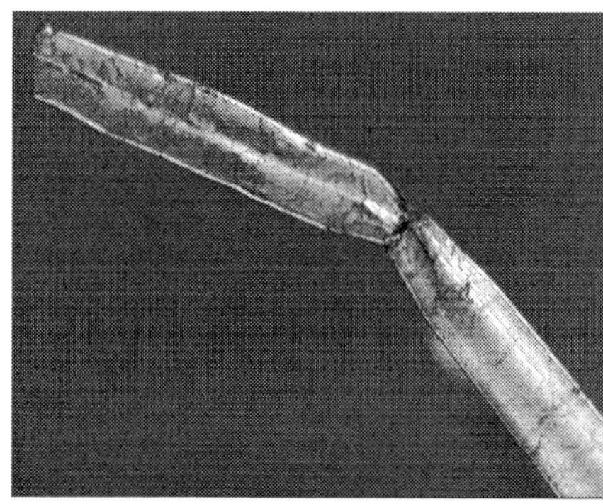

C.

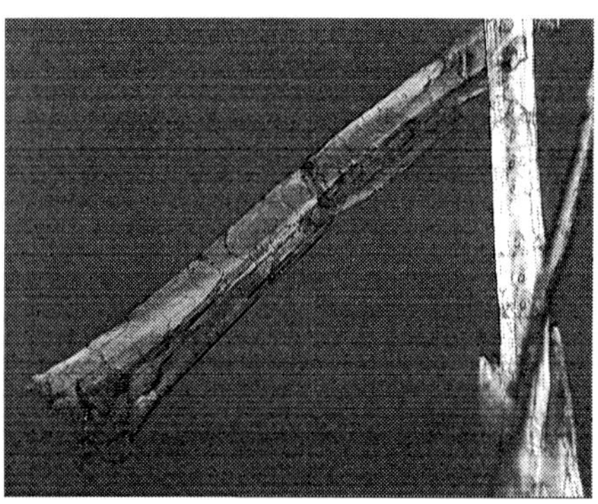

D.

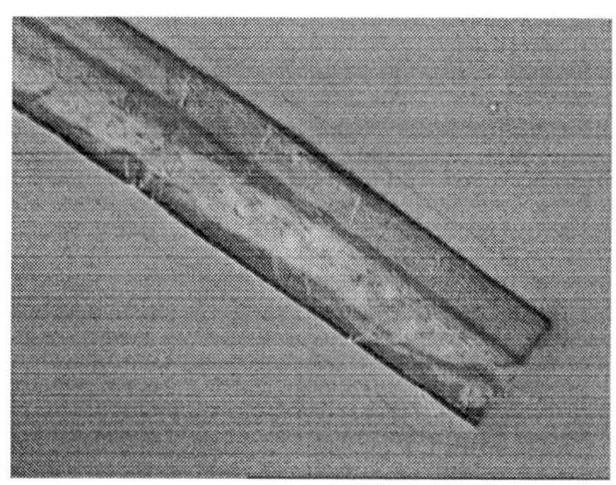

Figure 4A-D. Cleavage and degradation of pulp fibres by cellulase (EG2+CBH1). Scale bars: A-C 50 μm, D 25 μm.

3.4. Effect of xylanase

Pulps WU1, WU4, WU5 and pulp A with slightly different chemical composition were chosen and incubated under non-shearing conditions for 4 and 22h with endoxylanase. After xylanase treatment the fibres did not sediment faster than the water control and did not change much regarding dislocation structure. Some delamination was apparent as shown in *Figures 5A-B*. Such an effect of xylanase was earlier shown by Ander et al. (1996). The fibres in *Fig. 5B* contain many smaller dislocations/slip planes, which could be induced as a result of xylanase treatment.

3.5. Effect of laccase/HBT

Pulps WU4, WU6, WU10 (high lignin) and pulp A were incubated with or without oxygenation at high and low laccase activity in the presence of 10 mM HBT to observe the possible effect on dislocations. Since brown colouration started after 30 min, the HBT was oxidized by the laccase, and probably modifying lignin (Ander and Messner 1998). Despite this action on the lignin structure no visible change in dislocation pattern was obtained. One explanation may be that lignin is not present in dislocations or that NaOH-extraction

A. *B.*

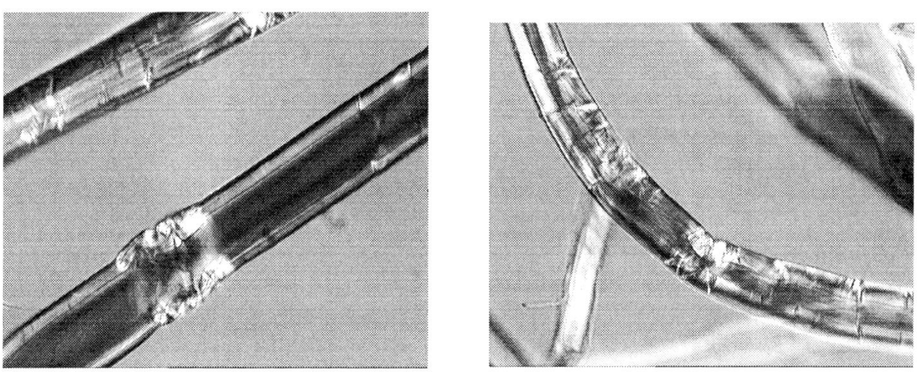

Figure 5. Part A. Fibre WU4 with a very large dislocation and fibre wall delamination.
Part B. Fibre WU1 with small and large dislocations inside the curve region. Xylanase 4h, objective 40x. Scale bars 25 μm.

of the fibres after the laccase reaction was not performed. The preferential staining with the lignin stain iodine/sulphuric acid in dislocations as reported by Robinson (1920) indicates that laccase/HBT should affect dislocations. The lignin modification performed here may be limited since Call and Mücke (1997) stated that laccase/HBT treatments should be done at 5 atmospheres of oxygen pressure.

4. FIBRE BALLOONING

Balloon swelling may be induced by cupriammoniumhydroxide, cupriethylenediamine, Fe(III)-tartrate/NaOH, LiCl/DMAc, phosphoric acid and other solvents or chemicals (Ander and Nyholm 2000; Hortling et al. 2001; Jayme and Harders-Steinhäuser 1964; Pionteck et al. 1996; Stawitz and Kage 1959; Steenberg 1947) and is a way to study the ultrastructure of the different cell walls (Nyholm and Daniel, to be published). In our studies, LiCl/DMAc (Westermark and Gustafsson 1994) was first used. This swelling agent is however, water sensitive with 1.5% water decreasing and 3% totally restricting balloon formation. Furthermore LiCl/DMAc remains on the fibre surface after washing and obscures the fibre details in SEM. Therefore, phosphoric acid (Steenberg 1947) was tested and found to be an excellent agent for balloon swelling.

In a number of publications swelling and ballooning of fibres in Fe(III)-tartrate/NaOH was reported (Stawitz and Kage 1959; Jayme and Harders-Steinhäuser 1964). Ballooning starts with a swelling solution penetrating the S1 layer in dislocations and pores (*Figure 6*). Then the S2 swells under ballooning and S1 is rolled off forming restriction rings between the balloons. The S1 microfibrillar angle, which is almost perpendicular to the fibre axis, facilitates this effect. The strong and compressed S1 is often seen as spiral bands around the balloons. Continued swelling and dissolution of the fibres create cigar-formed swellings, and the fibre is more and more dissolved. Often the S3 microfibrils are clearly visible showing a large microfibril angle (*Figures 6 &7*). In *Figure 7* it is also shown that in some fibres, the swelling stops in the balloon state with typical polarised light patterns (cf. Wardrop and Dadswell 1950). *Figure 8* shows fibre balloons as "string of pearls" when the balloons are stable. The cause of this remarkable regularity is unknown although dislocations may be involved.

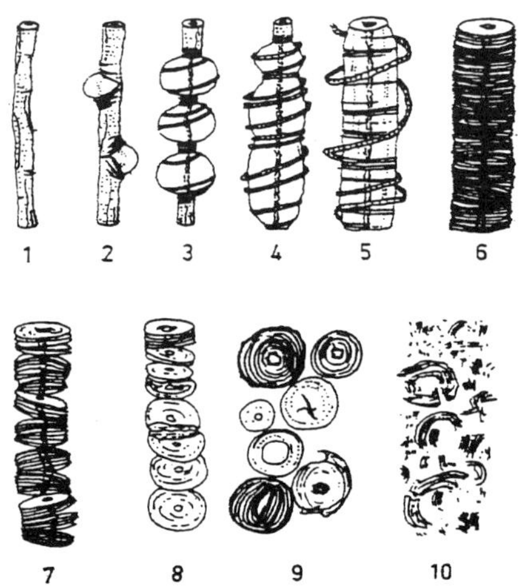

Figure 6. Different steps (1-10) in swelling, ballooning, spiral band formation and disintegration of a carboxymethylated cellulose fibre (Stawitz and Kage 1959; Jayme and Harders-Steinhäuser 1964).

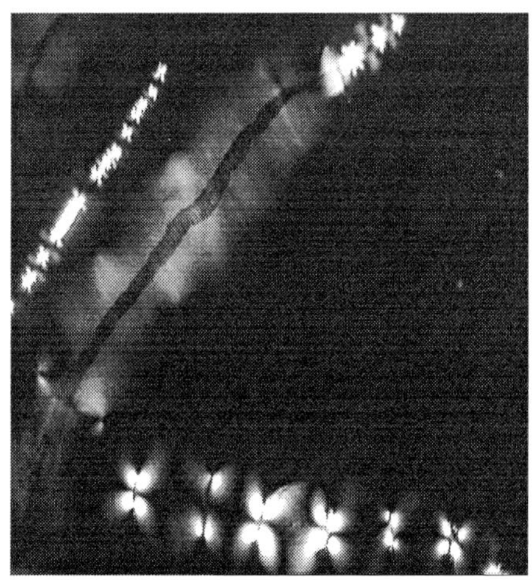

Figure 7. Balloons with typical polarised light pattern and S3 microfibril inside the swollen S2. Pulp fibre WU1 swollen in phosphoric acid. Above right. Scale bar 50 μm.

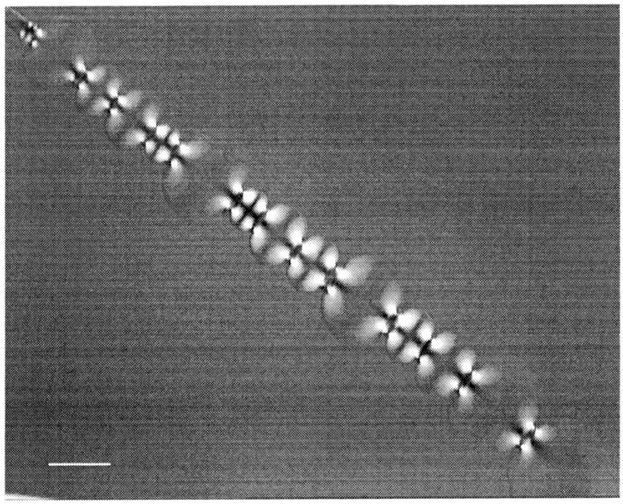

Figure 8. Balloons as "string of pearls" in pulp fibre WU6 swollen in phosphoric acid.
Scale bar 50 μm.

4.1 Balloon testing

During balloon testing it was found that in "blunt steel fibres", ballooning and the following dissolution of fibres was more rapid than in "sharp steel fibres". Cellulase treatment and bleaching also increased ballooning speed and dissolution. Most stable balloons were obtained with lignin-containing pulps (WU1, WU4, WU6 and pulp A), although with a high lignin pulp (WU10 with 10% lignin), almost no ballooning was obtained after 1.5h. Latewood fibres gave more stable balloons, compared with earlywood fibres (unpublished results). Stirring with HCl at 80°C, xylanase or laccase/HBT treatment of fibres did not affect ballooning.

5. DISCUSSION

In stirring experiments under a number of conditions at low consistencies (0.1-0.25%) it was found that laboratory stirring did not induce dislocations, and under shearing conditions they disappeared. During industrial chipping, pulping and bleaching dislocations have been found to increase (Allison et al. 1998; Ellis et al. 1997; Nyholm et al. 2001). Acid sulphite pulping or HCl (gas-phase or liquid) can lead to fibre cleavage (Gurnagul 1992; Hartler 1995) but this was not the case with HCl at 80°C as used here. The effect of dislocations on pulp and fibre quality, such as flexibility and strength, is under way during more industrial conditions or with industrial fibre samples (Ander and Nyholm, unpublished results).

Cellulases cleave fibres at dislocations to give shorter fragments and probably also act on the whole fibre surface indicating that the number of dislocations do not alone determine the release of sugar. Instead lignin content regulates cellulase attack. Xylanase had little effect on the four fibre types investigated, except for delamination in dislocations as described earlier (Ander et al. 1996).

Biotechnology in the Pulp and Paper Industry
L. Viikari and R. Lantto (Editors)

Recent Developments in Biopulping Technology at Madison, WI

G.M. Scott[a], M. Akhtar[b], R.E. Swaney[c], and C.J. Houtman[d]

[a]Faculty of Paper Science and Engineering, State University of New York, College of Environmental Science and Forestry, One Forestry Drive, Syracuse, NY 13224, USA.

[b]BioPulping International, Inc., 2403 Industrial Drive, , Madison, WI, 53713, USA.

[c]Department of Chemical Engineering, University of Wisconsin, 1415 Engineering Drive, Madison, WI 53706, USA.

[d]USDA Forest Service, Forest Products Laboratory, One Gifford Pinchot Drive, Madison, WI-53705, USA

Biopulping is defined as the treatment of wood or other lignocellulosics with a "natural" lignin-degrading fungus prior to pulping. Research consortia made up of the USDA Forest Service, Forest Products Laboratory (FPL) in Madison and the Universities of Wisconsin and Minnesota have evaluated biopulping from a small laboratory scale to a 50-ton semi-commercial scale over the past 12 years. The investigations were supported in part by 23 pulp and paper and related companies and the Energy Center of Wisconsin. The State University of New York, College of Environmental Science and Forestry has also joined as a partner in this research. The research established that biopulping substantially lowers the electrical energy required for mechanical pulping (or increases mill throughput), improves certain strength properties (reducing the need to augment with chemical pulps), and reduces environmental impact. Biopulping also reduces the pitch content of the pulp.

At a pilot scale, we have developed methods for decontamination of wood chips, cooling, and fungal inoculation sequentially in screw conveyers, and controlling temperature and moisture throughout the chip pile. Mill-scale refining of fungus-treated chips gave results similar to those obtained using the laboratory-scale bioreactors. With this information, a complete process flowsheet has been established for the commercial operation of the process. Based on the electrical energy savings and the strength improvements, the process economics looks very attractive. Several independent economic evaluations of biopulping have now been completed by both university and industry economists and engineers and are in agreement. Based on energy savings and reduction in kraft pulp in the final product, substantial savings can be realized. The additional benefits of increased throughput, and reduced pitch content and environmental impact improve the economic picture for this technology even further.

1. INTRODUCTION

Mechanical pulping accounts for about 25% of the wood pulp production in the world today. This volume is expected to increase in the future as raw materials become more difficult to obtain. Mechanical pulping, with its high yield, is viewed as a way to extend these resources. However, mechanical pulping is electrical energy-intensive and yields paper with lower strength than chemical pulps. Kraft pulp is often added to mechanical pulps to impart strength, but it is much more expensive than mechanical pulps. These disadvantages limit the use of mechanical pulps in many grades of paper. Biomechanical pulping, defined as the treatment of lignocellulosic materials with a natural lignin-degrading fungus prior to mechanical pulping, has the potential to ameliorate some of these problems. The reader is referred to several much more comprehensive summaries of the research and development leading to the current status of biomechanical pulping (1-5). Citations to the original papers and patents can be traced through these summaries.

2. PILOT SCALE EQUIPMENT

Once the biological variables were optimized, we turned to engineering process development and scale-up, focusing mainly on chip pretreatment, incubation conditions, and economics. Commercial biopulping will involve treatment of about 200 to 2,000 tons of wood chips per day. The gap between laboratory scale and the much larger commercial scale was bridged through a series of experiments, culminating in two 50-ton trials. The scale-up studies demonstrated that: (a) chips can be decontaminated and inoculated on a continuous basis rather than a batch process as was done on the laboratory scale and (b) the process can be scaled as expected from an engineering standpoint.

A treatment system was built based on two screw conveyers that transported the chips and acted as treatment chambers (Fig. 1). Steam was injected into the first screw conveyer, which heated and decontaminated the wood chip surfaces. A surge bin was located between the two screw conveyers to act as a buffer. From the bottom of the surge bin, a second screw conveyer removed the chips, which were subsequently cooled with filtered air blown into the second conveyer. In the second half of the second conveyer, the inoculum suspension was applied and mixed thoroughly with the chips through the tumbling action in the screw conveyer. From the screw conveyer, the chips fell into a pile for the 2-week incubation. Equipment of this design was used to treat 50 tons of spruce chips (dry weight basis) with *Ceriporiopsis subvermispora* at FPL at a throughput of 2 tons per hour (dry weight basis). During the subsequent 2-week incubation, the chip pile was ventilated with conditioned air to maintain the proper growth temperature (27-32°C) and chip moisture (50-60% on a wet weight basis) throughout the pile. Fig. 2 is a photo of one of the 50-ton trials with the equipment beside it.

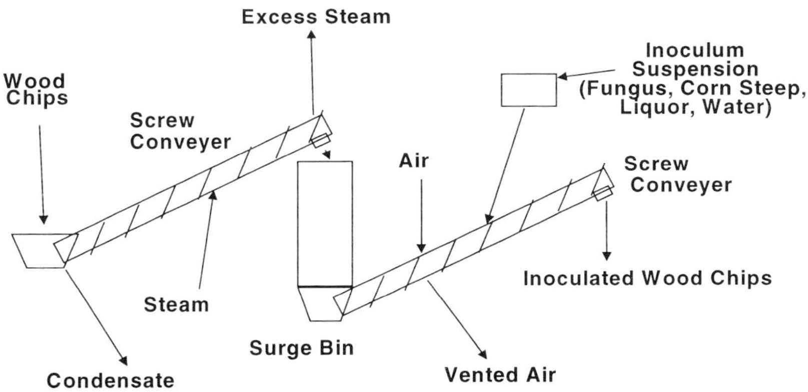

Figure 1. Overview of a continuous treatment process for decontaminating, cooling, and inoculating wood chips. The system is based on two screw conveyers with a surge bin between them.

Figure 2. Overview of a 50-ton trial held at the Forest Products Laboratory in Madison, Wisconsin. To the left of the pile are the aeration units, and to the right is the treatment apparatues.

3. EVALUATION OF FUNGUS-TREATED CHIPS IN PAPERMAKING

For many grades of paper, papermakers blend different pulps—mechanical and chemical, softwood and hardwood—to obtain the paper properties specific to that grade. In general, pulps such as softwood kraft are added to enhance the strength properties of the final sheet while mechanical pulps are added for their optical properties and lower costs. Since biopulping of the mechanical portion of the blend improves the strength properties of that fraction, significant economic savings can be realized since less of the more expensive chemical pulp can be used while still maintaining the sheet specifications.

The following detail two different case studies of blended pulps: One for lightweight coated paper and the other for a eucalyptus tissue paper. In each case, the kraft fraction was reduced resulting in significant annual savings.

3.1. Lightweight coated paper

Control and fungus-treated spruce chips (from one of the 50-ton trials) were refined through a commercial TMP mill producing lightweight coated paper (4,6). The fungal pretreatment saved 33% electrical energy (Fig. 3) and improved paper strength properties significantly compared to the control (Table 1). Yield loss for the biotreatment was approximately 2%. Since biomechanical pulp fibers were stronger than the conventional TMP fibers, we were able to reduce the amount of bleached softwood kraft pulp in the final product (7).

Table 1 shows that even with a 5% increase in the proportion of TMP in the blended pulp, the strength properties were still slightly improved over the control which consisted of a 50%/50% blend of TMP and softwood kraft. The optical properties (except for brightness) were also essentially the same. As has been noted many times with biopulping, there is a darkening of the biotreated pulp. However, as the table shows, this brightness can be recovered through an additional application of 1% hydrogen peroxide bleach on wood. This represents a 60% increase in the bleaching chemical used as compared with the control. Even with this additional bleaching, the process is economically feasible as will be discussed below.

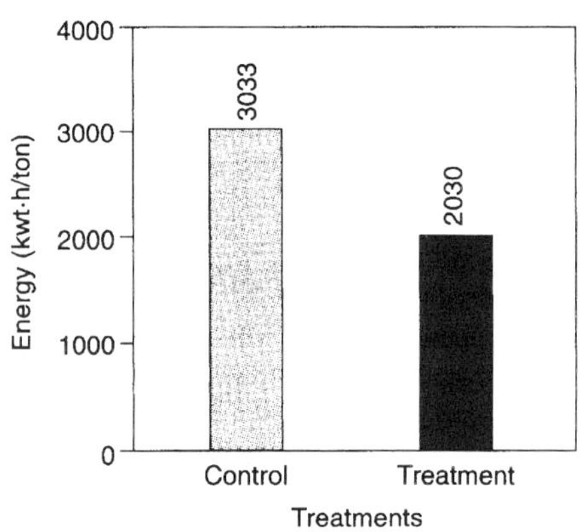

Figure 3. Energy requirements for the control and the fungus-treated chips from the 50-ton trial during thermomechanical pulping process to produce pulps at about 50 CSF.

Table 1. Strength and other physical properties comparison for blended pulp with spruce.

Parameters	Control (50%TMP + 50% kraft)	Treatment (55%TMP + 45% kraft)	
Burst index (kN/g)	2.29	2.39	
Tear index (mNm^2/kg)	9.92	10.0	
Tensile index (Nm/g)	38.9	41.2	
Brightness (%)	73.0	69.8[a]	73.0[b]
Opacity (%)	85.2	86.4	
Light scattering coefficient (m^2/kg)	46.2	47.7	
Drainage time (sec)	8.5	9.9	
Density (kg/m^3)	689	610	

[a]Same amount of hydrogen peroxide was applied in the control and treatment.
[b]Sixty percent more hydrogen peroxide was applied on the treatment. Strength and optical properties were not affected by the use of additional hydrogen peroxide (data not shown).

Mills can realize economic savings in several ways with the incorporation of biopulping into the process. The obvious one is the reduced electrical costs at the refiner. Savings can also be realized due to the increase in mill throughput as well as the savings due to the lower kraft pulp requirements. Although more difficult to quantify, the reduced pitch content of the pulp and the reduced environmental impact can also have economic benefits to the mill. Of course with the addition of biopulping to the process, there are some added costs. There is the additional cost of steam,power, and labor to operate the biopulping equipment as well as the cost of nutrients and inoculum. Additional bleaching chemicals may also be needed. Finally, there is a small amount of wood loss through the bioulping process, which needs to be accounted for.

Table 2 summarizes an example for a lightweight coated mill that produces 800 tons per day of paper that is a blend of kraft, TMP, and groundwood. The sheet is 50% softwood kraft, 27.5% TMP, and 22.5% groundwood. Table 3 summarizes the biotreatment operating costs. Additional electrical power is needed to operating the ventilation equipment and steam is used for the decontamination of the chips and for humidifying the ventilation air. A total cost of approximately US$15 is estimated. However, the 33% energy savings realized at the refiner result in a reduction in electrical costs of approximately US$40 per ton. Furthermore, as electrical energy costs increase, the savings become even more dramatic.

A complete economic analysis needs to take into account several other factors which are summarized in Table 4. As can be seen, while the biopulping process reduces the electrical energy requirements, there are increases in the wood, bleaching chemicals, and of course the biotreatment costs. However, there is still a net savings of approximately US$17 per ton of pulp for an annual savings of approximately US$1.3 million. Note that this analysis only takes into account the savings due to electrical energy savings. Other factors, such as increased mill throughput or kraft pulp reduction, would increase the savings. For lightweight coated paper, the increased TMP mill throughput and kraft pulp reduction could increase the annual savings to over US$11 million per year (7). Compared to the capital investment, these savings make the implementation of biopulping very attractive.

Table 2. Summary of economic assumptions for the economic analysis of an 800 ton/day lightweight coated paper mill.

Parameter	Value
Total production	800 ton/day
TMP production	220 ton/day
Kraft requirements @ US$650/ton	400 ton/day
Groundwood requirements @ US$233/ton	180 ton/day
Wood costs	US$80/ton
Production	350 days/year
TMP refiner energy	3033 kWh/ton
Electricity costs	US$0.04/kWh
TMP yield	95%
Treatment wood loss	2%
Additional bleaching chemicals	60%

Table 3. Biotreatment operating costs per ton of TMP pulp produced.

Operating parameter	Cost (US$/ton of pulp)
Steam	US$ 2.50
Electricity	US$ 6.16
Inoculum and nutrients	US$ 2.00
Labor, maintenance, taxes, overhead	US$ 4.10
TOTAL	US$14.76

Table 4. Comparison of conventional and biopulping TMP pulp manufacturing costs (US$/ton)

Cost	Conventional TMP	Biopulping TMP
Energy	121	81
Wood	84	86
Bleaching chemicals	10	16
Other costs	60	60
Biotreatment costs	-	15
TOTAL	275	258

3.2. Eucalyptus tissue paper

A tissue sheet is produced from a blend of 50% TMP and 50% hardwood bleached kraft. Again, as shown in Table 5, the biomechanical pulping of eucalyptus results in energy savings. While the savings in this case are not as dramatic as in the previous example, they are still significant. Furthermore, there is a very significant increase in the strength properties of the resulting pulp, with the tear and tensile indices more than doubling at a comparable freeness. The greatly improved strength properties of the biotreated TMP allow the kraft in the blended sheet to be reduced from 50% to 20% while still maintaining comparable properties (Table 6). While there is a decrease in the overall brightness, we expect that an increase in the bleaching chemicals used will recover this brightness loss as in the previous example.

Table 7 summarizes the economic assumptions for this particular mill producing a sheet that is a 50%/50% blend of eucalyptus TMP and hardwood kraft. Compared to the previous example, the electrical and wood costs are lower and the biotreatment energy savings are only 17%. Because of the lower energy savings, the process does not look economical based on energy savings along (Table 8). The electrical energy savings amount to only US$6 per ton, while the treatment costs are US$11 per ton. However, the much greater strength of the biotreated TMP pulp allows the amount of kraft to be reduced from 50% to 20% (Table 5). In essence, the biotreatment allows the substitution of kraft pulp at US$620 per ton with the biotreated TMP at US$195 per ton. The resulting savings is over US$11 million per year (Table 9). Although not considered in this analysis, additional bleaching (at a cost of US$10 per ton) would only reduce the annual savings to US$10.7 million per year.

Table 5. Energy savings and physical properties for the biomechanical pulping of eucalyptus.

Parameter	Control	Treatment
Freeness (ml)	402	390
Energy requirements (kWh/ton)	1005	833
Energy savings	-	17%
Burst index (kN/g)	0.20	0.34
Tear index (mNm^2/g)	1.03	2.93
Tensile index (Nm/g)	5.16	11.35

Table 6. Strength and other physical properties comparison for blended pulp with eucalyptus

Parameters	Control (50%TMP + 50% kraft)	Treatment (80%TMP + 20% kraft)
Burst index (kN/g)	0.35	0.38
Tear index (mNm^2/kg)	1.69	2.92
Tensile index (Nm/g)	9.40	11.26
Brightness (%)	79.8	74.5
Opacity (%)	88.5	88.8
Drainage time (sec)	5	5
Density (kg/m^3)	310	307

Table 7. Summary of economic assumptions for the economic analysis of a 260 ton/day eucalyptus mill.

Parameter	Value
Total production	260 ton/day
TMP production	130 ton/day
Kraft requirements @ US$620/ton	130 ton/day
Wood costs	US$20/ton
Production	350 days/year
TMP refiner energy	1400 kWh/ton
Electricity costs	US$0.025/kWh
TMP yield	95%
Treatment wood loss	2%

Table 1. Weight and component losses of *Pinus taeda* wood chips biotreated by *Ceriporiopsis subvermispora*

Component losses (%)	Biopulping time (days)			
	15	30	60	90
Weight	2.3 ± 0.5	3.0 ± 0.4	9 ± 2	13.8 ± 0.7
Glucan	0.9 ± 0.6	2 ± 1	1 ± 1	2 ± 1
Polyoses	1.4 ± 0.3	1.6 ± 0.1	7 ± 3	31 ± 3
Lignin	9.6 ± 0.6	10.7 ± 0.3	16.6 ± 0.1	22 ± 1
Extractives: Ethanol-soluble	28 ± 2	32 ± 2	48 ± 1	65 ± 4
Dichloromethane-soluble	22 ± 4	51 ± 3	64 ± 8	70 ± 10
Aryl-ether linkages	27 ± 3	53 ± 2	50 ± 1	56 ± 1

3. WOOD COMPONENTS IN SAMPLES BIOTREATED BY *C. SUBVERMISPORA*

3.1. Structural characteristics of residual lignin

Evaluation of the structural changes occurring in the wood components at the initial stages of biodegradation could help to elucidate whether these changes are related or not to the biopulping benefits. The DFRC and the CuO-oxidation methods have been reported to be appropriate to evaluate structural characteristics of lignin, because they are suitable for *in situ* characterization, lignin extraction from the wood matrix becoming unnecessary.

As earlier described by Chen (18), vanillin is the most abundant monomer found in the products of the CuO-oxidation of softwoods, while acetoguaiacone, vanillic acid and *p*-hydroxybenzoic acid are produced in small amounts. In *P. taeda* wood samples biotreated by *C. subvermispora*, the yield of vanillin resulting from CuO-oxidation decreases only after long biodegradation periods (9.5% in the undecayed control and 8.1% and 5.6% in samples biodegraded for 60 and 90 days, respectively) (19). The decrease in vanillin yield in this case has been attributed to an accumulation of condensed structures in lignin, since some carbon-carbon linkages, such as 5-5 biphenylics, are not cleaved during the oxidation by CuO (18). Such decrease may also result from oxidative ring opening, which normally occurs during lignin biodegradation by white-rot fungi (20).

The DFRC method has been reported to be specific for aryl glycerol β-O-aryl ether cleavage (21), whereas the yield of the degradation monomers can be used to estimate the aryl-ether linkage losses during wood biodegradation. Unlike the vanillin yield, the yield of DFRC-monomers decreased drastically at the initial stages of *P. taeda* biodegradation by *C. subvermipora* (27% and 53% after 15 and 30 days, respectively), while weight and lignin losses were very low in the same biodegradation periods (Table 1). These data illustrates well that an extensive lignin depolymerization occurs before its mineralization becomes significant. From these data it is also evident that the residual lignin in *C. subvermispora*-decayed samples, although depolymerized, is enriched in carbon-carbon substructures, corroborating CuO-oxidation data.

The decrease in the aryl-ether linkage contents after the first steps of wood biodegradation might be one of the explanations for the benefits of the biotreatment. A residual lignin extensively depolymerized should be easily solubilized in the kraft liquor. Gellerstedt and Lindfors (22) have shown that the initial delignification phase of kraft pulping mainly involves the release of low molecular weight lignin from the wood chips. This suggests that the bio-depolymerized lignin is easily released during the first stages of cooking, resulting in a faster and shorter initial delignification phase.

3.2. Molecular weight distribution of residual cellulose

Biodegradation of cellulose during biopulping by *C. subvermispora* is often considered low, because glucan (or cellulose) losses are very low even after long biodegradation periods (Table 1). However, the glucan loss measurement is based on the glucose released after acid hydrolysis of the wood samples (11, 23). This means that glucan loss represents the polymer degraded to carbon dioxide and water. Depolymerization of cellulose obviously starts before the polymer is mineralized, and a better estimation of the cellulose loss during biopulping needs a direct evaluation of the cellulose DP in biotreated samples. This has been evaluated by determining both, the alpha-cellulose contents in biodegraded wood samples and the molecular weight distribution of the residual cellulose. Figure 3 shows that, at the beginning of *P. taeda* biodegradation by *C. subvermispora*, neither glucan loss nor alteration in the yield of alpha-cellulose occurs. However, long biodegradation periods bring about a considerable decrease in the yield of alpha-cellulose, while the glucan loss remains low. In the case of *Eucalyptus grandis* hardwood biodegradation, the alpha-cellulose loss is already significant after two weeks of biotreatment (24). This suggests that degradation reactions start at the cellulose backbone, even when no significant glucan loss is detected. This kind of degradation reactions can generate low molecular mass glucans, which are soluble in the alkaline solution used for alpha-cellulose preparation, resulting in a decrease in the alpha-cellulose yield. The assimilation rate of these low-molecular-mass fragments by *C. subvermispora* seems to be very slow, since no significant increase in the glucan loss is detected even after long biodegradation periods. The degradation of the cellulose backbone can also be evidenced by size exclusion chromatography. Figure 4 shows the HPSEC of alpha-cellulose tricarbanilates prepared from *P. taeda* samples biodegraded by *C. subvermispora*. The average DP of cellulose was almost the same during the first 30 days of biodegradation, when wood weight losses were no higher than 3%. After this period, alpha-cellulose DP starts to decrease, which shows that although glucan losses are negligible in periods no longer than 90 days, after 30 days of biotreatment, the polymer loses its integrity. These data are in accordance with a report that shows a poor cellulolytic complex in *C. subvermispora*. Despite having some endo-glucanases activity (which might be responsible for cellulose depolymerization), *C. subvermispora* is not able to produce significant amounts of cellobiohydrolases responsible for releasing cellobiose from glucans produced by the endo-cellulases (25).

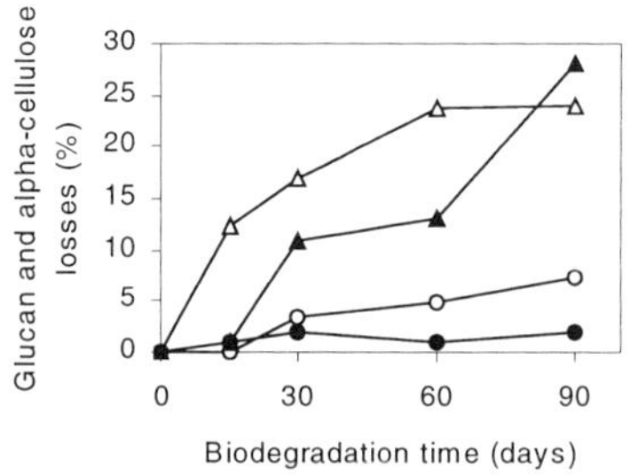

Figure 3. Glucan (-●-) and alpha-cellulose (-▲-) losses during biopulping of *P. taeda* (filled symbols) and *E. grandis* (open symbols) by *C. subvermispora* (24).

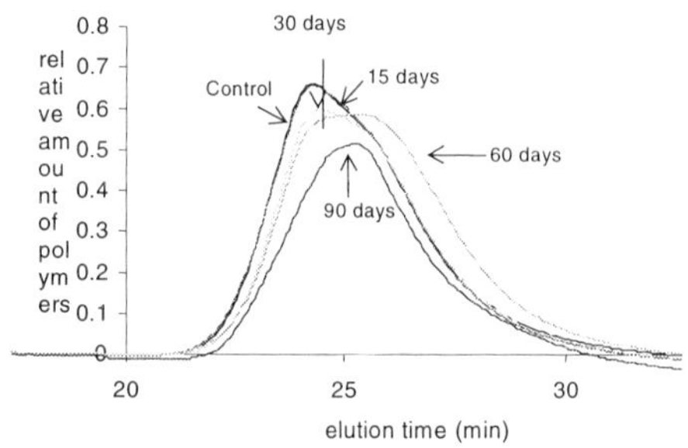

Figure 4. HPSEC of alpha-cellulose tricarbanilates prepared from *P. taeda* biotreated by *C. subvermispora*

4. ENZYMES PRODUCED DURING BIOPULPING BY *C. SUBVERMISPORA*

C. subvermispora is known to lack LiP activity in liquid cultures (26), despite the fact that LiP-like genes have been detected in this fungus (27). The production of this enzyme complex in cultures grown in solid-state fermentation, like biopulping, cannot be easily evaluated, because the aromatic compounds present in the extracts interfere with the veratryl alcohol method. However, it has been often assumed that this fungus lack LiP and that the main enzymes related to lignin degradation are MnP and laccases. The occurrence of lignin degradation inside the wood matrix during the first stages of wood decay, when the low cell wall permeability does not permit enzyme diffusion, has clearly indicated that some low molecular mass agents act in the initial stages of lignin biodegradation (28). Recently, MnP-induced lipid peroxidation has been proposed for lignin biodegradation by *C. subvermispora* (29-31). Under solid-state fermentation, it has been shown that at the initial stages of biodegradation, *C. subvermispora* produces unsaturated fatty acids that are transformed into hydroperoxides (32). Evaluation of the enzymes produced during biopulping of *P. taeda* by this fungus has shown that peroxidases are produced in high amounts (Table 2). These peroxidases are clearly dependent on Mn^{2+}, since the enzymatic activity of the extracts is significantly increased by the addition of this ion. Interestingly, laccases have been detected in solid-state wood cultures of *C. subvermispora* only when a co-substrate (malt-extract or glucose) is added to them (24, 32). This shows that MnP plays a key role in lignin degradation during biopulping, either by direct action on the lignin contained in the wood cell wall surfaces or by some mediated mechanism such as lipid peroxidation.

The total filter paper activity (33) produced by *C. subvermispora* under biopulping conditions was very low over all the biodegradation periods (Table 2).

Table 2. Enzymatic activities detected in aqueous extracts prepared from solid-state fermentation* (biopulping) of *Pinus taeda* by *Ceriporiopsis subvermispora*

Biopulping time (days)	Total enzymatic activities recovered from the cultures (IU/kg of wood chips on dry basis)		
	Manganese peroxidase		
	Syringaldazine as substrate	Phenol red as substrate	Filter paper activity (33)
15	702 ± 156	264 ± 78	76 ± 9
30	575 ± 31	246 ± 23	146 ± 7
90	503 ± 44	558 ± 29	64 ± 11

(*) The cultures did not contain co-substrates. Oxidation reactions to which H_2O_2 was not added (laccase activity) were negligible in all the cases.

5. CONCLUSIONS

Biopulping is reaching pulp-mill implementation, but little is known about the chemical and biochemical basis of the biopulping benefits. The rapid decrease in the aryl-ether linkage contents (easily estimated by the DFRC-method) found in residual lignins contained in biotreated samples relates well, at least, to the efficiency of the fungal pretreatment combined with high-yield kraft pulping. As the biopulping benefits are obtained within short biotreatment periods (low values of weight loss), it is possible that some biochemical mechanism based on the action of low molecular weight compounds might be involved in the lignin biotransformation. Considering that MnP is the main oxidative enzyme produced after short biodegradation periods (laccase is produced only if a co-substrate is added to the culture), the MnP-induced lipid peroxidation and the consequent degradation of lignin probably constitute the biochemical mechanism accounting for the biopulping benefits.

The measurement of glucan loss, which is usually based on glucose release from the wood samples after acid hydrolysis, is not a good procedure to evaluate the fungal selectivity during biopulping. Measuring the losses of alpha-cellulose content or directly determining the residual cellulose DP in biotreated samples seem to be more appropriate methods for evaluating biopulping selectivity.

REFERENCES

1. M. Akhtar, R.A. Blanchette, G. Myers and T.K. Kirk, An Overview of Biomechanical Pulping Research, in *Environmentally Friendly Technologies for the Pulp and Paper Industry*, R.Young and M. Akhtar (eds.), John Wiley & Sons, New York, 309-340, 1998.
2. M. Akhtar, M.C. Attridge, G.C. Myers and R.A Blanchette, Holzforschung, 47 (1993) 36.
3. G.M. Scott, M. Akhtar, M.J. Lentz and R.E. Swaney, Engineering, scale-up, and economic aspects of fungal pretreatment of wood chips, in *Environmentally Friendly Technologies for the Pulp and Paper Industry*, R.Young and M. Akhtar (eds.), John Wiley & Sons, New York, 341-383, 1998.
4. G.F. Leatham, R.A. Blanchette, G.C. Myers and T.H. Wegner, Tappi J., 73(3) (1990) 249.
5. G.F. Leatham, G.C. Myers and T.H. Wegner, Tappi J., 73(5) (1990) 197.

6. M. Akhtar, T.K. Kirk and R.A. Blanchette, Fungal delignification and biomechanical pulping of wood, in *Advances in Biochemical Engineering/Biotechnology* vol 57, K.E.L. Eriksson (ed.), Springer-Verlag, Berlin, 159-195, 1997.
7. A. Ferraz, R. Mendonça and F.T. Silva, J. Chem. Technol. Biotechnol., 75 (2000) 18.
8. A. Ferraz, J. Rodríguez, J. Freer and J. Baeza, J. Chem. Technol. Biotechnol., 75 (2000) 1190.
9. A. Ferraz, A. Guerra and R. Mendonça, *TAPPI Pulping/Process & Product Quality Conference CD-ROM*, Boston, USA, 2000.
10. A. Ferraz, J. Rodríguez, J. Freer and J. Baeza, W. J. Microbiol. Biotechnol., 17 (2001) 31.
11. R.A. Blanchette, T.A. Burnes, F.L. Gary and M.J. Effland, Biomass, 15 (1988) 93.
12. K. Messner, K. Koller, M.B. Wall, M. Akhtar and G.M. Scott, Fungal treatment of wood chips for chemical pulping, in *Environmentally Friendly Technologies for the Pulp and Paper Industry*, R.Young and M. Akhtar (eds.), John Wiley & Sons, New York, 385-419, 1998.
13. T.P. Oriaran, P. Labosky and P.R. Blankenhorn, Tappi J., 73(6) (1990) 147.
14. C.J. Biermann (ed.), Essentials of Pulping and Papermaking, Academic Press, New York, 1993.
15. V.L. Chiang, H.J. Cho, R.J. Puumala, R.C. Eckert and W.S. Fuller, Tappi J., 70(2) (1987) 101.
16. K. Fischer, M. Akhtar, R.A, Blanchette, T.A. Burnes, K, Messner and T.K. Kirk, Holzforschung, 48 (1994) 285.
17. L.J.F. Kohler, R.J. Dinus, E.W. Malcolm, A.W. Rudie, R.L. Farrell and T.S. Brush, Tappi J., 80(3) (1997) 135.
18. C. Chen, Nitrobenzene and cupric oxide oxidations, in *Methods in Lignin Chemistry*. Y.L. Lin and C.W. Dence (eds.), Springer-Verlag, New York. 301-321, 1992.
19. A. Guerra, R. Mendonça and A. Ferraz, Holzforschung (2001 in press).
20. T. Higuchi, J. Biotechnol. 30 (1993) 1.
21. F. Lu and J. Ralph, J. Agric. Food Chem., 45 (1997) 4655.
22. G. Gellerstedt and E-L. Lindfors, Holzforschung, 38 (1984) 151.
23. A. Ferraz, J. Rodriguez, J. Freer and J. Baeza, Bioresource Technol., 74 (2000) 201.
24. A.M. Córdova, M.Sc. Dissertetion, Faculdade de Engenharia Química de Lorena 112pp, 1999.
25. A. Sethuraman, D.E. Akin and K.E.L. Eriksson, Biotechnol. Appl. Biochem., 27 (1998) 37.
26. S. Lobos, J. Larraín, L. Salas, D. Cullen and, R. Vicuña, Microbiology, 140 (1994) 2691.
27. S. Rajakumar, J. Gaskell, D. Cullen, S. Lobos and R. Vicuña, Appl. Environ. Microbiol., 62 (1996) 2660.
28. R.A. Blanchette, E.W. Krueger, J.E. Haight, M. Akhtar and D.E.J. Akin, J. Biotechnol., 53 (1997) 203.
29. K. Jensen Jr., W. Bao, S. Kawai, E. Srebotnik and K. Hammel, Appl. Environ. Microbiol., 62 (1996) 3679.
30. E. Srebotnik, K. Jensen, S. Kawai and K. Hammel, Appl. Environ. Microbiol., 63 (1997) 4435.
31. A.N. Kapich, K.A. Jensen and K. Hammel, FEBS Letters, 461 (1999) 115.
32. M. Enoki, T. Watanabe, S. Nakagame, K. Koller, K. Messner, Y. Honda and M. Kuwahara, FEMS Microbiol. Letters, 180 (1999) 205.
33. M. Mandels, R. Andreotti and C. Roche, Biotechnol. Bioeng. Symp., 6 (1976) 2.

Biotechnology in the Pulp and Paper Industry
L. Viikari and R. Lantto (Editors)

Fungi as potential assisting agents in softwood pulping

A. Hatakka[a], P. Maijala[a], A. Mettälä[a], T. Hakala[a], L. Hauhio[b] and J. Ellmén[b]

[a]Department of Applied Chemistry and Microbiology
PO Box 56 (Viikki Biocenter), 00014 University of Helsinki, Finland
[b]Laboratory of Pulping Technology, Department of Forest Products Technology, Helsinki University of Technology, PO Box 6300, 02015 HUT, Finland

Biopulping, i.e. treatment of wood chips by a suitable lignin-selective white-rot fungus, has been successfully combined with mechanical pulping in the USA, and the process is now under commercalization. In contrast, biopulping combined with chemical pulping has been much less studied. The key factor for successful biopulping is to find a suitable fungus for the wood species of interest. More than 300 wood-rotting fungi, mostly newly isolated from old forests, were screened on plate tests, and the degradation of spruce wood blocks were tested by about 90 fungi. Suitable testing methods were developed and adopted to evaluate fungi that facilitate pulping. Most promising fungi were cultivated on spruce wood chips and the chips were then Kraft pulped. Some fungal treated samples were also mechanically pulped. White-rot fungi usually produce organic acids such as oxalic acid during the degradation of lignin in wood. Therefore the compatibility of the fungal treatment with the present Kraft pulping practice is difficult to achieve, although the best fungi clearly degraded lignin and also improved alkali solubility of lignin.

1. INTRODUCTION

Treatment of wood chips by a lignin degrading white-rot fungus combined with mechanical pulping is a process known as biopulping or biomechanical pulping. The process has proven its potential in large studies in the USA and is now under commercalization [1]. Biopulping combined with chemical pulping, i.e. successful fungal pretreatment of wood chips prior to chemical pulping is expected to improve penetration of cooking chemicals into wood chips and thus shorten the cooking time. During cultivation times of several weeks fungi substantially degrade lignin, which can be seen in the decrease of Kappa number. Fungal pretreatment of wood chips has also been studied in combination with sulfite [2] and Kraft pulping of wood [3, 4], and soda pulping of gramineous plants [5], but much less is known about the compatibility of the fungal treatment with the present Kraft pulping practice, compared with biomechanical pulping. The compatibility of the fungus with the selected wood is very important for the attack on lignin vs. wood carbohydrates. Also the conditions of wood chip treatment, e.g. temperature, are critical. The treatment time has a strong influence on the final result.

Fungi suitable for the treatment of hardwood species have been studied in most cases. Because in Northern countries softwood is industrially important, in this work the focus has been in identifying suitable fungi for softwood pulping. Norway spruce (*Picea abies*) was

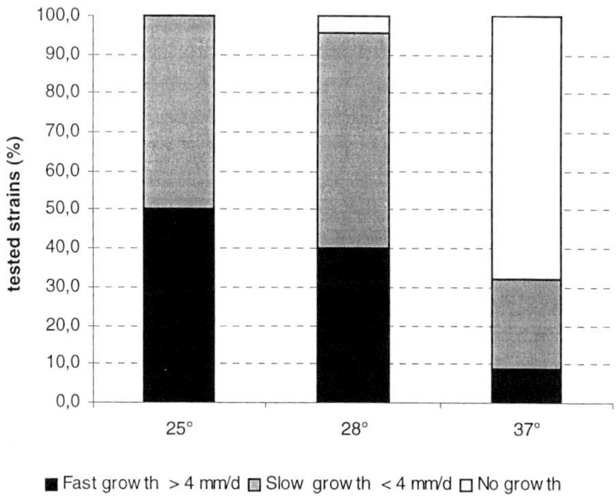

Figure 1. Growth rates of isolated wood rotting fungi in different temperatures on malt agar plates

The results showed that ca. 1-min steaming decreased 50% of the activity, with additional 2-min steaming resulting in about 1/10 of the activity of untreated small wood chips. Interestingly, some species that were poor colonizers of sterilized wood showed efficient colonizing ability only on steamed or untreated wood. Previously uncharacterized white-rot fungi having a good competitive ability were found [19, 20].

The ability of fungi to grow in higher temperatures would be beneficial in wood chip treatment since the chip piles become rather hot due to microbial activity. Figure 1 shows the relative amounts of fungi capable of fast growth also in relatively high temperature, 37°C. Fewer than 10% of the isolates could grew rapidly in this temperature.

The selectivity of fungi on wood block test is shown in Figure 2. The fungi have been grouped according to the weight loss they cause during 10 weeks cultivation time. Some of the new isolates showed high degradation of lignin compared to cellulose while the weight loss was relatively low. *C. subvermispora* CZ-3 was used as a reference fungus.

3. PULPING OF SPRUCE WOOD CHIPS TREATED BY WHITE-ROT FUNGI

3. 1. Fate of supplemented alkali in fungus-treated wood

Results from alkali impregnation tests [20] and Kraft pulping experiments showed that selectively lignin-degrading fungi such as *C. subvermispora* did degrade lignin in wood chips. Both the amount of alkali soluble compounds, as determined spectrophotometrically at A_{280} using Indulin AT as reference (Figure 3), and the small molecular weight lignin fraction in black liquor, as analyzed by gel permeation chromatography (results not shown), increased in fungal-treated samples. In these experiments the effects of *Phlebia gigantea*, a primary colonizing white-rot fungus, and Cartapip™ (colorless form of *Ophiostoma piliferum*), an ascomycete used for removal of extractives (pitch) from wood chips, were also studied.

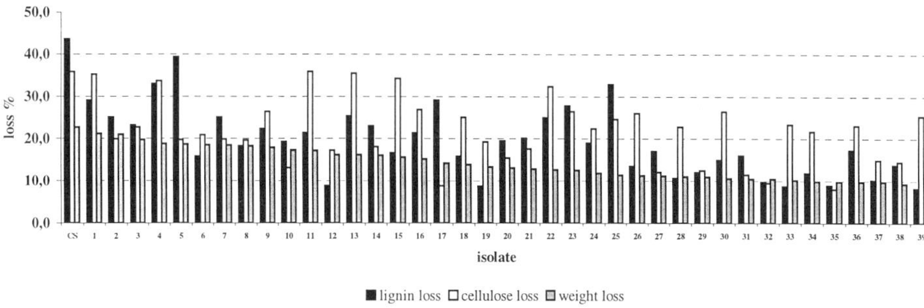

Figure 2. Weight, lignin and cellulose losses of spruce wood blocks by different white-rot fungi in the order of decreasing weight losses. Cultivation time 10 weeks. CS, *Ceriporiopsis subvermispora* CZ-3.

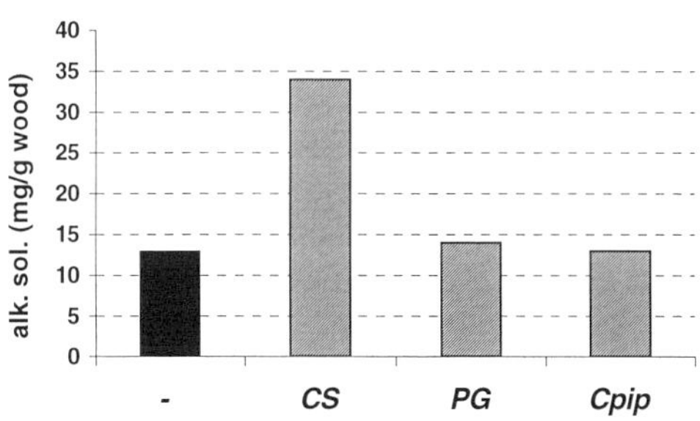

Figure 3. Alkali soluble (alk. sol.) compounds in spruce heartwood chips after pretreatment with fungi for 21 days. No fungus (-), *Ceriporiopsis subvermispora* (CS), *Phlebia gigantea* (PG), Cartapip (Cpip).

3.2. Chemical pulping of fungal treated spruce wood chips

Spruce sap- and heartwood were used in Kraft pulping experiments. Pulps were prepared in an air bath digester, hosting in a rotating frame six autoclaves of 2.5 litre each. For each treatment 300 g of screened oven dry chips were used. The conditions in pulping were as follows: sulfidity, 30 %, temperature, 170°C, time to 170°C (80->170°C), 40 min, and effective alkali (EA), typically 20-22%. Liquor to wood ratio was mostly 4. H-factor and EA were varied to produce pulps with a range of Kappa numbers. Lower Kappa number after

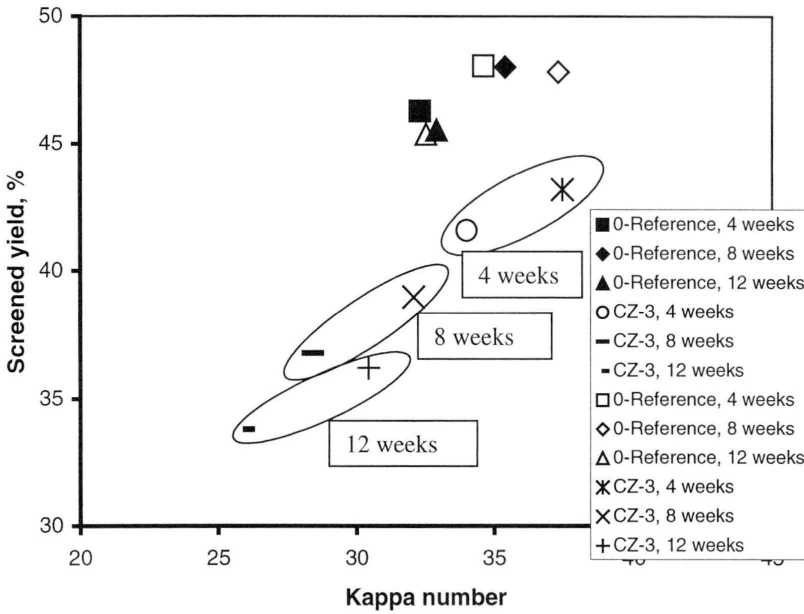

Figure 4. *Ceriporiopsis subvermispora* CZ-3 treatment of spruce chips: screened yield of pulp vs. Kappa number. Data points represent two pulping experiments. Growth times 4, 8 and 12 weeks.

pulping was found when the chips had been treated by *C. subvermispora*, especially after longer cultivation time, 8 or 12 weeks (Figure 4). However, the yields were lower and the consumption of alkali higher already after 2 weeks fungal treatment, the latter phenomenon probably due to the production of organic acids (oxalic acid) by the fungus [23].

3.3. Mechanical pulping of fungal treated spruce wood chips

Mechanical pulps were prepared from spruce wood chips treated with *C. subvermispora* CZ-3 for 14 days. For preparation of pulp, a wing defibrator was used in batchwise operation with 125 g dry chips. The refiner is a low intensity type, allowing high repeatability at constant refining consistency, adjustable temperature, pressure and refiner speed. Energy consumption could be measured precisely as a function of refining time. Fungal treatment caused energy savings of more than 20% compared to untreated chips and 10-12% compared to non-inoculated chips (Figure 5). Energy taken in the beginning of refining was clearly lower, and the effect remained throughout the refining (data not shown).

Tensile index in relation to energy consumption was improved, and density was higher in the same tensile index, freeness and energy consumption. As earlier reported from in the case of fungal treatments [1], ISO brightness was clearly lower in the fungal treatment of spruce wood chips, i.e. the pulps were darker than controls. Opacity was very high, up to 98%, with fungal treatments.

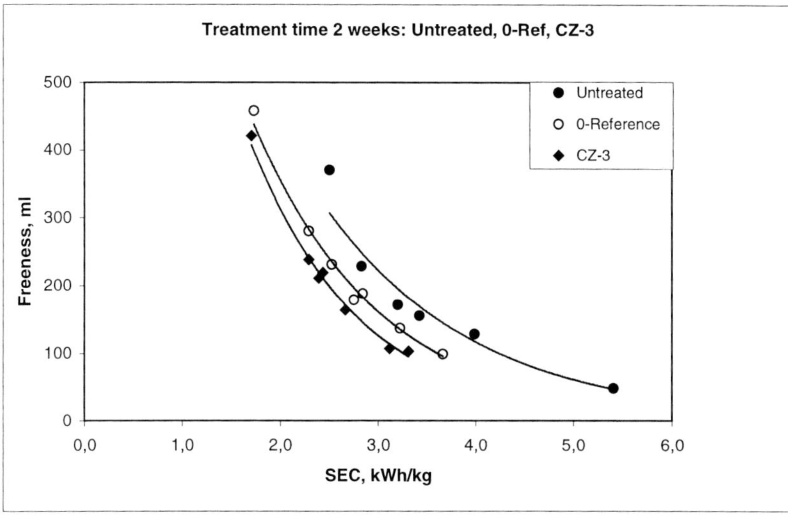

Figure 5. Energy consumption in refining of Norway spruce wood chips treated by *Ceriporiopsis subvermispora* CZ-3 for 14 days. &, untreated wood chips; o, autoclaved wood chips incubated 14 d without fungus (= 0-reference), ♦, autoclaved wood chips inoculated with *C. subvermispora* for 14 d. Aeration of wood chips during incubation, no additional nutrients.

4. CONCLUSIONS

For the treatment of spruce wood chips, about 300 wood rotting fungi were screened on plate tests and the most promising fungi were tested on spruce wood blocks to reveal their selectivity for lignin vs. wood polysaccharides. Very selective new fungi were found. However, the number of industrially interesting strains was low, which showed that fungi having good growth characteristics in addition to lignin-selectivity are difficult to find. Using appropriate methods the competitivity of fungi on wood chips and the behaviour of alkali in wood were studied by representative fungi. Results from Kraft pulping experiments showed that prolonged cultivation time (8 to 12 weeks) did decrease Kappa number but the yields were also lower. The compatibility of the fungal treatment with the present Kraft pulping practice is difficult to achieve because fungi produce organic acids while they degrade lignin, and this causes increased need of alkali. The advantages of fungal treatment, i.e. increased alkali solubility and removal of lignin by fungi, and the disadvantages, i.e. increased consumption of alkali and the lower yields obtained, point to that without modification of Kraft pulping process, the fungal treatment is not compatible with Kraft pulping. However, only a few white-rot fungi have been studied so far, and different types of fungi may be found for the treatment of wood chips prior to Kraft pulping. Fungal treatment combined with mechanical refining appeared to be more promising, and the potential of newly isolated lignin-selective fungi in biomechanical pulping will be studied in future.

5. ACKNOWLEDGEMENTS

The work was funded by the National Technology Agency of Finland, grants no. 40358/98, 40468/99 and 40600/00.

REFERENCES

1. M. Akhtar, R.A. Blanchette, G.C. Myers and T.K. Kirk, Environmentally Friendly Technologies for the Pulp and Paper Industry, R.A.Young and M. Akhtar, eds., pp. 309-419. J.Wiley & Sons, New York, 1998.
2. K. Messner and E. Srebotnik, FEMS Microbiol. Rev. 13 (1994) 351.
3. T.P.Oriaran, P. Labosky and P.R. Blankenhorn, Tappi J. 73 (1990) 147.
4. A. Ferraz, A. Guerra and R. Mendonca, Proc. 2000 TAPPI Pulping Conference, Boston, MA, November 5-9, 2000.
5. A. Hatakka, A. Mettälä, T. Härkönen and L. Paavilainen, Biotechnology in the Pulp and Paper Industry, E. Srebotnik and K. Messner, eds., pp. 229-232, Facultas-Universitätsverlag, Vienna, Austria. 1996.
6. T.K. Kirk and D. Cullen, Environmentally Friendly Technologies for the Pulp and Paper Industry, R.A. Young and M. Akhtar, eds., pp. 273-307. J.Wiley & Sons, New York, 1998.
7. A. Hatakka, Biopolymers. Vol 1. Lignin, Humic Substances and Coal, M. Hofrichter and A. Steinbüchel, eds., pp. 129-180. Wiley-VCH, Weinheim, 2001.
8. R.A. Blanchette, Can. J. Bot. 73 (1995) (Suppl. 1), S999-S1010
9. K.-E.L. Eriksson, R.A. Blanchette and P. Ander, Microbial and Enzymatic Degradation of Wood and Wood Components, Springer-Verlag, Berlin Heidelberg, 1990.
10. L. Otjen and R. Blanchette, Holzforschung 41 (1987) 343.
11. R.A. Blanchette, T.A. Burnes, M.M. Eerdmans and M. Akhtar, Holzforschung 46 (1992) 109.
12. P. Ander and K.-E. Eriksson, Physiol. Plant. 41 (1977) 239.
13. A.T. Martínez, S. Camarero, F. Guillén, A. Gutiérrez, C. Muñoz, E. Varela, M.J. Martínez, J.M. Barrasa, K. Ruel and J.M. Pelayo, FEMS Microbiol. Rev. 13 (1994) 265.
14. J. Dorado, G. Almendros, S. Camarero, A.T. Martinez, T. Vares, and A. Hatakka, Enzyme Microb. Technol. 25 (1999) 605.
15. V. Burlat, K. Ruel, A.T. Martinez, S. Camarero, A. Hatakka, T. Vares and J.-P. Joseleau, Proc. of the 7th Intl. Conf. on Biotechnol. in the Pulp and Paper Industry, pp. A75-A78. Vancouver, 1998.
16. R.A. Blanchette, Phytopathology 74 (1984) 725.
17. A. Hatakka, FEMS Microbiol. Rev. 13 (1994) 125.
18. T. Hakala, P. Maijala, B. Kluczek-Turpeinen, J. Konn and A. Hatakka, Abstracts, the 8th Intl. Conf. on Biotechnology in the Pulp and Paper Industry, pp. 161-162, Helsinki, 2001.
19. P. Maijala, M. Kalsi and A. Hatakka, Abstracts, the 8th Intl. Conf. on Biotechnology in the Pulp and Paper Industry, June 4-8, 2001, Helsinki, pp. 163-164, 2001.
20. A. Hatakka, A. Mettälä, J. Ellmén and P. Maijala, 2000. Proc. 2000 TAPPI Pulping Conference, Boston, MA, November 5-9, 2000.
21. T. Vares and A. Hatakka, Can. J. Bot. 75 (1997) 61.
22. M. Hofrichter, T. Vares, M. Kalsi, S. Galkin, K. Scheibner, W. Fritsche and A. Hatakka, Appl. Environ. Microbiol. 65 (1999) 1864.
23. S. Galkin, T. Vares, M. Kalsi and A. Hatakka, Biotechnology Techn. 12 (1998) 267.

Biotechnology in the Pulp and Paper Industry
L. Viikari and R. Lantto (Editors)

N-Hydroxy mediated laccase biocatalysis: Recent progress on its mechanism and future prospect of its application

Feng Xu[a], Kaichang Li[b], and Thomas J. Elder[c]

[a] Novozymes Biotech, Inc., Davis, CA 95616, USA

[b] Department of Forest Products, Oregon State University, Corvallis, OR 97331, USA

[c] School of Forestry and Wildlife Sciences, Auburn University, Auburn, AL 36849, USA

N-OH-containing compounds play important roles in many biological, pharmacological, and industrial processes. The traditional emphasis on the metal ion chelating property of the compounds has recently been shifted to the redox chemistry of the N-OH site, which is of great interest in developing mediated oxidoreductase-based biocatalyses, such as laccase-catalyzed delignification, decontamination, and organic synthesis. In an N-OH-mediated laccase biocatalysis system, N-OH is first oxidized into N-O• by laccase. As shown by a comparative study of 33 N-OH-containing compounds and seven fungal laccases, the oxidation is controlled by the electron-transfer from N-OH to laccase whose rate depends on the redox potential difference between laccase and N-OH. Higher redox potential tends to reduce the oxidation rate of N-OH, similar to the cases of other laccase substrates such as phenols. The redox potential of N-OH is related to the frontier molecular orbital energy, which is proportional to electron-withdrawing ability of *N*-phenyl substituents. Using cyclic and differential pulse voltammetry, the N-O• decay can be quantitated. The stability of N-O• varies, depending on the heat of formation of the radical. Lower redox potential, better delocalization/resonance, or more extensive steric effect tend to make N-O• more stable. However, stabilization of N-O• mitigates against its oxidation of the target molecule of the biocatalysis. Balancing the reactivity and stability of N-O• is key to the catalytic efficiency. The prospect of N-OH mediated laccase biocatalysis is discussed in terms of applying quantum calculation, rational design, and methodology development.

1. LACCASE-BASED BIOCATALYSIS

Following hydrolases or acyl/glycosyl transferases (such as proteases, lipases, carbohydrate hydrolases, lyases, and isomerases), oxidoreductases are emerging as the next generation of biocatalysts useful for a wide range of industrial/medicinal applications [1]. Among the promising oxidoreductases, laccase (EC 1.10.3.2), a multi-Cu oxidase that catalyzes the oxidation of phenol, aniline, and other reducing substances with the concomitant reduction of O_2, has an excellent potential as the biocatalyst for wood fiber modification, biosensor, and water/soil remediation [2-6]. Recently laccase-based biocatalysis has been expanded, with the aid of small "mediators", to pulp delignification [7-20], textile dye bleaching [21,22], polycyclic hydrocarbon degradation [23-27], destruction of pesticide, insecticide, or chemical warfare [28-30], and organic synthesis [28-36]. In these biocatalyses, a redox-active mediator shuttles electrons between laccase (acceptor) and target molecule (donor) that is either insoluble or a poor laccase substrate. The electron-transfer from mediator to laccase, the stability of oxidized mediator active species, and the oxidation of target molecule by the oxidized mediator (leading to mediator regeneration) can all be vital for such biocatalyses.

One of the most promising laccase mediator families contains the *N*-hydroxy (N-OH) functional group. These compounds, particularly hydroxamic acids (N(OH)C(O)), play an important role in a broad range of biological, pharmacological, and industrial processes [37-39]. Their *in vivo* functions include microbial metal ion acquisition, antibiotic-antifungal-antitumor activities, and activation of cytochrome P450 for amide-based drug metabolism. Traditionally, the metal ion-chelating property of these compounds has made them useful for extractive metallurgy, corrosion prevention, and nuclear fuel processing. Recent studies have been focused on the redox property of NOH, which is of great interest for the development of laccase-catalyzed pulp delignification, water/soil decontamination, chemical synthesis, and other biocatalyses [7-10, 12, 13, 15-19, 21-24, 27, 31, 34, 40].

2. GENERAL REDOX CHEMISTRY OF N-OH

N-OH-containing compounds, including *N*-substituted *N*-hydroxyamines and oximes shown in Fig. 1, are weak acids ($pK_a \sim 5$ to 6) and can undergo both reduction and oxidation because of the intermediate valency of the N atom (-1) [39]. During oxidation of N-OH (by PbO_2, for instance), the first product is often the nitroxide radical N-O•, which can be further oxidized to -NO and -NO_2. The N-O• can be detected by EPR or UV-visible spectrum, which is generally red-shifted in comparison to the corresponding N-OH [41,42]. The single-electron redox potential (E°) of the N-O•/N-OH couple, spin density, stability, and electrophilicity of N-O• can be affected by the *N*-substituents [41-51].

In general, the planar N-O• is in equilibrium with the dipolar, slightly pyramidal mesomeric form N^+•-O^-, leading to an apparent 3-electron N-O bond unfavorable to any dimerization via either N or O. The extent of pyramidalization is dependent on the energy difference between the lowest unoccupied molecular orbital (LUMO) and the singly occupied molecular orbital (SOMO), which is enhanced by π-electron-accepting/σ-electron-donating substituents [41]. The unpaired electron in the nitroxide radical is in a π molecular orbital, in contrast to the iminoxyl radical =NO• (derived from oxime) in which the unpaired electron is in a σ molecular orbital with a strong s character. An N-O• is only stable in the absence of α-C-H; otherwise the -CH(R)-N(O•)- would convert into a nitrone -C(R)=N^+(O^-)- [41].

Some N-O• with significantly steric structure around the radical site, such as 2,2,6,6-tetramethyl-1-piperidinyloxyl (TEMPO) radical, are very stable and can be readily generated in an aqueous solution. Other N-O• have much shorter half-lives and inert solvents (such as $CHCl_3$) are often needed to generate them by oxidants such as PbO_2 or peroxide-peracid with metallo-catalysts. To study N-O• in an aqueous solution, transient methods are required, and one of the mostly applied methods is the cyclic voltammetry (CV) [43, 45, 47-49, 51]. Four important sets of information about N-OH redox chemistry can be obtained from CV: E°,

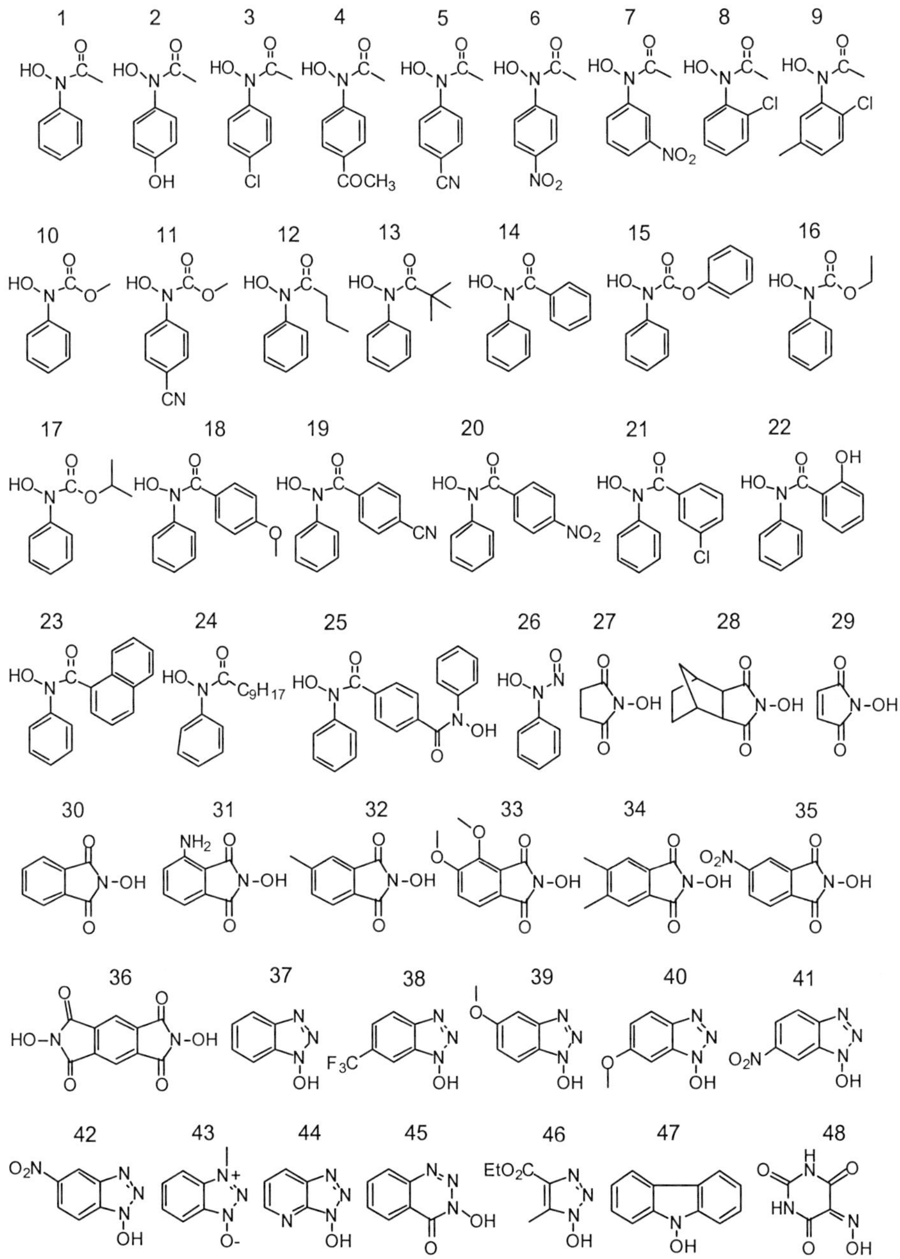
1
2
3
4
5
6
7
8
9
10
11
12
13
14
15
16
17
18
19
20
21
22
23
24
25
26
27
28
29
30
31
32
33
34
35
36
37
38
39
40
41
42
43
44
45
46
47
48

Figure 1. Structure of various N-OH-containing compounds. Reference: **21**, [47]; **22**, [49]; **31** to **36**, [31]; **39** to **42**, **46**, **47**, [44]; **43**, [43]; the rest, [56]. Other similar N-OH-containing compounds have also been reported [46-49, 61].

oxidation valency change (or number of electron exchanged), stability of oxidized/reduced product, and reactivity of oxidized/reduced product with other molecules (in a coupled homogenous reaction). For instance, hydroxamic acids often show three redox reactions on their voltammogram, at neutral pH and from -0.5 to 1.4 V range, as follows (E° referred to the Normal Hydrogen Electrode (NHE)) [47, 48]:

$\phi N(OH)COX - e^- \leftrightarrows \phi N(O\bullet)COX + H^+$ (E° = 0.7 to 0.8 V) (1)
$\phi N(O\bullet)COX - e^- \rightarrow [\phi N^+(O)COX] \rightarrow \phi NO + XCO^+$ (E° ≈ 1.1 to 1.3 V) (2)
$\phi NO + 2e^- + 2H^+ \leftrightarrows \phi N(OH)H$ (E° = 0.1 to 0.4 V) (3)

To mediate laccase catalysis, reaction (1) is the most important. From the anodic and cathodic peak potentials and currents one can extract information on E° and N-O• stability, as well as reactions between the N-O• and a target molecule or between the N-OH and laccase, should the target molecule or laccase be present, by coupling the homogenous reaction to the heterogeneous electron-exchange between the N-OH and electrode [45]. Subjecting a series of N-OH-containing compounds to CV would allow the elucidation of various electronic and steric effects from the *N*-substituents.

3. GENERAL MECHANISM OF N-OH-MEDIATED LACCASE BIOCATALYSIS

The early study on 1-hydroxybenzotriazole (HBT; **37** in Fig. 1), violuric acid (VA; **48** in Fig. 1), and *N*-hydroxyacetanilide (NHA; **1** in Fig. 1) have indicated a general mechanism as shown in Fig. 2 [7-10, 12, 13, 15-19, 21-24, 27, 31, 34, 40, 45]. First, an N-OH is oxidized by the type 1 (T1) Cu in laccase, whose E° ranges from 0.5 to 0.8 V [2-6, 52]. Second, the T1-Cu transfers electrons to the type 2/type 3 (T2/T3) trinuclear Cu cluster. Third, an O_2 molecule bound at the T2/T3 cluster accepts the electrons and is reduced to H_2O. Based on the transient UV-visible spectrum, the stoichiometrical measurement of O_2 consumption, and the dependence of the oxidation rate on the E° of N-OH-containing compounds, the N-O• is most likely the active species mediating the indirect oxidation of a target substance (such as a lignin model molecule) by laccase [7, 9, 19, 42, 53-55].

The mechanism outlined in Fig. 2 involves four major reactions for an N-OH mediator: N-OH oxidation by laccase (***O***), N-O• decay (***D***), N-O• reduction by target molecule (***R***), and laccase inactivation by N-O• (***I***). In general, an N-OH with lower E° tends to have a faster reaction ***O*** to generate a more stable N-O•, which would be adverse for reactions ***D***, ***R***, and ***I***. The steric hindrance around N-OH tends to slow down all four reactions. Hence, the balance of the reactions is vital for the overall efficiency of the N-OH-mediated laccase biocatalysis. To better understand the electronic and steric effects from the N-OH structure, a comparative study of a series of NHA and HBT derivatives has recently been made [56].

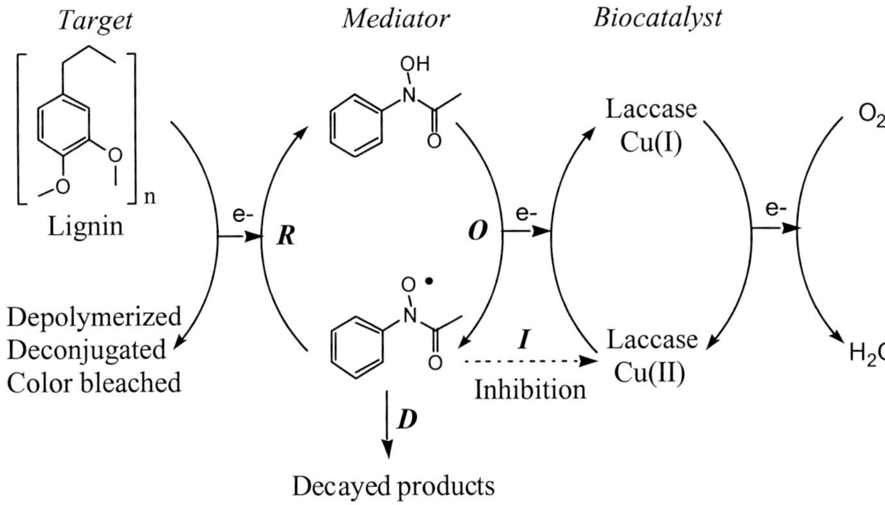

Figure 2. Generalized mechanism for mediated biocatalysis of laccase. During the catalysis, the mediator is first oxidized by laccase (reaction ***O***), then the oxidized mediator is reduced by target molecule (reaction ***R***) to be regenerated for next round mediation. The oxidized mediator may decay into inactive byproducts (reaction ***D***) or inactivate laccase (reaction ***I***), diminishing the overall mediation/catalysis efficiency.

4. OXIDATION OF N-OH TO N-O•

4.1. Redox potential E°

Using CV and differential pulse voltammetry (DPV), the E° of numerous N-OH-containing compounds have been measured [43-45, 48, 49, 51, 54, 56]. In general NHA (**1**) derivatives and VA (**48**) have a reversible or quasi-reversible anodic/cathodic peaks, taking place near 0.7 to 0.9 V, for the N-O•/N-OH couple (1). For substituted *N*-phenyl-*N*-hydroxyacetamides (**1** to **9**), *N*-phenyl-*N*-hydroxycarbamic acid methyl esters (**10**, **11**), and *N*-phenyl-*N*-hydroxybenzamides (**14**, **18** to **20**), the E° correlates to the conjugative Hammett σ constant of the substituents, similar to phenols [57-59]. Substituents with stronger electron-withdrawing ability (larger σ) result in lower electron density on N-OH, via π orbital interaction, and consequently higher E°. For substituted *N*-phenyl-*N*-hydroxyamines (**1**, **10**, **14**, **20**), *N*-phenyl-*N*-hydroxyamides (**1**, **10**, **12** to **15**, **23** to **25**), and *N*-phenyl-*N*-hydroxycarbamic acid esters (**10**, **15** to **17**), however, the E° does not correlate to the inductive Taft σ* constant. Since the unpaired electron in N-O• is in a π orbital, its energy (related to E°) is not perturbed much by inductive substituents via σ orbital interaction.

Unlike NHA derivatives, HBT derivatives (**37**, **38**, **44**, **45**) undergo the first oxidation near 1.2 V, as shown by an irreversible single-electron anodic peak [43-45, 54, 56]. Nevertheless, their E° still correlate to the σ constant of the substituent on the benzene ring, but not to the σ* of the groups directly linked to N-OH, similar to *N*-hydroxydiacetamides (**27** to **30**).

Because of the concomitant H^+ release when N-OH is oxidized (1), E° of an N-OH in general decreases as pH increases by a rate of ~ 59 mV/pH at 25 °C [54].

4.2. Molecular energetics

Using semi-empirical UHF/AM1 quantum chemical methods (the Spartan software from Wavefunction), we have calculated the molecular orbital energies of N-OH compounds (Table 1). The geometries of N-OH and N-O• are almost identical [51], indicating a small entropy change (ΔS) for (1). Thus the enthalpy change (ΔH) contributes the most to the Gibbs free energy change (ΔG) for the oxidation ($\Delta G \approx \Delta H$). This is consistent with the observation that E°, which equals to $-F\Delta G$ (F: Faraday constant), correlates with the highest occupied molecular orbital (HOMO) energy, equivalent to the ionization energy or ΔH of the reaction N-OH → $N\text{-}OH^+ + e^-$ (Fig. 3). The E° also correlates with the bond dissociation energy (BDE) or ΔH of the reaction N-OH → N-O• + H•, since the ΔG for the reactions $N\text{-}OH^+$ → N-O• + H^+ and $H^+ + e^-$ → H• are essentially identical for the N-OH compounds.

As shown in Fig. 4, the atomic orbitals of the *N*-phenyl, particularly at the *p*- and *o*-sites, contribute significantly to the SOMO of N(phenyl)(O•) (**1** to **20**, **23** to **26**). This is consistent with the observation that substituents on the *N*-phenyl can significantly affect the E° as well as the reactivity of N(phenyl)(OH) [56]. In comparison, the contribution from *N*-C=O or the *C*-phenyl to the SOMO is less significant in N(OH)C(O)(phenyl) (**14**, **18** to **20**). The ester O in N(OH)COO phenyl further reduces the resonance of the phenyl, or other substituents, with N(OH)C(O) (**10**, **11**, **15** to **17**) [56]. A similar effect also exists for the HOMO [51].

For *N*-OH-phthalimide (**30**), the phenyl group contributes little to the SOMO. For HBT derivatives (**37**, **38**, **44**), the 5- and 7-sites of the phenyl group show significant contribution to SOMO. For VA (**48**), the α-C hosts the major portion of SOMO, indicating that the σ-like =N(O•) has a significant super-exchange with the α-C via the σ C-N bond.

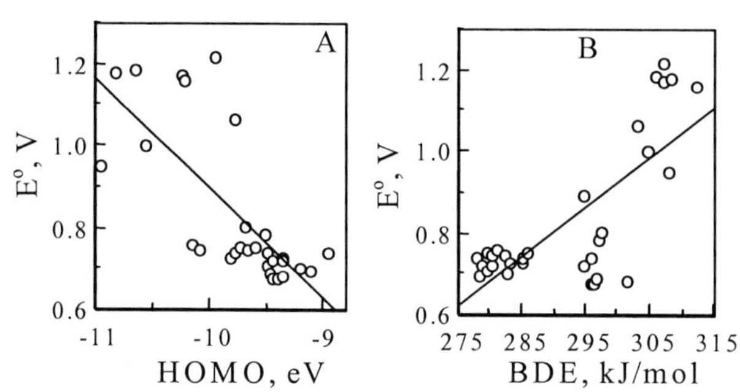

Figure 3. Correlation of the measured E° (in solution) with the calculated HOMO energy and BDE (in gas phase). N-OH-containing compounds examined: **1** to **20**, **23**, **25** to **30**, **37**, **38**, **44**, **45**, and **48** (Fig. 1). E° values are from [56]. VA (**48**), whose radical is of σ type, is not included in (A), which covers π type N-O• only. Regression line: (A) E° = -0.27HOMO - 1.8, $r^2 = 0.554$; (B) E° =0.012BDE - 2.7, $r^2 = 0.553$. From the equation E° = $-F\Delta G = -F(\Delta H - T\Delta S)$ = $-F(\text{BDE} - T\Delta S)$, a slope of 0.0104 (= 1000/F) is expected for the correlation line in (B).

Table 1
Molecular orbital energy, heat of formation, spin density, and mediated phenol red oxidation rate for 33 N-OH/N-O•[1].

#	HOMO	H_{-N-OH}	SOMO	$H_{-N-O\bullet}$	BDE	Spin-N	Spin-O	Rate[2]
1	-9.4	-49	-9.3	13	280	0.39	0.54	11.1 ± 0.1
2	-9.1	-234	-8.93	-173	278	0.40	0.53	
3	-9.4	-78	-9.33	-17	279	0.39	0.55	11.5 ± 0.2
4	-9.7	-203	-9.53	-140	280	0.39	0.55	18.3 ± 0.7
5	-9.7	85	-9.63	146	279	0.38	0.55	22.5 ± 0.4
6	-10.2	-31	-10.03	32	281	0.38	0.56	16.5 ± 0.9
7	-10.1	-28	-9.9	36	282	0.39	0.559	17.5 ± 1.6
8	-9.7	-70	-9.7	10	297	0.39	0.59	14.6 ± 0.4
9	-9.5	-102	-9.7	-23	297	0.38	0.59	13.3 ± 0.6
10	-9.5	-212	-9.5	-133	296	0.41	0.53	12.2 ± 0.2
11	-9.8	-78	-9.8	0.2	296	0.40	0.55	28.0 ± 0.3
12	-9.5	-103	-9.2	-41	279	0.40	0.54	10.3 ± 0.1
13	-9.4	-258	-9.3	-181	294	0.41	0.52	10.1 ± 0.1
14	-9.4	101	-9.2	166	283	0.40	0.54	
15	-9.5	-36	-9.4	42	296	0.40	0.53	
16	-9.4	-237	-9.4	-158	296	0.40	0.53	14.0 ± 0.3
17	-9.4	-254	-9.3	-176	295	0.40	0.53	10.5 ± 0.4
18	-9.2	-60	-9.1	4.1	282	0.41	0.53	
19	-9.6	238	-9.3	306	286	0.42	0.52	
20	-9.8	125	-9.6	192	285	0.39	0.54	8.5 ± 2.4
23	-9.0	187	-9.2	247	278	0.40	0.53	
24	-9.5	-274	-9.2	-213	279	0.39	0.54	1.9 ± 0.6
25	-9.5	111	-9.3	178	285	0.40	0.54	3.4 ± 0.1
26	-9.4	150	-9.5	233	301	0.32	0.55	0.16 ± 0.02
27	-10.8	-308	-10.4	-218	308	0.30	0.64	0.95 ± 0.02
28	-10.6	-280	-10.2	-192	306	0.31	0.64	0.98 ± 0.01
29	-11.0	-132	-10.4	-43	307	0.33	0.63	2.40 ± 0.04
30	-10.6	-99	-10.2	-12	305	0.33	0.63	5.47 ± 0.03
37	-9.8	444	-9.6	528	303	0.32	0.56	21.1 ± 0.1
38	-10.3	-200	-10.1	-111	307	0.30	0.57	2.34 ± 0.03
44	-10.2	501	-10.1	595	312	0.28	0.59	1.05 ± 0.09
45	-10.0	226	-10.1	315	307	0.26	0.61	1.3 ± 0.1
48	-11.4	-370	-11.5	-293.1	295	0.55	0.63	15.6 ± 0.1

[1] Unit: HOMO and SOMO, eV; H and BDE, kJ mol^{-1}; v, μmol(phenol red oxidized) min^{-1} .
[2] For experimental details of N-OH-mediated, *Trametes villosa* laccase-catalyzed Phenol Red oxidation, see Fig. 8.

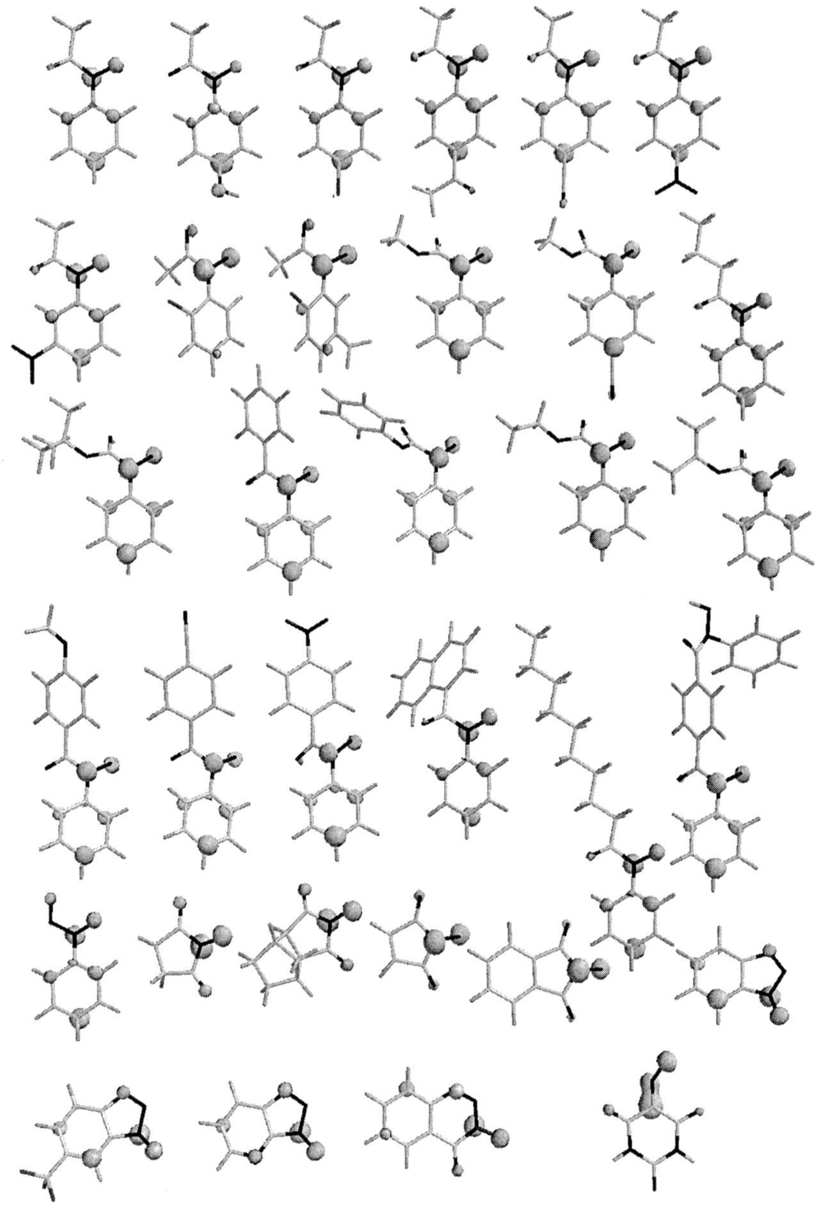

Figure 4. SOMO of N-O•. N-OH-containing compounds examined: **1** to **20**, **23** to **30**, **37**, **38**, **44**, **45**, and **48** (Fig. 1), arranged from top left to top right to bottom right.

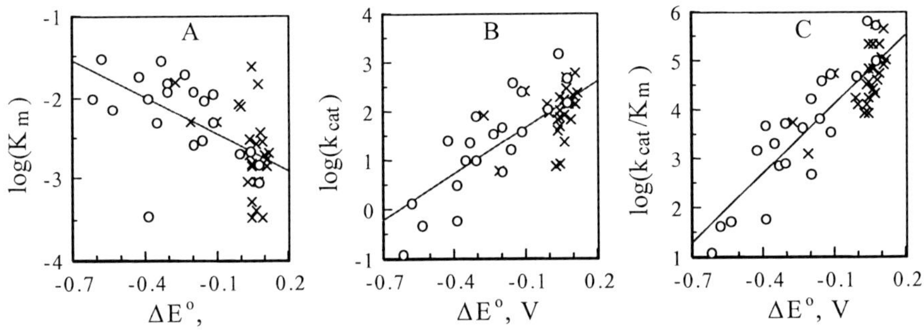

Figure 5. Correlations of K_m, k_{cat}, and k_{cat}/K_m to E° for laccase-catalyzed N-OH oxidation. K_m, k_{cat}, and ΔE° (= $E^\circ_{laccase}$ - $E^\circ_{N\text{-}OH}$) are from [54, 56]. Symbol: (x) oxidation of 25 N-OH-containing compounds (**1** to **20**, **23**, **26**, **30**, **37**, **48**) by *Trametes villosa* laccase-1 whose E° value is 0.79 V, (o) oxidation of three N-OH compounds (**1**, **37**, and **48**) by seven laccases whose E° values range from 0.48 to 0.79 V. Regression line: (A) $\log(K_m) = -1.5\Delta E^\circ - 2.6$, $r^2 = 0.333$; (B) $\log(k_{cat}) = 3.1\Delta E^\circ + 2.0$, $r^2 = 0.572$; (C) $\log(k_{cat}/K_m) = 4.7\Delta E^\circ + 4.6$, $r^2 = 0.761$. Unit: K_m, M; k_{cat}, min^{-1}; k_{cat}/K_m, $M^{-1}min^{-1}$; ΔE°, V. For phenols, phenazines, and other laccase substrates, the correlation $\log(k_{cat}/K_m) = 5.8\Delta E^\circ + 6.5$, $r^2 = 0.659$ has been found [54].

4.3. Oxidation by laccase

Like phenols and phenazines, a linear correlation between $\log(k_{cat})$ or $\log(k_{cat}/K_m)$ and ΔE°, the difference between E° of laccase and E° of N-OH, exists for the oxidation of N-OH by laccase (Fig. 5) [54, 56]. The higher the E° of laccase or the lower the E° of N-OH is, the faster the oxidation rate tends to be, consistent with the Marcus "outer-sphere" mechanism in which the thermodynamic "driving-force" ΔE°, together with reorganization energy and transmission coefficient, dominates the electron-transfer from the N-OH to the T1 Cu of laccase. The variation of k_{cat} among different *N*-substituted N-OH containing compounds is attributable to the substitutional effect on the E° of N-OH as discussed above [56]. However, the non-negligible scattering of the data shown in Fig. 5 indicates that other factors, such as the proper docking of N-OH molecule or its interaction with the amino acid residues surrounding the T1 Cu, can also play important roles in the fine-tuning of the oxidation rate. The negative correlation between $\log(K_m)$ and ΔE° (Fig. 5A) could result from the effect of ΔE° on the interaction between the filled (valence) molecular orbitals of N-O and the half- or unoccupied molecular orbitals of T1 Cu [54]. *N*-substituents in N-OH can also impact K_m. For instance, the observed high K_m of **24** and **25** could be caused by the steric hindrance of the bulky group attached to N-OH, whereas the high K_m of **27** to **29** and **37**, **38**, **44**, and **45** might be due to the lack of a freely rotating phenyl group attached to N-OH, which may disfavor the interaction of the substrate with the amino acid residues in the T1 pocket.

52. F. Xu, Biochemistry, 35 (1996) 7608.
53. K. Li, F. Xu, and K.-E. L. Erikssen. Appl. Environ. Microbiol., 65 (1999) 2654.
54. F. Xu, J. Kulys, K. Duke, K. Li, K. Krikstopaitis, H.-J. Deussen, E. Abbate, V. Galinyte, and P. Schneider, Appl. Environ. Microbiol., 66 (2000) 2052.
55. K. Li, N. G. Prabhu, D. A. Cooper, F. Xu, T. Elder, and K.-E. L. Eriksson, in Fundamentals and Catalysis of Oxidative Delignification Process (D. S. Argyropoulos ed.), American Chemical Society, 2000.
56. F. Xu, H.-J. Deussen, B. Lopez, L. Lam, and K. Li, Eur. J. Biochem., 268 (2001) 4169.
57. S. Steenken and P. Neta, J. Phys. Chem., 86 (1982) 3661.
58. J. Lind, X. Shen, T. E. Eriksen, and G. Merenyi, J. Am. Chem. Soc., 112 (1990) 479.
59. S. V. Jovanovic, M. Tosic, and M. G. Simic, J. Phys. Chem., 95 (1991) 10824.
60. A. Majcherczyk and C. Johannes, Biochim. Biophys. Acta, 1474 (2000) 157.
61. J. Sealey, A. J. Ragauskas, and T. J. Elder, Holzforschung, 53 (1999), 498.
62. I. B. Afanasev, in Chemical Kinetics of Small Organic Radicals (Z. B. Alfassi, ed.), pp. 97-160, CRC Press, Boca Raton, 1988.

ABBREVIATIONS

HBT, 1-hydroxybenzotriazole; VA, violuric acid; NHA, *N*-hydroxyacetanilide; T1 Cu, type 1 Cu; NHE, Normal Hydrogen Electrode; CV, cyclic voltammetry; DPV, differential pulse voltammetry; k_{decay}, apparent decay rate of N-O•; E°, single-electron redox potential; ΔE°, $E^\circ_{laccase}$ - $E^\circ_{N\text{-}OH}$ for reaction ***O*** or $E^\circ_{N\text{-}OH}$ - E°_{target} for reaction ***R***; LUMO, lowest unoccupied molecular orbital; SOMO. singly occupied molecular orbital; HOMO, highest occupied molecular orbital; BDE, bond dissociation energy; H, heat (or enthalpy) of formation; IE, ionization energy.

ACKNOWLEDGMENT

This research was supported, in part, by the National Initiative Competitive Grants Program of the United States Department of Agriculture (Award No. 2001-35103-10149) to K. Li, and by the United States Department of Energy (Award DE-FC36-99GO10374) to T. J. Elder. F. Xu thanks Novozymes colleagues whose contributions are reflected by the cited references.

Biotechnology in the Pulp and Paper Industry
L. Viikari and R. Lantto (Editors)

Metal-activated laccase promoters

V. Faraco[a], P. Giardina[a] ,G. Palmieri[b] and G. Sannia[a],

[a]Dep. Chimica Organica e Biochimica, Complesso Universitario Monte Sant'Angelo, Via Cinthia, 80126 Napoli, Italy

[b]IABBAM Consiglio Nazionale delle Ricerche, Via Argine 1085, 80147 Napoli, Italy

Production of laccase isoenzymes POXC and POXA1b in the white rot basidiomycete *Pleurotus ostreatus* is induced by copper at the level of gene transcription. *poxc* and *poxa1b* promoter regions, extending 200 nucleotides upstream of the ATG, contain multiple putative metal responsive elements (MREs) similar to the core MRE *consensus* sequence identified in the promoters of metallothionein genes from higher eukaryotes. It has been demonstrated that one or more proteins specifically bind *poxa1b* and *poxc* promoter regions when the fungus is grown in the presence of added copper. On the other hand, some factors interact with the analysed *poxc* promoter region also when the fungus is grown without copper addition. Furthermore, the presence of copper is needed to form specific complexes.

1. INTRODUCTION

Pleurotus ostreatus is a basidiomycete that secretes several laccase isoenzymes, four of which, named POXC (1), POXA1w (2), POXA2 (2), and POXA1b (3) have been purified and characterised. Three different *P. ostreatus* laccase genes and their corresponding cDNAs have been cloned and sequenced: *poxc* (1) (previously named *pox2*), *pox1* (4) (coding for a not yet identified laccase isoenzyme), and *poxa1b* (3). Recently, we tested several putative inducers of *P. ostreatus* laccase production and demonstrated that the addition of copper sulphate to *P. ostreatus* growth medium causes a strong increase of total laccase activity and induces the production of POXC and POXA1b isoenzymes. We have also demonstrated, by Northern blot analysis, that copper has a marked effect on the induction of *poxa1b* and *poxc* gene transcription (5).

Nucleotide sequences of *poxc* and *poxa1b* promoter regions, extending about 400 nucleotides upstream of the ATG, have been analysed, allowing the identification of multiple regulatory sites, such as MREs (Metal Responsive Elements), XREs (Xenobiotic Responsive Elements) and HSEs (Heat Shock Elements) (Fig. 1). The sequence of all MREs is similar to the core MRE *consensus* sequence identified in the promoters of metallothionein (*mt*) genes, TGCPuCNC, with that of cMRE3 being identical. Other laccase promoters have already been reported to possess multiple putative MRE sites: *Ceriporiopsis subvermispora* laccase promoter contains five putative MREs (6), and laccase1 and laccase3 promoters from the basidiomycete fungus I-62 contain several putative MREs (7). However, no further information about the functioning of these putative laccase promoter elements has been reported so far.

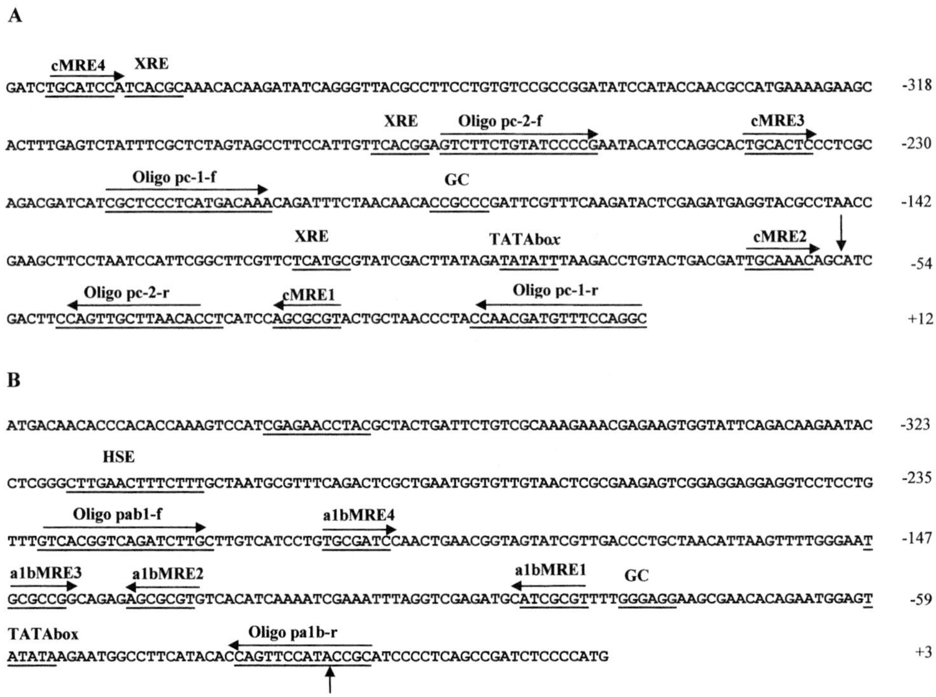

Figure 1 Nucleotide sequences of *Pleurotus ostreatus poxc* (A) and *poxa1b* (B) promoters. The region extending about 400 nucleotides upstream of the ATG in each gene is shown. Transcription-initiation sites are indicated by an arrow. Putative TATAbox, GCbox, MREs, HSEs, and XREs are underlined. *poxc* MREs are denoted as cMREs, whilst *poxa1b* MREs as a1bMREs. In each promoter the MRE sites are numbered according to their proximity to the ATG. The orientation of MREs is indicated by arrows. Oligonucleotides used for probe amplifications are also indicated.

In order to shed light on the mechanism involved in copper regulation of *poxc* and *poxa1b* transcription, we examined the role of putative *poxc* and *poxa1b* MREs in metal-induction of laccase gene transcription, with the aim of identifying *cis*-acting elements in this regulatory system. We tested the ability of endo-cellular proteins to bind these promoters in copper supplemented cultures and investigated the ability of copper to activate metal-responsive transcription factors and/or to induce their synthesis.

2. MATERIALS AND METHODS

2.1 Organism and culture conditions

White rot fungus, *P. ostreatus* (Jacq.: Fr.) Kummer (type: Florida), was maintained through periodic transfer at 4 °C on potato-dextrose agar plates (Difco) in the presence of

0.5% yeast extract (Difco). The mycelium was grown both in liquid basal medium (24 g/l of potato dextrose broth, 5 g/l of yeast extract) and in the same medium supplemented with 150 μM $CuSO_4$ at the time of inoculation. The mycelium was grown at 28 °C for 48 hours after inoculation.

2.2 Preparation of mycelia crude extract

Total protein extract from *P. ostreatus* mycelium was prepared as follows. Lyophilised cells were ground in a mortar with a pestle. The ground material was resuspended in cold extraction buffer [200 mM Tris-HCl pH 8.0, 400 mM $(NH_4)_2SO_4$, 10 mM $MgCl_2$, 1 mM EDTA, 10% glycerol, 1mM PMSF (Phenylmethylsulfonylfluoride), and 7 mM ß-mercaptoethanol] and then centrifuged at 4 °C for 1 hour at 15,000 x g. $(NH_4)_2SO_4$ (70% w/v) was added to the supernatant and material was centrifuged for 1 hour at 15,000 x g. The pellet was resuspended in a minimal volume of protein buffer (20 mM HEPES pH 8.0, 5 mM EDTA, 7 mM ß-mercaptoethanol, 1mM PMSF, 20% glycerol) and dialysed extensively against the same buffer. Protein concentration was determined using the Bio-Rad Protein Assay with bovine serum albumin as standard.

2.3 Probes

Recombinant plasmids containing *poxc* and *poxa1b* promoters extending about 1400 nucleotides upstream of the ATG had been previously selected from the genomic *P. ostreatus* DNA library (1, 3, 4). These plasmids were used as templates for PCR reactions to amplify probes to be used in electrophoretic mobility shift assays. Two different fragments, pc1 and pc2, covering +12/ –219 and -32/ -273 regions of the *poxc* gene respectively were generated by PCR using the oligonucleotide primers pc-1-f (-219 to -203) and pc-1-r (+12 to -6) for the amplification of pc1 fragment and the oligonucleotide primers pc-2-f (-273 to -258) and pc-2-r (-32 to -48) for the amplification of pc2 fragment respectively. Only one fragment, pa1b, consisting of the region –22/-231 of the *poxa1b* gene was generated by PCR using the oligonucleotide primers designated pa1b-f (-231 to -214) and pa1b-r (-22 to -35). Amplified fragments were purified on agarose gels using the QIAquick Nucleotide Removal kit (Qiagen); the fragments were radioactively end-labelled using T4 Polynucleotide kinase (Boehringer) and [γ-^{32}P] ATP (Amersham) and separated from free [γ-^{32}P]ATP with QIAquick Nucleotide Removal kit.

2.4 Electromobility shift assay

Electromobility shift assays were performed in 20 μl reaction volume, in binding buffer (20 mM HEPES pH 7.6, 100 mM NaCl, 2mM EDTA, 0.5 mM $MgCl_2$, 0.7 mM ß-mercaptoethanol, 0.1mM PMSF, 10% glycerol) using 1.0-10 μg of poly(dI-dC) to minimise non-specific binding. Total extract (20-50 μg of proteins) was pre-incubated in binding buffer containing poly(dI-dC) for 20 min at room temperature. After addition of the ^{32}P-labelled DNA fragment (0.5-2 ng, 50,000-100,000 cpm) the reaction was incubated for 20 min at the same temperature. Mixtures were then analysed by electrophoresis on 5% native polyacrylamide gel (29:1 cross-linking ratio) in 0.5x TBE (45 mM Tris-HCl pH 8.0, 45 mM boric acid, 1 mM EDTA). Electrophoresis was performed at room temperature at 200 V (20 V/cm). The gels were dried and analysed by autoradiography.

For competition experiments either non specific competitor (pGEM 7Zf(+)) or unlabeled specific competitor DNA (100-3,000 fold molar excess) was added to the pre-incubation mixture. For chelation inactivation experiments, specified amounts of EDTA or Cyclam (1, 4, 8, 11-Tetraazacyclotetradecane) were added to the pre-incubation mixture. The effect of

EDTA was analysed using final concentrations of the chelating agent in the 1-250 mM range, whilst the effect of Cyclam was tested using final concentrations in the 5-90 mM range.

3. RESULTS

3.1 Binding of specific proteins to poxc and poxa1b promoters.

Proteins extracted from both mycelium grown in the basal medium (control) and from mycelium grown in the same medium supplemented with 150 μM $CuSO_4$ were tested in electromobility shift assays with pc1, pc2 probes (regions +12/ -219 and –32/ -273 of the *poxc* gene respectively) and with pa1b probe (region –22/-231 of the *poxa1b* gene).

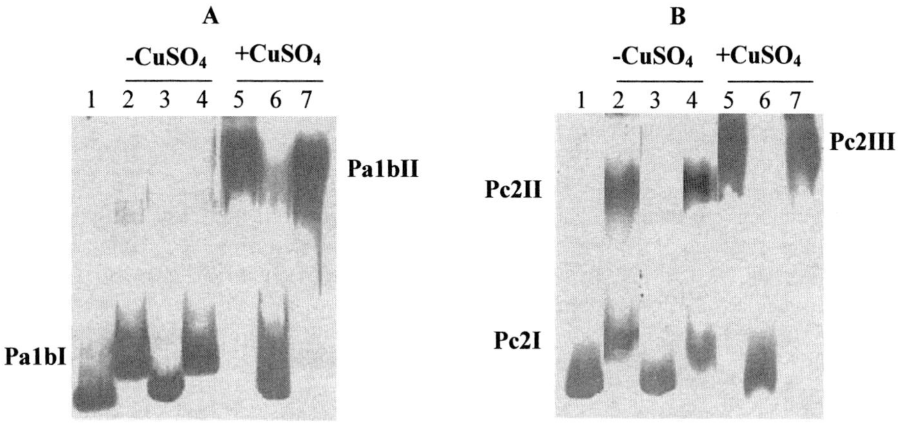

Figure 2 Electrophoretic mobility shift analysis of pa1b (A) and pc2 (B) probes incubated with proteins extracted from *P. ostreatus* grown in the absence (-$CuSO_4$) and in the presence (+$CuSO_4$) of added $CuSO_4$. Lane 1: probe; lane 2, 3, 4: probe incubated with proteins extracted from fungus grown in the absence of inducer (lane 2); and with specific competitor (lane 3) or non-specific competitor, (lane 4); lane 5, 6, 7: probe incubated with proteins extracted from fungus grown in the presence of inducer (lane 5), and with specific competitor (lane 6) or non-specific competitor (lane 7).

As shown in Fig. 2A, incubation of radio-labelled pa1b probe with proteins extracted from *P. ostreatus* grown in the absence of added $CuSO_4$ produced only one complex, Pa1bI. On the other hand, incubation of the radio-labelled pc2 probe with the same protein extract resulted in the formation of two complexes, Pc2I and Pc2II (Fig. 2B).

Incubation of each radio-labelled probe with proteins extracted from mycelium grown in the presence of 150 μM $CuSO_4$ resulted in the formation of only one low-mobility complex. As far as pc2 probe is concerned, this large complex was denoted as Pc2III, its electrophoretic mobility being lower than that of Pc2II (Fig. 2B). On the other hand, incubation of pa1b probe with proteins extracted from mycelium grown in the presence of copper produced another complex, Pa1bII (Fig. 2A). Pc2III and Pa1bII complexes display similar electrophoretic mobility.

The specificity of the binding was demonstrated by competing the binding of radio-labelled pc2 and pa1b probes to intracellular proteins with the same unlabeled DNA fragments, whilst non-specific DNA fragments were unable to compete. Pa1bI and Pa1bII complexes are displaced by a 100-fold molar excess of unlabelled probe; formation of Pc2I and Pc2II requires a 1000-fold molar excess to be competed, whilst a 3000-fold molar excess is necessary to compete the formation Pc2III complex.

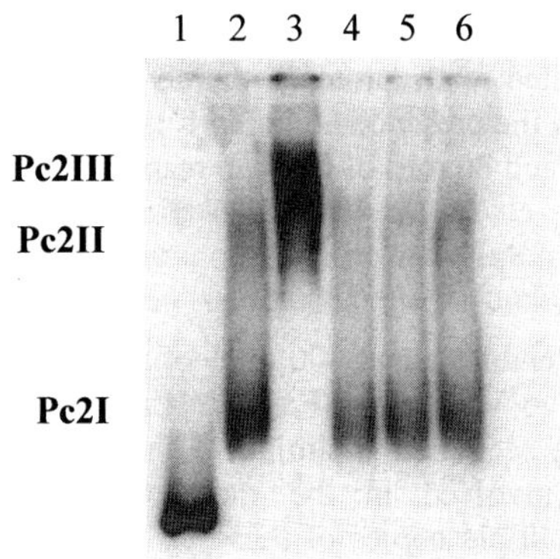

Figure 3 Effect of copper sulphate addition on electrophoretic mobility shift assay of pc2 probe. Lane 1: probe; lane 2: probe incubated with proteins from fungus grown in the absence of added $CuSO_4$; lane 3: probe incubated with proteins from fungus grown in the presence of $CuSO_4$; lane 4, 5 and 6: probe incubated with proteins from fungus grown in the absence of added $CuSO_4$ after pre-incubation with the specified amount (0.5, 0.7 and 1 mM respectively) of $CuSO_4$.

Fig. 3 reports electromobility shift assays in which different amounts of copper sulphate were added to proteins extracted from fungus grown in the absence of added copper, before incubation with the pc2 radio-labelled fragment. No formation of Pc2III complex was observed after copper sulphate addition (0.5, 0.7 and 1 mM final concentrations), thus indicating that this complex can only be formed if copper is present during fungal growth. Therefore the metal should induce *de novo* synthesis and/or post-translational modification of the involved proteins. The same behaviour was observed for Pa1bII complex.

3.2 Metal-dependence of binding of specific proteins to *poxc* and *poxa1b* promoters.

Radio-labelled pc1, pc2, and pa1b probes were incubated with protein extracts in the presence of either EDTA or Cyclam. Results obtained using pc2 probe and Cyclam are reported in Fig. 4. Similar results were obtained with pc1 and pa1b incubation products; EDTA caused the same effect for all the analysed DNA fragments. Formation of complexes Pc2II, Pc2III, and Pa1bII was found to be sensitive to both EDTA and Cyclam, whereas Pa1bI and Pc2I complex formation was unaffected by both chelating agents. This finding suggests that formation of Pa1bI and Pc2I complexes is presumably due to the presence of elements such as TATA box and GC sequence in the analysed fragments.

Pc2II complex, which is formed in non induced conditions and is displaced by the presence of EDTA and Cyclam (100 and 40 mM respectively) could be due to the presence of metal-responsive factors present in basal protein extract.

Formation of Pc2III complex is inhibited by 250 and 80 mM final concentration of EDTA and Cyclam, respectively, whereas formation of Pa1bII complex is inhibited by a final concentration of 150 and 70 mM of the chelating agents, respectively.

Biotechnology in the Pulp and Paper Industry
L. Viikari and R. Lantto (Editors)

Protease mediated processing of a Cu-induced laccase in *Pleurotus ostreatus*: a natural approach to improve protein stability[1]

P. Giardina[a], G. Cennamo[a], C. Greco[a], G. Palmieri[b] and G. Sannia[a]

[a]Dipartimento di Chimica Organica e Biochimica, Università di Napoli "Federico II", Complesso Universitario Monte S. Angelo, via Cinthia, 80126 Napoli, Italy

[b]IABBAM Consiglio Nazionale delle Ricerche, Via Argine 1085, 80147 Napoli, Italy

Laccase isoenzymes from *Pleurotus ostreatus*, a white rot basidiomycete fungus, have been extensively studied by our research group. More recently we have reported that the addition of $CuSO_4$ to culture broth results in a large increase of the total laccase activity and in the production of a new isoenzyme: POXA1b.

POXA1b secreted in the culture broth exhibits interesting properties with regard to pH stability. In fact the enzyme shows an increase of $t_{1/2}$ from pH 3.0 to 10.0 and, surprisingly, displays a $t_{1/2}$ value at pH 10.0 of about 100 days. Furthermore POXA1b is partly secreted, in fact analysis of proteins from cellular extract showed the presence of a larger amount of POXA1b in this extract than in the culture broth. The enzyme purified from cellular extract (POXA1b-I) shows some differences respect to the secreted enzyme (higher molecular mass, slightly different catalytic constants and lower pH stability).

Extra-cellular POXA1b, named POXA1b-P, has been also purified from fungal culture in the presence of a serine protease inhibitor PMSF, in order to prevent protease action that may affect this isoenzyme. Comparison of properties of the three forms make evident significant similarity between POXA1b-P and POXA1b-I respect to POXA1b. A marked difference can be observed with regard to pH stability, as a fact POXA1b is the most stable form at alkaline pHs.

1. INTRODUCTION

White rot fungi play a central role in pulping biotechnology: by using the extracellular ligninolytic enzyme systems they can selectively remove or alter lignin polymers allowing cellulose fibres to be obtained. These fungi produce different extracellular lignin-modifying enzymes: manganese peroxidases, lignin peroxidases and laccases (1). Laccases (E.C. 1.10.3.2) are multinuclear copper-containing oxidases which exploit the full oxidising power of O_2 to perform one-electron oxidation of a large number of organic (notably phenols) and inorganic substrates (2).

[1] This work was supported by grants from the Ministero dell'Università e della Ricerca Scientifica (Progetti di Rilevante Interesse Nazionale grant PRIN 98 and PRIN 2000), and from Consiglio Nazionale delle Ricerche (Progetto Finalizzato Biotecnologie).

a colorimetric reaction. Immunological detection was performed according to the manufacturer's instructions.

2.7. Western blots.

Proteins were separated by SDS-PAGE according to Laemmli (9) and electroblotted on to PVDF ProBlott™ Membrane (Applied Biosystems). Electroblotting was performed in 10 mM CAPS (3-[Cyclohexylamino]-1-propanesulfonic acid) pH 11, 10% (v/v) methanol at 50 V for 180 min at room temperature. Blocking solution was 5% (w/v) dried milk in PBS (phosphate-buffered saline solution) supplemented with 0.2% (v/v) Triton X-100 (washing buffer). The membrane was washed and incubated with the primary antiserum diluted 1:100 in washing buffer, at room temperature for 1 h, under continuous shaking. Subsequently the membrane was washed and incubated as above with Anti-rabbit IgG, peroxidase conjugate (Sigma), diluted 1:2000 in washing buffer. Blots were visualised with 100 mM Tris-HCl pH 7.5, 0.5 mg/ml 3,3'-diaminobenzidine, 0.03% (w/v) $NiCl_2$, 0.006% (v/v) H_2O_2.

2.8. Protein sequence analysis.

Automated N-terminal degradation of the electroblotted protein or of purified peptides was performed using a Perkin Elmer-Applied Biosystems 477A pulsed-liquid protein sequencer equipped with a model 120A PTH-analyzer for the on-line identification and quantification of PTH-amino acids.

2.9. Mass spectrometry analysis.

MALDI mass spectra were recorded using a Voyager DE MALDI-TOF mass spectrometer (Perkin Elmer-Perseptive Biosystem); a mixture of analyte solution, α-ciano-4-hydroxy-cinnamic acid or 2,5-dihydroxy-benzoic acid as matrices, bovine insulin and horse heart myoglobin were applied to the sample plate and dried in vacuo. Mass calibration was obtained using as internal standards horse myoglobin (16,952.50 *m/z*), bovine insulin (5,734.59 *m/z*) and α-ciano-4-hydroxy-cinnamic acid (379.06 *m/z*). Raw data were analysed by using a computer software provided by the manufacturer and are reported as average masses.

3. RESULTS AND DISCUSSION

3.1. Induction and purification of POXA1b-I and POXA1b-P.

Copper addition in the culture broth lead to a significant increase in laccase activity; the extracellular POXA1b produced in these conditions was purified and characterised (5, 8). In order to verify copper induction of POXA1b present in cellular extract, Western blot analysis was performed on proteic extract from cells grown in the presence or absence of copper. Figure 1 shows that a protein of about 63 kDa is recognised by antiPOXA1b antibodies in the extract from cultures supplemented with copper.

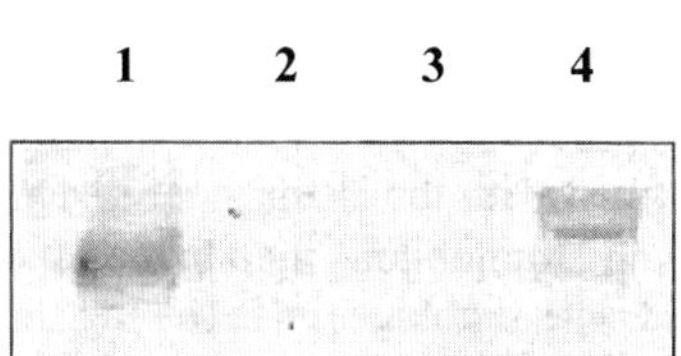

Figure 1. Western-blot analysis. Lane 1: protein sample from culture broth in the presence of copper; lane 2: protein sample from culture broth in the absence of copper; lane 3: protein sample from cellular extract of culture in the absence of copper;: lane 4: protein sample from cellular extract supplemented with copper.

Therefore POXA1b was purified from cellular extract (POXA1b-I) with the aim to investigate on the molecular properties of this isoform. A summary of the purification procedure is shown in Table 1.

Table 1
Purification of intracellular laccase POXA1b-I

Purification step	Total activity (U)	Total protein (mg)	Specific activity (U/mg)	Recovery (%)
Extract	8,200	476	17	100
$NH_4(SO_4)_2$ prec.	6,500	11	590	80
S-Sepharose	2,600	1.5	1,700	32
Superdex 75	2,500	1	2,500	30

SDS-PAGE, isoelectrofocusing and gel filtration chromatography showed that the purified protein is homogeneous. The purified enzyme corresponds to 30% of the cellular laccase activity and has a specific activity of 2,500 U/mg, a slightly lower value than that determined for the extracellular POXA1b. On the other hand, the amount of the protein from cellular extract is about twofold that of the protein purified from the corresponding culture broth. This sharing has not been observed for other *P. ostreatus* laccase enzymes and could be explained either by an inefficient secretion mechanism for POXA1b or by the occurrence of a physiological role of this isoenzyme inside the cell or on the cell wall.

The addition of a protease inhibitor cocktail to the culture broth lead to an increase of POXA1b activity and allowed persisting of the extracellular isoenzyme for prolonged time (Figure 2). When the serine protease inhibitor PMSF was added, a less marked effect was observed, thus suggesting that other proteases than serine ones are involved in POXA1b degradation (8).

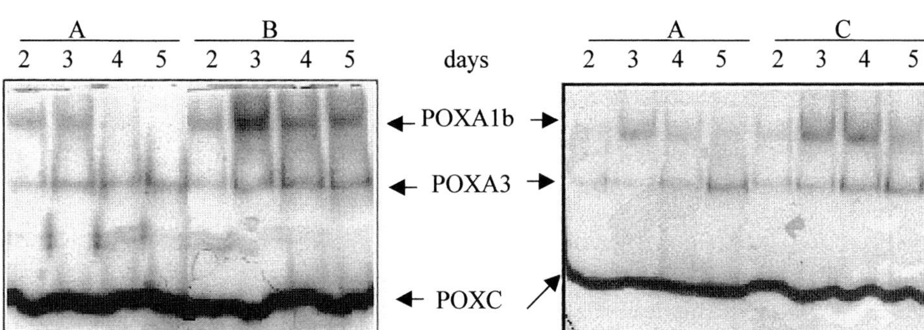

Figure 2 Zymograms of laccase isoenzymes in the absence (A) and in the presence of protease inhibitor cocktail (B) or in the presence of PMSF (C). Samples collected at different times, containing 0.015 U of laccase activity, were loaded.

POXA1b-P has been purified from the extracellular broth of cultures supplemented with PMSF (less expensive than the inhibitor cocktail) and grown for three days (Table 2).

Table 2
Purification of laccase POXA1b-P from PMSF-containing medium

Purification step	Total activity (U)	Total protein (mg)	Specific activity (U/mg)	Recovery (%)
Broth	59,000	135	437	100
$NH_4(SO_4)_2$ prec.	52,000	34.5	1,507	88
S-Sepharose	5,800	1	5,800	10

The specific activity of POXA1b-P is about twofold respect to that of POXA1b.

3.2. Structural characterisation of POXA1b-I and POXA1b-P

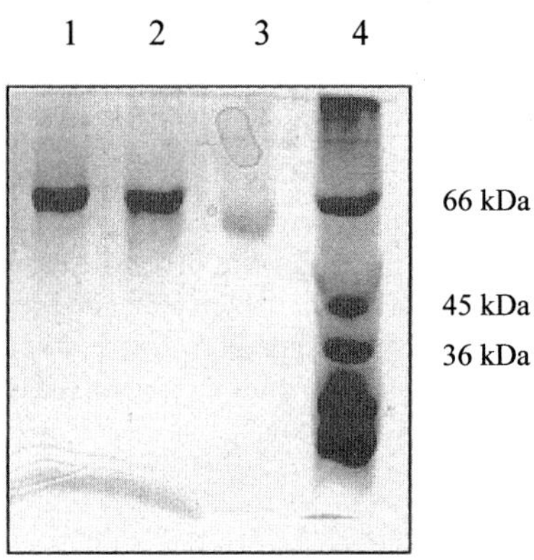

Figure 3. SDS PAGE of the three purified proteins. Lane 1 POXA1b-I, lane 2 POXA1b-P, lane 3 POXA1b, lane 4 molecular weight markers.

Figure 3 shows a SDS-PAGE analysis of the three purified proteins POXA1b-I, POXA1b-P and POXA1b. The former two proteins show an apparent molecular mass about 2000 Da larger than the latter. These results have been supported by MALDI mass spectrometry analysis. Peaks centred at 61,645, 61,540, and 60,410 kDa respectively, were observed. The monomeric nature of the three proteins was inferred by gel filtration experiments. The pI determined for both POXA1b-I and POXA1b-P was 6.9, the same value determined for POXA1b. In order to investigate on the molecular differences among these proteins, N-terminal sequences were determined. The first 10 aminoacids are identical for all the three isoforms, thus N-terminal processing of POXA1b can be excluded. Furthermore comparison of tryptic maps of POXA1b and POXA1b-P, obtained by MALDI MS, does not show any difference, even if the C-terminal peptides could not be identified in both maps. Therefore the observed structural differences could be due to a C-terminal processing which occurs after POXA1b secretion. As a matter of fact, alignment of deduced amino acid sequence of POXA1b with those of many other fungal laccases predicts a 16-amino acid C-terminal extension, whose hydrolysis could justify the molecular mass differences among the three forms. Furthermore, data have been reported supporting the hypothesis of a C-terminal processing for *N. crassa* and *C. coprinus* laccases, whose sequences show a C-terminal extension when aligned with those of other laccases (10, 11).

POXA1b, POXA1b-P and POXA1b-I samples were analysed for the presence of oligosaccharides by the lectin-binding assay. All the three isoforms were specifically recognised by *G. nivalia* agglutinin lectin which binds to terminal mannose residues and *D. stramonium* agglutinin specific for galactose β(1-4)-N-acetyl-glucosammine, thus excluding significant differences in the glycosylation pattern among the three isoforms. On the basis of these results further investigations are needed on the enzymes aimed to analyse in detail their structural differences.

3.3. Stability and catalytic properties

Table 3 shows the catalytic parameters of the three isoforms towards three different substrates. Moreover, all the isoforms are unable to oxidise guaiacol, a typical laccase substrate, and catalyse oxidation of the substrate analysed at the same optimal pH (2.5-4.0, 4.0-6.0, 4.0-5.0 for ABTS, syringaldazine and DMP, respectively).

Tab.3 Catalytic parameters of extra-cellular laccase POXA1b (A1b), extra-cellular laccase from PMSF-containing medium POXA1b-P (A1b-P) and intracellular laccase POXA1b-I (A1b-I) towards ABTS, syringaldazine (syr) and DMP as substrates.

Substrate	Km (mM)			k_{cat} (min^{-1})			k_{cat}/Km (mM^{-1} min^{-1})		
	A1b	A1b-P	A1b-I	A1b	A1b-P	A1b-I	A1b	A1b-P	A1b-I
ABTS	0.37	0.64	0.16	$1.5x10^5$	$3.4x10^5$	$1.0x10^5$	$4.1x10^5$	$5.3x10^5$	$5.0x10^5$
syr	0.22	0.03	0.03	$2.0x10^4$	$3.1x10^4$	$7.8x10^4$	$9.1x10^4$	$1.0x10^6$	$2.2x10^5$
DMP	0.26	0.52	0.35	$3.6x10^5$	$5.9x10^4$	$3.4x10^4$	$1.4x10^6$	$1.1x10^5$	$1.5x10^5$

POXA1b-I and POXA1b-P showed a significant increase in affinity (about tenfold-lower Km) towards syringaldazine than POXA1b and a lower k_{cat} towards DMP, whilst only slight differences in Km and k_{cat} of the three isoforms towards the substrate ABTS can be observed. Therefore the structural differences among the isoforms are reflected on their catalytic properties.

Tab. 4. $t_{1/2}$ values, at different temperatures and pHs, of extra-cellular laccase POXA1b; extra-cellular laccase from PMSF-containing medium, POXA1b-P; and intracellular laccase, POXA1b-I.

	POXA1b	POXA1b-P	POXA1b-I
Temperature (50°C)	8 h	13 h	10 h
Temperature (60°C)	3 h	8 h	9 h
pH 3	11 h	9 h	8 h
pH 5	38 h	14 h	6 h
pH 7	38 days	9 days	6 days
pH 9	33 days	13 days	4 days
pH 10	100 days	32 days	12 days

As far as the stability of the three isoforms is concerning, the resistance at 50 and 60 °C and the stability at various pH values have been determined. As shown in Table 4, both POXA1b-I and POXA1b-P are more stable at high temperature respect to POXA1b. On the contrary, POXA1b displays an unusual stability at alkaline pH whilst the same characteristic is not so marked for POXA1b-I and POXA1b-P.

In conclusion, POXA1b is a promising candidate for biotechnological applications requiring alkaline conditions. Noteworthy the improvement of enzyme stability is a result of a natural process which involves fungal secreted proteases. As a fact, the unprocessed isoforms display higher molecular weights and different catalytic constants. Further investigation are needed in order to recognise protein structural sites altered by the process and to understand how these modifications affect the structure-function relationship of the protein. Moreover the identification of the involved protease/s could elucidate a complex proteolytic system addressed to generate a more stable biocatalyst.

REFERENCES

1. Hatakka A., FEMS Microbiol. Rev. 13 (1994) 125-135
2. Thurston C.F., Microbiology 140 (1994) 19-26
3. Palmieri G., Giardina P., Marzullo L., Desiderio B., Nitti G. and Sannia G., Appl. Microbiol. Biotechnol., 39 (1993) 632-636.
4. Palmieri G., Giardina P., Bianco C., Scaloni A., Capasso A., and Sannia G., J. Biol. Chem. 272, (1997) 31301-31307.
5. Giardina P., Palmieri, G. Scaloni A., Fontanella B., FaracoV., Cennamo G. and Sannia G., Biochem. J. 341 (1999) 655-663
6. Giardina P., Aurilia V., Cannio R., Marzullo L., Amoresano A., Siciliano R., Pucci P., and Sannia G., Eur. J. Biochem. 235 (1996) 508-515
7. Palmieri G., Giardina P., Bianco C., Fontanella B. and Sannia G., Appl. Environ. Microbiol. 66 (2000) 920-924
8. Palmieri G., Bianco C., Cennamo G., Giardina P., Marino G., Monti M. and Sannia G., Appl. Environ. Microbiol. 67 (2001) 2754-2759
9. Laemmli, U. K. Nature 227 (1970) 680-685
10. Germann, U. A., Muller, G., Hunziker, P. E., and Lerch, K., J. Biol. Chem. 263 (1988) 885-896
11. Yaver, D. S., Overjero, M. C., Xu, F., Nelson, B. A., Brown, K. M., Halkier, T., Bernauer, S., Brown, S. H. and Kauppinen, S. Appl. Environ. Microbiol. 65 (1999) 4943-4948.

Biotechnology in the Pulp and Paper Industry
L. Viikari and R. Lantto (Editors)

Reactivity of high and low molar mass lignin in the laccase catalysed oxidation

M. L. Niku-Paavola[a], T. Tamminen[b], B. Hortling[b], L. Viikari[a] and K. Poppius-Levlin[b]

[a]VTT Biotechnology, P.O. Box 1500, FIN-02044 VTT, Finland
E-mail: marja-leena.niku-paavola@vtt.fi

[b]KCL, Science and Consulting, P.O. Box 70, FIN-02151 Espoo, Finland
E-mail: tarja.tamminen@kcl.fi

Residual lignins from pine kraft pulp (KraftRL), oxygen delignified pine kraft pulp (KraftOORL), and fractionated KraftRL were treated with laccase. The reactivity of lignin substrates in laccase reaction (oxygen uptake) and alterations in molar mass distribution and in content and type of phenolic hydroxyl groups were analysed.

KraftRL was fractionated according to the molar mass. Each fraction had an equal level of phenolic hydroxyl groups. The lower the molar mass, the higher was the reactivity and the degree of polymerization. The results indicate that the molar mass is critical for polymerization and not the content of free phenolic hydroxyl groups. Generally, the content of free phenolic hydroxyl groups was only slightly decreased in these polymerization and depolymerization reactions although they are the sites of primary oxidation.

Also the structure of lignin was shown to affect the reactivity. The reactivities of KraftRL and KraftOORL were similar when evaluated by oxygen uptake. Although both were polymerized, the polymerization of KraftOORL was not as strong as that of KraftRL. Obviously the competing depolymerization reactions were more pronounced in KraftOORL than in KraftRL.

1. INTRODUCTION

White-rot fungi are the most efficient lignin degraders in nature. During their growth on lignin containing materials they secrete several enzymes capable to modify lignin substructures (reviewed in [1]). The role of individual enzymes in lignin degradation has been studied intensively. Fungal cultures grown on lignosulfonates, lignosulfonate fractions [2], Kraft and Björkman-lignin [3] and sulfite-pulp effluent [4] have shown to secrete laccase and to polymerize lignin. Polymerization of macromolecular lignin has identified as the main reaction also by purified laccase enzymes separated from the cultures. However, minor degradation of specific lignin substrates has also been reported (reviewed in [5]).True delignification resulted finally due to laccase-mediator concept [6,7], where a proper primary substrate of laccase mediates the oxidation towards a secondary substrate, e.g. lignin. Recently it was shown that also laccase alone may modify lignin in a destructive manner. Maldi-Tof-MS-technique revealed that phenolic substances like creosol and syringaldehyde initially formed oligomers in the laccase reaction but were thereafter degraded to a small extent [8]. Slight depolymerization was also discovered when laccase treated synthetic guaiacyl lignin was analysed by GPC [9]. Further, the amount of biphenyl structures in residual lignin isolated from oxygen delignified pine kraft pulp was decreased by laccase treatment, indicating lignin degradation [10].

To get more information concerning the relationship between laccase function and lignin structure, residual lignins from pine kraft pulp and oxygen delignified pine kraft pulp were treated with laccase. The residual lignin from kraft pulp was also fractionated as a function of molar mass, and the fractions were treated with laccase. Reactivity of lignin substrates in laccase reaction was analysed as a function of oxygen consumption. Simultaneous alterations in molar mass distribution and in content and type of phenolic hydroxyl groups of lignin substrates were estimated.

2. MATERIALS AND METHODS

2.1. Lignin substrates

Residual lignins were isolated from pine kraft pulp(KraftRL) and from two-stage oxygen-delignified pine kraft pulp (KraftOORL) as described [11]. For the laccase treatments, the lignins were dissolved in 0.1M NaOH, pH was adjusted to 5 with citrate buffer and 1M HCl. The final lignin concentration (based on dry weight) was 0.15 mg/ml and 1.5 mg/ml.

Lignin fractions with low polydispersity were obtained from KraftRL by preparative gel permeation chromatography with Superdex 200 prep grade gel (Pharmacia Biotech) in 0.1 M NaOH [12]. The lignin sample, 100mg, was reduced with $NaBH_4$, dissolved in 5 ml of eluent and fractionated. The elution curve was obtained by measuring the content of lignin in each 10 ml fraction at 280 nm using an absorptivity of 20 l/g*cm. The fractions were collected as indicated in Figure 5. Fractions 24 and 25 were pooled to fr.1; fraction 26 was named fr. 2; fraction 27 fr.3; fraction 28 fr.4; fraction 29 fr.5; fraction 30 fr.6; fractions 31 and 32 were pooled to fr.7. These fractions were diluted with 0.1M NaOH, the pH was adjusted to 5 with citrate buffer and 1M HCl. The final lignin concentration was 0.13 mg/ml (based on UV absorption at 280 nm).

2.2. Laccase treatments

Laccase treatments were performed using partially purified laccase preparation from *Trametes hirsuta* strain VTT-D-443 prepared as described [13]. Enzyme doses for unfractionated lignins were 20nkat/mg (corresponding to *ca.* 740 nkat/g of pine kraft pulp and 260 nkat/g of two-stage oxygen-delignified pine kraft pulp) and 170nkat/mg (corresponding to 6300 and 2200 nkat/g of the two pulps, correspondingly). Dose 170 nkat/mg was used for the lignin fractions. High enzyme doses were used to allow a complete oxidation.

2.3 The reactivity of lignin

The reactivity of lignin was measured as oxygen consumption during the reaction with laccase in closed erlenmeyer flask using a Clark type oxygen electrode (Orion 97-08)[13].

After the measurement (15min) at room temperature the lignin samples were stored for additional 72h at 4°C and analysed directly from the solution.

2.4. Analysis methods

Phenolic hydroxyl groups, content and type, were determined using the modified differential ionization UV method [14].

Molar mass distributions were measured by analytical gel permeation chromatography with Superdex 200 prep grade gel in 0.5 M NaOH. Detection was at 280 nm and calibration with monodisperse Na-polystyrene sulfonates [12].

3. RESULTS

3.1. Reactions of KraftRL and KraftOORL, catalysed by laccase

The reactivity of the residual lignins in the laccase reaction was assayed by oxygen consumption. Only minor differences were observed between KraftRL and KraftOORL. Increase in laccase dose from 20 to 170nkat/mg increased oxygen uptake 2-3 fold. Ten-fold increase in lignin concentration at constant laccase dose (20 nkat/mg) increased oxygen uptake approximately ten-fold, indicating that the same oxidation level in lignin was reached even at higher lignin concentration (Figure 1).

The effect of laccase on the content of phenolic hydroxyl groups is shown in Figure 2. Slight reduction is seen for KraftOO RL, but not for KraftRL. The proportion of α-conjugated phenolic structures increased slightly in both series indicating formation of α-oxidized structures.

When lignin content in the sample was increased ten-fold (1.5mg/ml) the results were analogous (not shown). Phenolic hydroxyl groups in both lignins remained nearly constant, and the amount of α-conjugated phenols increased .

The effect of laccase treatment on the molar mass distribution was determined by analysing the samples containing lignin 1.5mg/ml. GPC-elution curves in Figure 3a and 3b show that untreated KraftRL contains a minor amount of high molar mass fraction eluted in the void volume of the GPC-column (Figure 3a). The amount of this high molar mass fraction increased remarkably by the laccase treatment. The KraftOORL before and after treatment is in the range of the column separation performance and the increase in the amount of large molecules is very modest (Figure 3b).

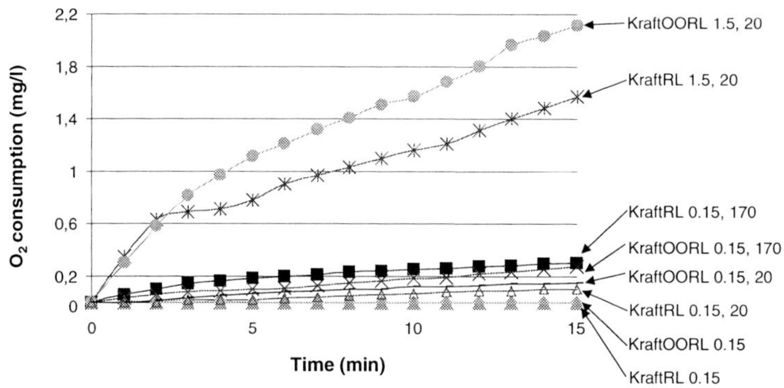

Figure 1. The reactivity of KraftRL and KraftOORL as a function of time, lignin concentration (0.15 and 1.5mg/ml) and laccase charge (20 and 170 nkat/mg).

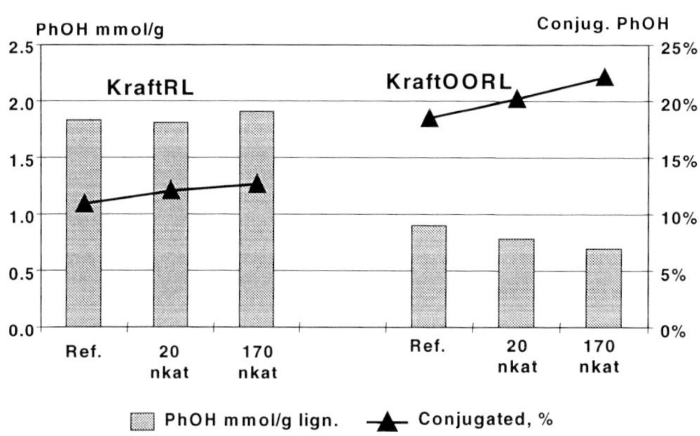

Figure 2. The effect of laccase on the content of phenolic hydroxyl groups. Lignin concentration 0.15mg/ml, laccase charge 20 and 170 nkat/mg.

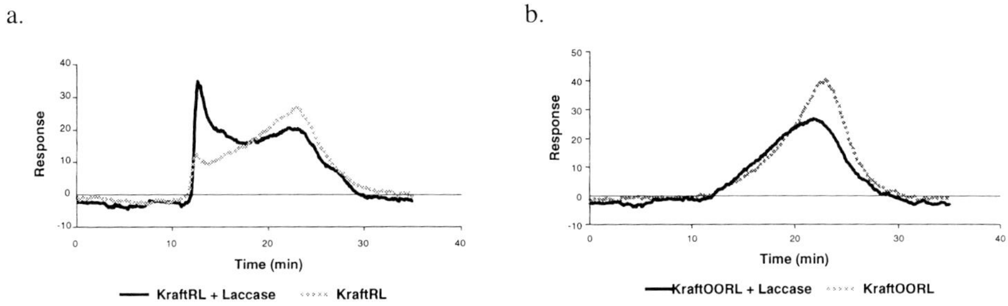

Figure 3. GPC elution curves of KraftRL (a) and KraftOORL (b) before and after laccase treatment. Lignin concentration 1.5 mg/ml, laccase charge 20 nkat/mg.

The weight average molar masses and polydispersity of the samples before and after laccase treatment are given in Figure 4. The molar mass and polydispersity of both lignin samples increased in the laccase treatment, that of KraftRL more prominently than that of KraftOORL.

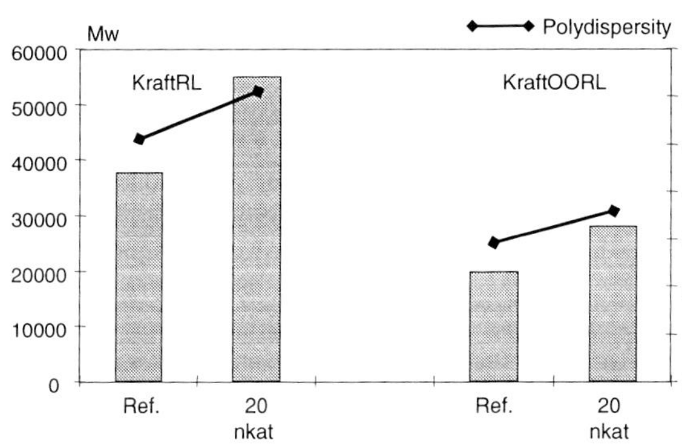

Figure 4. Changes in the weight average molar mass and polydispersity of KraftRL and KraftOORL in laccase treatments. Lignin concentration 1.5mg/ml, laccase charge 20 nkat/mg.

Oxygen uptake seems not to correlate directly with the increase in molar mass. The consumption of oxygen cannot be explained by the elimination of phenolic structures either, as their contents remained almost constant during the laccase treatment. Thus, oxygen was obviously consumed in other types of reactions, which could not be analysed with the methods applied.

3.2 Reactions of fractionated KraftRL catalysed by laccase

To get more information about the factors affecting the reactivity of lignin during laccase treatments, lignin substrates with low polydispersity were prepared by fractionation of KraftRL using preparative GPC.The fractions collected are indicated in Figure 5.

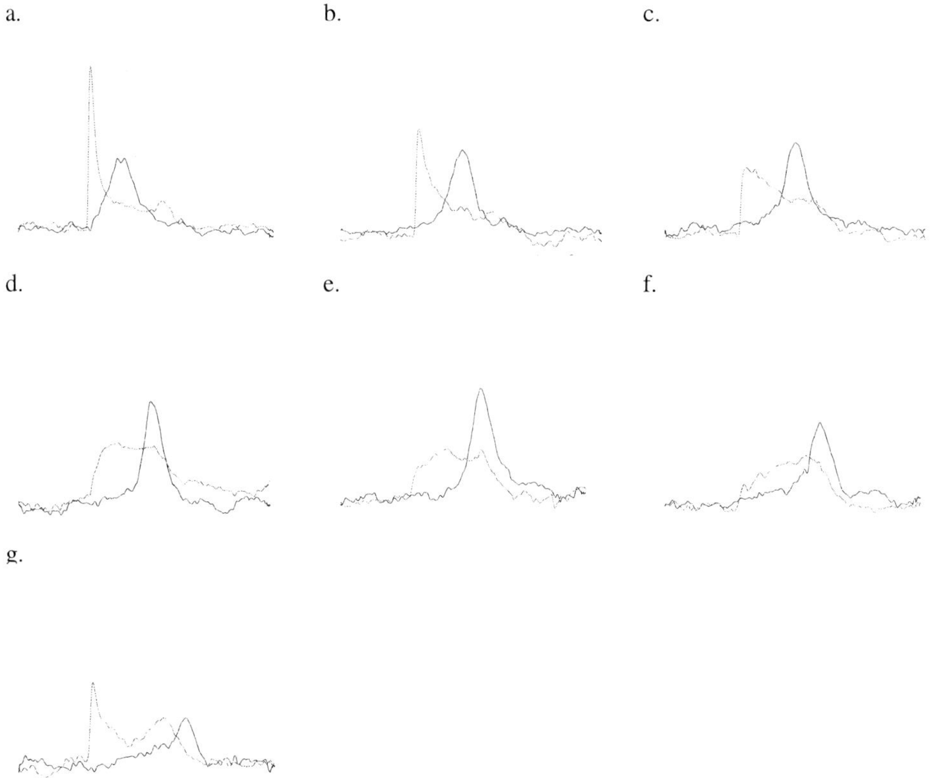

Figure 8. GPC elution curves of the KraftRL fractions before ------ and after - - -laccase treatment. .Lignin concentration 0.13 mg/ml, laccase charge 170 nkat/mg. Fr1 (a); fr2 (b); fr3 (c); fr4 (d); fr5 (e); fr6 (f) and fr7 (g).

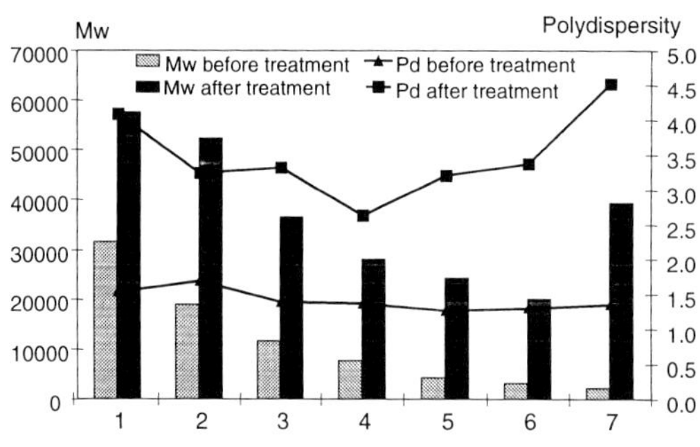

Figure 9. Changes in weight average molar mass and polydispersity of KraftRL fractions in the laccase treatments. Lignin concentration 0.13 mg/ml, laccase charge 170 nkat/mg.

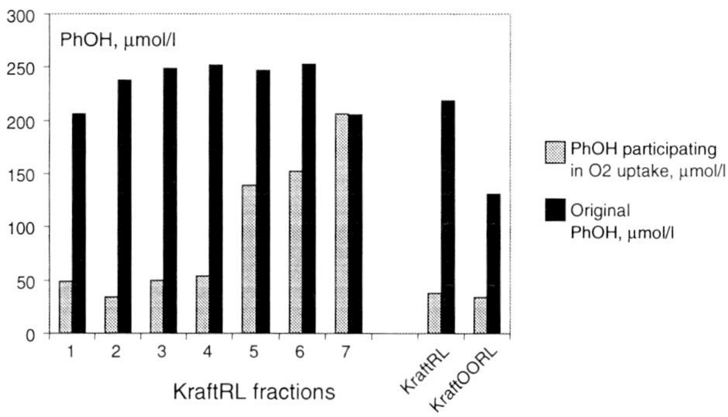

Figure 10. Comparison between the original, determined PhOH-group content and the theoretically oxidized content calculated from the oxygen consumption.

4. DISCUSSION

The reactions of unfractionated KraftRL and KraftOORL with laccase proceeded in different manners. The reactivities of KraftRL and KraftOORL were similar when evaluated by oxygen uptake. However, the treatment with laccase resulted in pronounced polymerization of KraftRL while the polymerization of KraftOORL was not as strong as that of KraftRL. The amount of phenolic hydroxyl groups remained practically constant in all cases, but the proportion of α-oxidized phenolic structures increased, indicating formation of this type of structures.

The molar mass of KraftRL affected the reactivity. The reactivities (oxygen uptake) of the high molar mass KraftRL fractions towards oxidation by laccase were on a similar level as those of the unfractionated KraftRL, while the reactivities of the low molar mass fractions were significantly higher with a consequent increase in molar mass. Generally, the amount of PhOH-groups decreased only slightly, but most significantly in the largest molar mass fraction. On the contrary, the proportion of α-oxidized phenolic structures increased most remarkably in the smallest molar mass fraction.

Oxygen seems to be consumed mainly in the polymerization reactions of low molar mass KratRL whereas in the high molar mass fractions oxygen is probably also consumed in degradative reactions. This is consistent with the observed more pronounced decrease in the content of phenolic hydroxyl groups in the high molar mass lignin fractions.

The oxidation of phenolic compounds by laccases is known to start by oxidation of PhOH-groups [15]. The primarily formed phenoxyl radical is resonance stabilized, and therefore the reaction may proceed further via any of the reactive sites in the aromatic ring or side chain. As the content of phenolic hydroxyl groups was practically unchanged in most cases, the main reaction route seemed not to involve oxidation of PhOH or polymerization *via* PhOH. The reaction routes observed were formation of carbonyl groups in the benzylic position, polymerization and depolymerization (disproportionation). The content of PhOH was not decisive, based on the equal oxygen uptake by KraftOORL and KraftRL despite of the higher PhOH content of the latter.

However, in other respects the structure of lignin was crucial. The residual lignin in oxygen delignified pulp is known to be condensed, which includes tertiary carbons. This kind of structure may provide stabilization of the intermediate radical at the cross-linking sites, leading favourably to depolymerization. Another specific feature of KraftOORL is its higher content of carboxylic acid groups compared to KraftRL.This may lead to more open and flexible macromolecular configuration in KraftOORL compared to KraftRL [16]. This may be one reason for the observed better accessibility of the phenolic hydroxyl groups of KraftOORL with laccase, even if their frequency is lower than in Kraft RL.

5. CONCLUSIONS

The molar mass and structure of lignin was shown to affect the reactivity of lignin in oxidation by laccase. These results add to the general understanding of the enzymatic modification of lignin and help to develop enzyme aided depolymerization or polymerization methods.

REFERENCES

1. A. Hatakka, FEMS Microbiol. Rev. 13 (1994) 125.
2. J.-F. Selin, V. Sundman and R. Räihä, Arch. Microbiol. 103 (1975) 63.
3. A. Huttermann, C. Herche and A. Haars, Holzforschung 34 (1980) 64.
4. X. Geng, R. Kondo, K. Sakai and Z. Lee, J. Wood Sci. 45 (1999) 143.
5. S. Sarkanen, R.A. Razal, T. Piccariello, E. Yamamoto and N.G. Lewis, J. Biol. Chem. 266 (1991) 3636.
6. R. Bourbonnais and M.G. Paice, Appl. Microbiol. Biotechnol. 36 (1992) 823.
7. H-P. Call, Process for Modifying, Breaking down or Bleaching Lignin, Materials Containing Lignin or Like Substances WO 94/29510 (1994).
8. A. Potthast, T. Rosenau, H. Koch and K. Fischer, Holzforschung 53 (1999) 175.
9. E. Strebotnik and K.E.Hammel, J. Biotechnol. 81 (2000) 179.
10. T.Tamminen, T. Elder, W. Wang, and K. Poppius-Levlin. Proceedings of the 7th International conference on biotechnology in the pulp and paper industry, Vancouver, BC, Canada, June 16-19, 1998. Vol.A, p A107.
11. K. Poppius-Levlin, W. Wang, T.Tamminen, B.Hortling, L.Viikari and M-L.Niku-Paavola, J. Pulp Pap. Sci. 25 (1999) 90.
12. B. Hortling, E.Turunen and P. Kokkonen, Proceedings of the 10th International Symposium on Wood and Pulping Chemistry, Yokohama, Japan, June 7-10, 1999. Vol 1, p. 48.
13. M-L. Niku-Paavola, W. Wang, K. Poppius-Levlin, H. Anke and L.Viikari, Proceedings of the 9th International Symposium on Wood and Pulping Chemistry, Montreal, Canada, June 9 - 12, 1997. 80-1.
14. T.Tamminen and B.Hortling, In: Progress in Lignocellulosics Characterization. Ed. D. Argyropoulos, TAPPI Press, Atlanta, 1999, Chapter 1, p. 1.
15. T.K. Kirk and M. Shimada, M, In: Biosynthesis and Biodegradation of Wood Components. Ed. T. Higuchi, Academic Press, Inc., Orlando, 1985, p. 579.
16. M. Norgren and B. Lindström, Holzforschung 54 (2000) 528.

Biotechnology in the Pulp and Paper Industry
L. Viikari and R. Lantto (Editors)

The effect of oxidation with single electron oxidants compared to a laccase treatment of TMP wood fibres

S. Barsberg and J. Hassingboe

Plant Fibre Laboratory, Royal Veterinary and Agricultural University, Taastrup, Denmark, sbar@kvl.dk, jeh@kvl.dk

Previous work has shown that laccase is able to catalyse auto adhesion of lignin rich wood pulp. This promises the replacement of environmentally hazardous glue in composites with a laccase treatment of the pulp. In the present work we compare spectroscopic results of this treatment with those of other better defined oxidative treatments to examine the origin of this auto adhesion. The results of the laccase treatment is also achieved by laccase-ABTS or –HBT treatment. This points to the fact that the auto adhesion depends on general features of solid state lignin oxidation.

1. INTRODUCTION

The utilisation of enzymes related to biosynthesis and -degradation of lignin for, e.g., bleaching of wood fibre pulps has in recent years been the subject of much research. Laccases are known to catalyse the degradation of lignin phenolic model compounds (albeit some laccases degrade non-phenolic lignin models as well [1]) reducing oxygen to water. In order to catalyse the degradation of lignin of wood pulps, which is poorly accessed by the enzyme active site, the enzyme needs a low molecular weight substrate to mediate the oxidation between the lignin and the laccase-oxygen complex. Such a substrate is commonly referred to as a *mediator*.

The laccase catalyses the oxidation of the inactive mediator into an active radical state. In this state the mediator oxidises lignin (or other substrates) by which it returns to its inactive reduced state – a mechanism proposed already in 1985 [2]. A non-natural mediator can thus be added together with laccase to bleach pulp fibres. Such biobleaching systems typically employ ABTS (2,2-azinobis-3-ethylbenzthiazoline-6-sulfonate) or HBT (1-hydroxybenzo-triazole) as mediator and numerous other compounds will function as well [3-7]. The oxidation mechanisms have been studied by the replacement of fibres with lignin model compounds [8-12]. A notable observation is the fact that the substrate range of laccase is extended by the mediator to include, e.g., non-phenolics.

Oxidation of solid state lignin may give rise to phenomena not captured by studies on model compounds. In our laboratory we have examined the effect of a laccase treatment of wood TMP (thermo mechanical pulp) on properties of boards made from these treated fibres [13, 14]. A significant increase in board strength was obtained without addition of glue. This promised the perspective of replacing environmentally hazardous glue in, e.g., MDF (medium density fibre) boards, with a laccase pre-treatment. This treatment did *not* depend on the addition of non-natural mediators.

The fibres obtained from the treatment were examined by spectroscopic means [15]. Radicals, with lifetimes of *weeks*, were generated in the fibres, together with spin-less species which caused fluorescence quenching. An excess UV-VIS absorption band appeared with these radicals with a corresponding reddening of the fibres. It was suggested that (1.) lignin cation radicals and/or neutral (phenoxyl) radicals, as well as their decay products, were generated and that long lived lignin cation radicals caused most of the UV-VIS absorption, whereas lignin decay products caused fluorescence quenching, and that (2.) these species were generated by the action of one or several unknown mediators already present in the fibre samples.

Future improvement and utilisation of laccase catalysed wood pulp adhesion depends on a better understanding of the basic mechanisms of the treatment with respect to both the solution phase as well as the solid (fibre) phase.

In this work we examine the spectroscopic modifications of beech TMP fibres resulting from laccase-mediator treatments (ABTS & HBT) and treatment with the mild one-electron oxidant HCF (hexacyanoferrat – $Fe(CN)_6$) in its two oxidation states (HCFII/HCFIII). The fibres have been previously thoroughly extracted to reduce the effect of the already present unknown mediators. These modifications are compared with those achieved by the less well defined laccase treatment.

2. MATERIALS AND METHODS

2.1. Materials

Beech (*Fagus sylvatica*) TMP fibres were supplied by the MDF-board plant of Junckers Industries, Køge, Denmark, where fibration is carried out using an Asplund process. The fibres were frozen after fibration, thawed and conditioned to the ambient environment just before usage. A *Trametes villosa* laccase, SP504 (EC 1.10.3.2), was supplied by Novo Nordisk A/S, Bagsværd, Denmark. Laccase activity (LACU) was measured using a syringaldazine assay [13-15]. ABTS, HBT, (Potassium-)HCFII and HCFIII were used as obtained from Sigma-Aldrich A/S.

2.2. Pre-treatment/Extraction

20 g (dry matter) of fibres were immersed into 1 l, pH = 4.5, T = 40°C, 0.1 M acetate buffer. After 30 min's the solution was removed through a filter funnel. 1 l of new buffer solution at the same conditions was added to the drained fibres and the extraction procedure repeated in this manner all-in-all 10 times. A ready-to-use extracted fibre batch resulted upon drying the fibres (overnight in an oven at 40°C). An additional batch of fibres was obtained by subjecting 10 times extracted fibres to 5 additional extractions each of which lasted 2h (but otherwise the same conditions).

2.3. Laccase-mediator / HCF treatments

Pre-treated samples of 1.00 g dry material was immersed in 50 ml, pH = 4.5, T = 40°C, 0.1 M acetate buffer containing 10^{-3} M ABTS/HBT or 10^{-3} M HCF. Controls were made (1.) without mediator, with laccase, (2.) without laccase, with mediator and (3.) with pure buffer. Previous work [13-15] showed that treatment with heat inactivated laccase was equivalent to the pure buffer (control) treatment. Time *zero* was defined as the moment when 260 LACU of laccase solution (~ 0.2 ml) was added (or for treatments without laccase as the moment when the solution was added to the dry fibres). The treatment finished when the fibre suspension was removed through a filter funnel at time *t*, the fibres thoroughly rinsed with de-ionised water, and finally thoroughly drained and transferred to an oven where they were dried overnight at 40ºC. The rinsing lasted typically 5-10 min's.

Fibres extracted 15 times were subjected to laccase-ABTS treatment for 10, 20, 30, 45 min's, 1h and 6h, laccase –, laccase-HBT –, HCFII – and HCFIII treatments for 1h, an ABTS treatment for 6h. Pure buffer control treatments for 1h and 6h were also made. Fibres extracted 10 times were subjected to a 1h laccase treatment and a 1h control (pure buffer) treatment. In the remaining part of the paper we refer to fibres extracted 15 times if nothing else is mentioned.

2.4. Spectroscopic measurements

Electron Spin Resonance (ESR) measurements were done with a Bruker EMS 104 spectrometer. Each fibre sample (~50 mg) was densely packed within a Wilmad glass tube such that the bottom ~10 mm of the tube contained the sample. The sample was placed in the centre of the 35 mm long instrument cavity. UV-VIS measurements were done in diffuse reflectance mode with a Cintra 40 Spectrometer equipped with an integrating sphere. Fluorescence measurements were done with a Perkin Elmer LS50B spectrometer using Front Face setup. All measurements were done under ambient conditions with (dry) conditioned fibre samples.

3. RESULTS

All results on pure buffer control fibres (1h and 6h treatment times) show no time dependency. It is therefore safe to extrapolate all results of the 1h treatment time to smaller times. All reported results are averages of five independent samples with standard deviation of the average as uncertainty. Decay behaviour of the results is not a subject of this work. The meaningfulness of these are, however, ensured by the fact that the least stable results, ESR excess signals generated by some of the treatments, have half lives exceeding two weeks (results not shown) and by the fact that all measurements were done within five days after a treatment.

3.1. ESR

The ESR intensity I_s, given in arbitrary units, is normalised to the dry matter content (weight) of each sample s. Results are, however, often most appropriately given *relative* to the intensity I_c of a pure buffer control as $\Delta I_s = I_s - I_c$.

The laccase-ABTS system shows two phases: A rapid initial production of radicals, taking place for $t < 10$ min's, followed by a slow production, see fig. 1. For $t = 6$h the excess signal has increased further to $\Delta I = 581\pm14$. The initial population of radicals is I_c (1h)= 268±8 and remains constant up to $t = 6$h where I_c (6h)= 262±6. ΔI of the laccase treatment (1h: 164±13) is less than half of that observed for the laccase+ABTS treatment (1h) and increases only very slowly (6h: 188±8). ΔI of the ABTS treatment is non-zero (6h: 33±7).

The laccase-HBT system as well as *the HCFIII system* also generate radicals albeit to a smaller extent with the HCFIII system as the weakest oxidising system.
The 1h laccase treatment (control) shows no dependency on the extent of pre-extraction with $\Delta I = 156 \pm 13$ (10 times extracted) and $\Delta I = 164 \pm 13$ (15 times extracted).

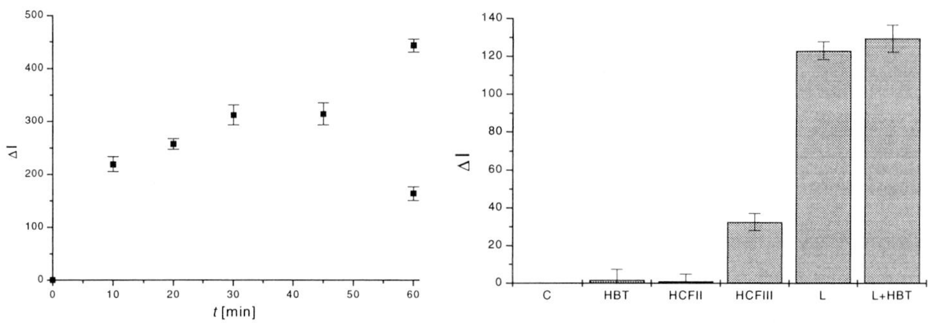

Figure 1. Left: The production of radicals by the laccase-ABTS system. The lowest ΔI (60 min) is for the laccase treatment. Right: Production of radicals by the other systems (t = 1h). The laccase treatment is not identical to that of the left part of the figure due to the fact that a different batch of fibres was used for these systems.

3.2. UV-VIS spectra

The laccase-ABTS system shows now three phases: An excess spectrum $\Delta F(\lambda)$, the *shape* of which is nearly time-invariant for $t <$ 1h, with maximum absorption $\Delta F(\lambda)_{max}$ for $\lambda = 510$ nm and a shoulder appearing at $\lambda = 400$ nm. The dependency of $\Delta F(\lambda)_{max}$ on t, for $t <$ 1h, roughly correlates with the ESR results $\Delta I(t)$, where it exhibits two phases, i.e., a rapid increase of $\Delta F(\lambda)_{max}$ followed by a slower additional increase (results not shown). For $t >$ 1h a *third* phase with spectral shape dynamics sets in as is evident when the spectral shape for t = 1h is compared with that of t = 6h, see fig. 2.

The laccase-HBT system is characterised by the same spectral shape as the laccase-ABTS system (t = 1h). The value of $\Delta F(\lambda)_{max}$ is, however, half of that of the laccase-ABTS system. *The HCFIII system* exhibits an even smaller $\Delta F(\lambda)_{max}$ value and the shoulder at $\lambda = 400$ nm does not appear, see fig. 2.

The 1h and 6h (results not shown) laccase treatments give the same spectral shapes as the 1h laccase-ABTS treatment.

The spectrum of the ABTS treated fibres (t = 6h) shows absorption for $\lambda < 400$ nm whereas for $\lambda > 400$ nm it is identical to that of the control fibres (results not shown). The spectra of HCFII and HBT treated fibres were identical to the spectrum of the pure buffer control.

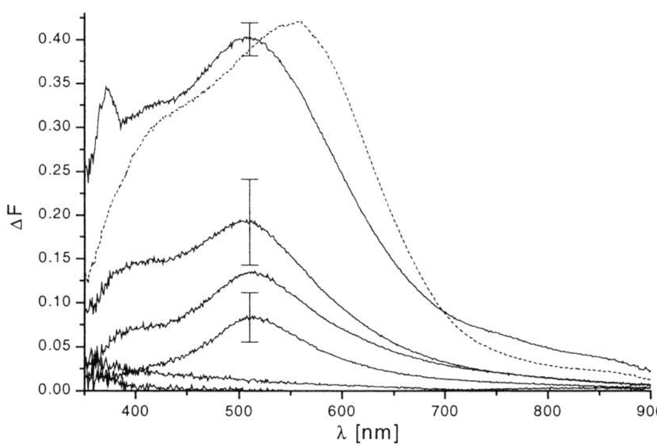

Figure 2. UV-VIS spectra in Kubelka-Munk units relative to the pure buffer control. $F=(1-R)^2/2R$ where R is reflectance. The top two spectra are laccase-ABTS for t = 1h (full line) and t = 6h (dashed line, *reduced* by a factor 3). In order of decreasing ΔF follows: Laccase-HBT, Laccase, HCFIII, HCFII and HBT (all t = 1h).

3.3. Fluorescence

All oxidative treatments lead generally to emission quenching for all excitation energies with the most pronounced quenching obtained for excitation at 350 nm [15]. The results are therefore expressed in terms of a quenching factor α, for excitation at 350 nm, where $\alpha = 1$ for the pure buffer control by definition. The emission spectrum $I_s(\lambda_{em})$ of any sample s is then expressed as $I_s(\lambda_{em}) = \alpha(s)\, I_c(\lambda_{em})$ where $I_c(\lambda_{em})$ is that of the control.

The laccase-ABTS system shows again two time domains: A rapidly increasing quenching (i.e., decreasing α) followed by a slower additional quenching, see fig. 3, reaching α = 0.329±0.012 for t = 1h. For t = 6h the quenching is pronounced with α = 0.177±0.007.

The laccase-HBT system (t = 1h) and *the HCFIII system* (t = 1h) also lead to quenching, but to a smaller degree, with α = 0.56±0.02 and α = 0.76±0.03, respectively.

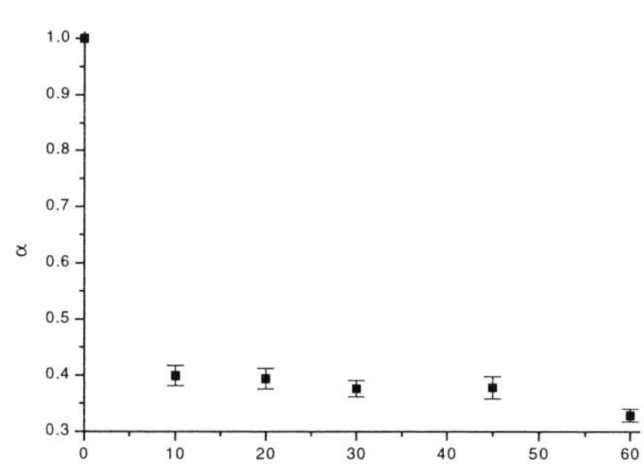

Figure 3. Development of quenching factor α with t for the laccase-ABTS system.

The 1h laccase treatments also give rise to quenching where the 10 times extracted fibres show a relatively larger degree of quenching (α = 0.59±0.02) compared to the 15 times extracted fibres ($\alpha = 0.72 \pm 0.03$). The laccase treatment causes additional quenching for $t > 1$h with $\alpha = 0.64 \pm 0.02$ for $t = 6$h.

The other treatments also exhibit quenching where $\alpha = 0.79 \pm 0.03$ (6h ABTS), $\alpha = 0.86 \pm 0.03$ (1h HBT) and $\alpha = 0.89 \pm 0.02$ (1h HCFII).

Table 1. Quenching factor α. The table is divided into a left and a right part after the two different batches of extracted fibres used. The laccase treatment was applied to both 10 and 15 times extracted fibres symbolised by "L_{10}" and "L", respectively.

	L_{10}	L	ABTS	L-ABTS		L	L-HBT	HCFIII	HCFII	HBT
t =1h	0,59 ±0,02	0,72 ±0,03		0,330 ±0,012		0,653 ±0,012	0,56 ±0,02	0,76 ±0,03	0,89 ±0,02	0,86 ±0,03
t =6h		0,64 ±0,02	0,79 ±0,03	0,177 ±0,007						

4. DISCUSSION

Of the treatments examined the laccase-ABTS system shows the largest effects. These are beyond doubt due to the catalytic effect of ABTS in addition to the effect of the active enzyme alone. The catalytic effect of HBT is less pronounced but shows up clearly in the fluorescence results. The results of the HCFIII treatment show that the effects need *not* be tied to the presence of the active enzyme. It does, however, have a clear effect when present without mediator as is seen from the results of the laccase treatment. The fluorescence results show that the extend of pre-extraction (10 vs. 15 cycles) also matters which points to the fact that a catalytic effect of natural mediators may be present in all oxidising systems.

HBT and HCFII have in themselves no oxidising effect since they generate no excess radical or – UV-VIS signal. They do, however, generate weak fluorescence quenching. This we attribute to their sorption into the lignin matrix of the fibres. The results of the ABTS treatment may also be explained in terms of sorption of neutral ABTS, which leads to the small UV-VIS peak at $\alpha = 370$ nm superimposed on the laccase-ABTS spectrum (see fig. 2).

All oxidising treatments lead to the concomitant generation of excess long-lived radicals, - UV-VIS absorption and fluorescence quenching. This generation divides into clearly distinguished phases as functions of treatment time *t* and nature of the oxidative system.

The active one-electron abstractors have quite different reduction potentials (vs. NHE) with $E_{red} = 0.41$ V for HCFIII [16], $E_{red} = 0.79/1.09$ V for ABTS cation radical/dication and $E_{red} = 0.79$ V for the laccase-oxygen complex [8]. HCFIII is thus a relatively weaker oxidant. Quantitative comparisons between results are not possible due to recycling of the mediator oxidation state which depends on the uncontrolled oxygen availability. The results do, however, suggest that HCFIII is a relatively weak oxidant since the majority of ABTS is converted (*via* laccase) to its radical state(s) within a few minutes, and since only a small fraction of HCFIII is reduced within 1h (UV-VIS results not shown). Even though the initial basic mechanism connecting the solution phase with the solid state phase may be different for the laccase-HBT system (the HBT radical, $E_{red} = 1.08$ V [8], may abstract a lignin H atom, or a lignin electron and a solvent proton) the effects are qualitatively the same for $t < 1$h.

The laccase treatment of non-extracted TMP fibres has the same general effects as the oxidative treatments examined in this work [15]. Most notable is the fact that the UV-VIS excess spectrum of this treatment shows the *same* features as the spectra of the laccase-ABTS – and the laccase-HBT systems, i.e., strong absorption at α = 510 nm and a shoulder at α = 400 nm. More work including $t > 1$h may, however, discriminate between the various systems.

The action of HCFIII and ABTS cation radical/dication is selective electron abstraction from the lignin producing *cation radicals*. The electron transfer rate from a given lignin sub-structure to, e.g., ABTS cation radical may be orders of magnitude larger than that to HCFIII due to the difference in reaction free energy (i.e., difference in E_{red} of the electron acceptors) assuming the invariance of other electron transfer parameters (reorganisation free energy, electronic matrix element).

The selectivity is also controlled by this rate. Hence, the ABTS dication abstracts electrons from both phenolic as well as non-phenolic lignin albeit at very different rates [8]. HCFIII only oxidises phenolic lignin since its E_{red} value is too low relative to, e.g., E_{ox} = 1.38 V [8] of veratryl alcohol which is a typical non-phenolic lignin model.

The weak shoulder at $\alpha \sim 400$ nm in the UV-VIS absorption spectrum observed for all oxidative systems, *except* the HCFIII system, may result from such selectivity. Hence, spectral dynamics (as function of t and system type) may result from a combination of material (lignin) heterogeneity and localisation of radicals and their decay products. It is clear that as the extent of oxidation for the relatively powerful laccase-ABTS system increases ($t > 1$h) the UV-VIS absorption increases significantly (showing a shift towards the red end of the spectrum, see fig. 2) together with increased fluorescence quenching and – ESR signal intensity.

The initial optical darkening (i.e., the excess UV-VIS spectrum), which have been observed before during enzymatic bleaching of high lignin content Kraft pulp (see [7] and references herein), may be caused by long-lived cation radicals [15]. The stability of these may be tied to the fact that their decay pathways depend on the surrounding (lignin) medium such that, e.g., de-protonation is suppressed relative to a solution medium [17]. Phenoxyl radicals may also be produced, either from cation radical decay or directly (i.e., the laccase-HBT system). Both types of radicals are likely produced by all the oxidative systems and - for phenols - these radicals are part of a general "scheme of squares" characterising the initial oxidation [18]. Subsequent reactions of these radicals, which may involve other induced radicals, mediator radicals or molecular oxygen, lead to spin-less products as observed in solution studies. These typically contain carbonyl groups which are generally non-fluorescent. The generally observed emission quenching may be due to such products.

The production of long lived radicals is most likely tied to the mechanism of laccase catalysed bonding either directly, by causing fibre-to-fibre covalent bonding during hot pressing, or indirectly by their decay into oxidation products which may improve adhesion by increasing the (non-covalent) Lewis acid-base bonding. The bonding is likely to depend on the type of oxidative system as well as on the physico-chemical state of the fibre lignin, i.e., its solid state properties. These depend on wood origin, pulping methods and further pre- or post-treatments of the pulp.

Biotechnology in the Pulp and Paper Industry
L. Viikari and R. Lantto (Editors)

Studies on Inactivation and Stabilization of Manganese Peroxidase from *Trametes versicolor*

Hakan Bermek[a], Kaichang Li[b] and Karl-Erik L. Eriksson[c]

[a] Istanbul Technical University, Dept. of Molecular Biology and Genetics, Maslak, 80626 Istanbul – Turkey
[b] Department of Forest Products, Oregon State University, Corvallis, OR 97331, USA
[c] Center for Biological Resource Recovery, Dept. of Biochemistry and Molecular Biology, University of Georgia, Athens, GA 30602-7229, USA

High concentrations of H_2O_2 cause inactivation of manganese peroxidase (MnP). The mechanism of this inactivation has been investigated. A method for stabilization of MnP used for pulp bleaching in the presence of high concentrations of H_2O_2 has been developed. It was demonstrated that formation of the inactive form of MnP (Compound III) is the major reason for the inactivation in which hydroxyl free radicals or superoxide anion radicals are not involved. It was also found that 1-hydroxybenzotriazole, 2,2,6,6-tetramethyl-1-piperidinyloxy free radical, violuric acid, 3-hydroxy-1,2,3-benzotriazin-4-(3*H*)-one and chlorpromazine stabilized MnP. The stabilizing effect by 1-hydroxybenzotriazole is due to the conversion of the inactive Compound III to the native enzyme.

1. INTRODUCTION

Lignin is a complex and heterogenous aromatic biopolymer which comprises 20-30% of the woody plants. Extensive degradation of lignin is accomplished by white-rot fungi (1). These fungi secrete several extracellular enzymes for this degradation. MnP, which is a Mn^{2+} and hydrogen peroxide dependent heme-containing peroxidase, constitutes one of the most important lignin-degrading enzymes together with lignin peroxidase and laccase.

Complete removal of lignin from wood pulp is one important step in the production of high quality paper. Pulp bleaching is conventionally achieved by treatment of unbleached pulp with chlorine based chemicals. However, chlorinated organic compounds formed in the reaction of chlorine with lignin or other pulp components in this process are dispensed in the effluents which cause environmental problems (2).

Biobleaching using MnP and/or laccase, on the other hand, could be developed into environmentally benign bleaching processes, and could thus be a very promising way to replace conventional ones. It has been shown that MnPs from various white-rot fungi are capable of bleaching kraft pulp (3,4,5,6). The actual oxidant in MnP-based pulp bleaching is believed to be chelated Mn^{3+} that is able to diffuse into the pulp fiber walls to attack and degrade lignin. MnP could thus be an ideal candidate for use in pulp bleaching. However, one of the major hurdles in applying MnP to pulp bleaching is that MnP is inactivated very rapidly. This inactivation is especially related to high concentrations of H_2O_2 (7). In this paper, we have investigated the reasons for this inactivation, and found a way of stabilizing MnP in the presence of high concentrations of H_2O_2.

2. MATERIALS AND METHODS

2.1. Organism

Trametes versicolor strain 52J was kindly provided by M. G. Paice (Pulp and Paper Research Institute of Canada).

2.2. Pulp

An unbleached hard wood kraft pulp (HWKP) was obtained from a Georgia, USA pulp mill.

2.3. Chemicals

All chemicals were analytical grade and obtained from commercial sources such as Sigma or Aldrich chemical companies, USA. In the experiments, only double distilled deionized water was used.

2.4. MnP production

Cultivation of the fungus and purification of the enzyme was performed as described before. (8). The culture medium was modified to contain 10 g l^{-1} glucose instead of 40 g l^{-1}, and the Cu content in the trace metals solution was reduced from 1 mM to 0.1 mM.

2.5. Studies of MnP stability

A variety of compounds such as metal ions, various organic compounds, free radical scavengers and organic peroxides, and hardwood kraft pulp (HWKP) were incubated with 1 μM MnP for various times at 30°C in the absence or presence of excess H_2O_2 (250 μM or higher). In each set of the experiments, 100 μl samples were drawn after 0, 15, 30, 60, 120 and 180 min of incubation, and the remaining MnP activity in these samples was assayed by monitoring the increase of Mn(III)-malonate formation at 270 nm (ε_{max}= 11590 M^{-1} cm^{-1}) at 30°C (9). Assay mixtures (3 ml) contained malonate buffer (pH 4.5, 50 mM), $MnSO_4$ (1.0 mM), H_2O_2 (0.1 mM) and 100 μl of the samples containing the MnP to be assayed. Assays continued for 2 min, and 1.0 U of activity was defined as 1 μmol of product formed per minute.

2.5.1. Metal ions

To investigate the ability of metal ions to stabilize MnP, 3 ml of reaction solutions containing 50 mM malonate buffer (pH 4.5), 1 μM MnP, and 1 mM of compounds containing various metal ions ($Mn[CH_3COO]_3$, $MnSO_4$, $FeSO_4$, $FeCl_3$, $CaCl_2$, $CoSO_4$, $CuSO_4$, or $MgSO_4$) were incubated in the presence (or absence) of 250 μM H_2O_2 for 0, 15, 30, 60, 120 and 180 minutes at 30°C.

A control experiment was run without any metal ions, and another control was run in the presence of $MnSO_4$ in distilled water (replacing the malonate buffer).

2.5.2. Various organic compounds

1-hydroxybenzotriazole (1-HBT), 3-hydroxy-1,2,3-benzotriazin-4-(3H)-one (HBTO), 2,2,6,6-tetrametyl-1-piperidinyloxy free radical (TEMPO), violuric acid and chlorpromazine were added (10 mM) respectively to a solution of MnP (1μM), and malonate (50 mM, pH 4.5). The solution was incubated in the presence of 250 μM H_2O_2 at 30°C. MnP stability was determined as described in the previous section. Experiments were also performed with various concentrations of 1-HBT: 1, 2, 5 and 10 mM. In the experiments with 10 mM 1-HBT, remaining H_2O_2 in the incubation solutions was monitored at various time points according to an established method (10). Different concentrations of H_2O_2 (250, 500, 750, 1000, 1500, 2000, and 2500 equivalents of MnP) were incubated with MnP in the presence of 1-HBT (10 mM) to assess the efficacy of 1-HBT for protection of MnP.

2.5.3. Absorption spectra of native MnP, MnP compounds I-III and 1-HBT

Absorption spectra of the Soret bands of MnP (0.5 mM) compound III were recorded after its formation by the addition of 250 eq H_2O_2, and monitored after addition of 1-HBT (10 mM) to this mixture by spectroscopic scanning between 350-700 nm. The reference cuvette contained malonate buffer only. Reaction of 1-HBT with MnP was also investigated with UV-Vis spectroscopy. A solution of 1-HBT (0.1 mM) was scanned between the wavelengths 200-700 nm after addition of MnP (1 U), $MnSO_4$ (0.1 mM), and H_2O_2 (0.2 mM) in malonate buffer (50 mM, pH 4.5). The scans were recorded within 0, 10, 20 and 30 mins of incubation.

2.5.4. Free radical scavengers

Free radical scavengers such as sodium benzoate (10 mM), mannitol (10 mM) or 150 U of superoxide dismutase (SOD) were used under similar conditions as in the experiments described above. Controls were also run without the addition of the scavengers.

2.5.5. Organic peroxides

Sodium perborate or m-chloroperoxy benzoic acid (mCPBA) instead of H_2O_2 was added at a concentration of 0.25 mM under the same conditions as described above.

2.5.6. Stability of MnP in the presence of pulp

To investigate the ability of Mn(II) and 1-HBT to stabilize MnP in the presence of wood pulp, 3 ml of reaction solutions containing 50 mM malonate buffer (pH 4.5), 1 μM MnP, 0.04 g (oven-dried weight) HWKP and 1 mM of $MnSO_4$ or 10 mM 1-HBT were incubated in the presence of 250 μM H_2O_2 for 0, 15, 30, 60, 120 and 180 minutes at 30°C.

3. RESULTS

3.1. Impact of various metal ions on MnP activity at excess H_2O_2 concentrations

In the control experiments in which MnP and malonate were incubated together with 250 eq H_2O_2, MnP activity decreased very rapidly; after 30 min, no activity was detectable in the incubated solution (Table 1). However, both Mn(II) and Mn(III) were able to protect the MnP activity in the presence of H_2O_2. It appeared that Fe(II) slightly inhibited MnP in the absence of H_2O_2 while addition of Fe(II) to the MnP solution containing H_2O_2 was somewhat less inhibitory. Fe(III) alone showed virtually no inhibitory effect on the enzyme, but did not protect the MnP activity in the presence of excess H_2O_2. Ca(II) and Co(II) did not inhibit MnP while 25% of the activity was lost in the presence of Mg(II) (Table 1). Cu(II) was found to be a strong inhibitor of MnP even in the absence of H_2O_2. With 1 mM Cu(II), MnP activity decreased by 65% during the first 30 min, but remained the same thereafter (Table 1). In fact, Cu(II) concentrations at even 10 μM were inhibitory to the enzyme (Data not shown). None of the four metal ions i.e. Ca(II), Co(II), Mg(II) and Cu(II) was thus effective to protect MnP activity against inactivation by H_2O_2.

Table 1. Stability of MnP incubated with 250 eq. Hydrogen peroxide and 1 mM of various metal ions.

Metals	Remaining MnP activity (%) at various incubation times (min)*							
	Metal only				Metal+250 eq. H_2O_2			
	0	15	30	60-180	0	15	30	60-180
None	100	93	92	95	100	9.5	3.5	0
Mn(III)	100	104	116	110	100	105	112	107
Mn(II)	100	87	85	87	100	107	99	102
Mn(II)+pulp	-	-	-	-	100	103	93	77
Fe(II)	100	81	62	67	100	92	86	85
Fe(III)	100	93	93	95	100	24	16	6
Ca(II)	100	111	110	109	100	6.4	3.8	0
Co(II)	100	103	102	100	100	8.4	2.4	0
Mg(II)	100	78	80	77	100	7.7	2.3	0
Cu(II)	100	38	34	35	100	4.2	1.6	0

*Data are the means of values from three independent experiments, with a maximal sample mean deviation of ±5%.

3.2. Inactivation of MnP by free radicals and peroxides other than H_2O_2

To find out whether free radicals such as hydroxyl free radical (HO•) and superoxide (O_2^-•) are involved in the inactivation of MnP, free radical scavengers such as sodium benzoate, mannitol, and superoxide dismutase were used. None of them had any stabilizing effect on MnP in the presence of excess H_2O_2 (not shown).

To investigate if H_2O_2 could be replaced by other peroxides that are able to function as electron acceptors but might not inactivate the enzyme, two peroxides (sodium perborate and mCPBA) were studied. While addition of sodium perborate caused a direct and rapid decline in the MnP activity, similar to the effect of H_2O_2, addition of mCPBA caused a slower decrease in the activity than H_2O_2 during the 3 hrs incubation. However, the remaining MnP activity after 3 hrs was only 20% of the original value (Table 2).

Table 2. Stability of MnP in the presence of 250 eq. Hydrogen peroxide and 10 mM of various organic compounds, and MnP substrates. Various concentrations of 1-HBT at increasing concentrations of Hydrogen peroxide were also examined for stabilization of MnP.

Various organic compounds*	Remaining MnP activity (%) at various incubation times (min)**			
	0	15	30	60-180
TEMPO+250 eq H_2O_2	100	100	100	92
Violuric acid+250 eq H_2O_2	100	93	88	85
HBTO+250 eq H_2O_2	100	77	76	83
Chlorpromazine+250 eq H_2O_2	100	98	98	82
1 mM 1-HBT+250 eq H_2O_2	100	62	28	24
2 mM 1-HBT+250 eq H_2O_2	100	84	85	83
5 mM 1-HBT+250 eq H_2O_2	100	92	85	84
1-HBT+250 eq H_2O_2	100	103	103	102
1-HBT+250 eq H_2O_2+pulp	100	92	87	89
1-HBT+250 eq H_2O_2+pulp+0.5 mM Mn(II)	100	102	99	99
1-HBT+500 eq H_2O_2	100	97	98	98
1-HBT+750 eq H_2O_2	100	82	82	83
1-HBT+1500 eq H_2O_2	100	69	66	61
1-HBT+2000 eq H_2O_2	100	50	50	42
1-HBT+2500 eq H_2O_2	100	12	5.5	0
0.25 mM mCPBA	100	83	64	20
0.25 mM Sodium perborate	100	10	4	0

*Concentrations of the compounds used are 10 mM except where otherwise noted. **Data are the means of values from three independent experiments, with a maximal sample mean deviation of ±5%.

3.3. Stabilization of MnP activity by various organic compounds at excess H_2O_2.

It was found that 1-HBT could completely protect MnP from inactivation by H_2O_2 (250 eq.). When various concentrations of 1-HBT (1, 2, 5, 10 mM) were applied, it was observed that 10 mM 1-HBT was able to maintain the MnP activity at 100% during the time course studied (Table 2). Lower concentrations of 1-HBT (2 and 5 mM) did not protect the enzyme as effectively as did 10 mM 1-HBT. At 1 mM of 1-HBT, the enzyme activity decreased by 75% during a 3 hr incubation time. Table 2 also shows the range of H_2O_2 concentrations in which MnP is protected by the presence of 10 mM 1-HBT. In the presence of 10 mM 1-HBT and 0.25 mM or 0.5 mM H_2O_2 (250 or 500 eq of MnP, respectively), MnP was stable throughout the experiment. However, at a higher H_2O_2 concentration (2.5 mM, 2500 eq), a rapid decrease in MnP activity was observed after 30 min. To understand whether the stabilization of MnP by 1-HBT is due to the consumption of H_2O_2 by 1-HBT, the H_2O_2 concentration was measured in samples taken from the incubations of 1-HBT (10 mM) with H_2O_2 (250 μM) for 0, 5, 15, and 60 min (10). It was found that the H_2O_2 concentration remained essentially unchanged during the incubation (Data not shown). Other organic compounds such as TEMPO, violuric acid, HBTO and chlorpromazine also appear to have strong MnP activity protecting effects although none of them was as efficient as 1-HBT . At the end of the incubation (3 hrs) with 10 mM of these compounds, most of the MnP activity remained unchanged in the presence of H_2O_2 (250 eq to MnP) (Table 2).

In comparison to the experiments without pulp, Mn(II) alone in the presence of pulp, did not appear to be as efficient in maintaining MnP activity during 3 hrs of incubation. At the end of the 3 hrs only 77% of the activity remained (Table 1). When Mn(II) was not included, but 1-HBT was added, 89% of MnP activity remained (Table 2). In the presence of pulp and 0.5 mM Mn(II), 1-HBT stabilized MnP as efficiently as it did in the absence of pulp.

3.4. Changes in MnP absorption spectra in the presence of 1-HBT.

Native MnP showed a Soret absorption maximum at 407 nm, and small peaks were also found at 500 and 630 nm. Addition of 250 eq of H_2O_2 gave rise to a shift in the Soret absorption maximum to 418 nm, and the two other peaks were replaced by new ones at 545 and 580 nm which indicates the formation of the inactive form of MnP, Compound III (7). However, when 1-HBT was added to this mixture, the Compound III peaks disappeared and the 407 nm as well as 630 peaks re-appeared, which indicated that Compound III was transferred back to the native enzyme (Fig. 1). When a solution of 1-HBT only was scanned, 2 peaks were observed at 227 and 304 nm (Fig. 2). After addition of MnP to this solution, these peaks decreased gradually by time. At the end of 30 mins, the peaks had decreased significantly and also slightly shifted to 224 and 310 nm, respectively, indicating consumption of 1-HBT.

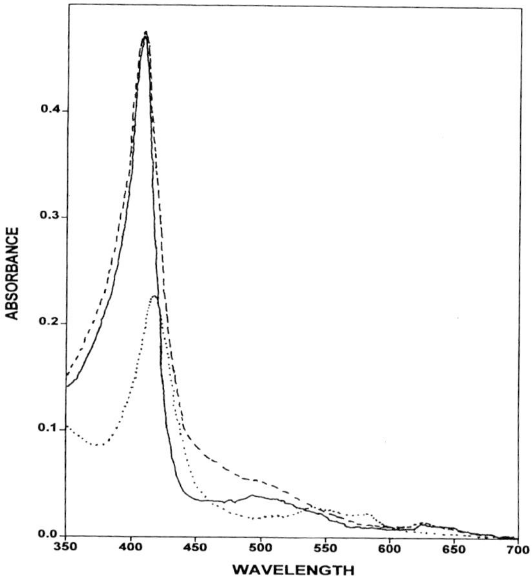

Figure 1. Changes in the absorption spectra of native MnP after the addition of excess H_2O_2 and 10 mM 1-HBT. Legends: — MnP alone; .... MnP+H_2O_2; ---- MnP+H_2O_2+1-HBT.

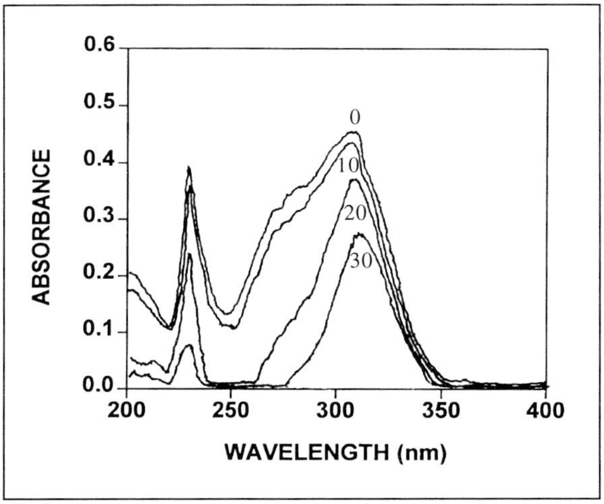

Figure 2. Absorption spectra of 0.1 mM 1-HBT after the addition of MnP, $MnSO_4$, and H_2O_2 in 50 mM malonate buffer. Legends: 1-HBT scanned after 0 min, 10 min, 20 min and 30 min. No absorbance was detected in the 400-700 nm range.

4. DISCUSSION

MnP activity is dependent on H_2O_2, but it is also inactivated by H_2O_2. Oxidation of the native enzyme by 1 equivalent H_2O_2 forms the active Compound I (7) that subsequently oxidizes Mn(II) or a phenolic substrate. The Compound II formed in this process is reduced back to the native enzyme, oxidizing an additional Mn(II) to Mn(III). Mn(II) is a mandatory substrate for Compound II (7). When excess concentrations of H_2O_2 are present, native MnP is directly converted to Compound III that is an inactive form of the enzyme. Compound III can also be formed from Compound II in the presence of excess H_2O_2 (7). The inactivation of MnP by high concentrations of H_2O_2 is one of the major obstacles in applying MnP in pulp bleaching. The understanding of the inactivation mechanism, especially the role of various metal ions in the inactivation, would greatly facilitate the development of a MnP-based pulp bleaching technique.

In fact, it has yet to be demonstrated that the formation of Compound III is the only reason for inactivation of MnP. It is well known that active oxygen species such as hydroxyl free radical formed by the reaction of transition metal ions with H_2O_2 have detrimental effects on the stability of proteins. Some metal ions may compete with the Fe ligand in heme resulting in the inactivation of the enzyme. According to our results, Mn(II) stabilizes the enzyme in the presence of excess H_2O_2. This conclusion is also in concert with the data in a recent paper (11).The stabilization of MnP by Mn(II) is enhanced by malonate. The reason for this stabilization is that Mn(II) competes with H_2O_2 for Compound II, thereby preventing the formation of the inactive Compound III. Malonate chelates and removes Mn(III) from the active site of MnP, thereby accelerating the catalytic cycle (12). In this way, Mn(II) competes more efficiently with H_2O_2 for Compound II. However, Mn(II) cannot stabilize MnP in the presence of an unbleached pulp.

Both Fe(II) and Mn(III) can destroy H_2O_2 according to the equations (1 and 2) taken from

Cheton et al. (13). The stabilizing effects of Fe (II) and Mn(III) on the MnP activity, therefore, could be due to their ability to destroy H_2O_2. Another reason for the stabilizing effect of Fe(II) could

$$Fe(II) + H_2O_2 \rightarrow HO\bullet + Fe(III) + OH^- \quad \text{(Fenton's Reaction)} \quad (1)$$

$$2Mn(III) + H_2O_2 \rightarrow 2Mn(II) + O_2 + 2H^+ \quad (2)$$

be due to its conversion to Fe(III) by H_2O_2 since Fe(II) itself inhibits MnP to some extent. This is not a desirable way of stabilizing MnP since a certain amount of H_2O_2 is needed for the function of the enzyme. However, stabilization of the enzyme by Fe(II) indicates that HO• formed from the Fenton's Reaction does not play an important role in the inactivation of MnP. In addition to these results, the inability of free radical scavengers to maintain MnP activity leads us to the conclusion that hydroxyl free radicals as well as superoxide anion radicals do not inactivate MnP and that the formation of compound III is the sole reason for its inactivation.

Studies on the crystal structure of MnP have revealed two Ca(II) binding sites in the enzyme (14). It was also demonstrated that low amounts of Ca(II) were able to efficiently prevent the thermal inactivation of *Phanerochaete chrysosporium* MnP (15). We found that addition of Ca(II) had no stabilizing effect on the enzyme in the presence of H_2O_2 while Ca(II) alone somewhat enhanced MnP activity. This implies that inactivation of MnP in the presence of H_2O_2 has nothing to do with the release of Ca(II) from MnP. Similarly, neither Mg(II) nor Co(II) stabilized MnP. While 0.1 mM Co(II) alone inhibited 67% of *P. chrysosporium* MnP activity (16), no inhibition of *T. versicolor* MnP was observed even at a 1 mM Co(II) concentration.

The effects of the Cu(II) and Fe(II) ions on MnP may have an important implication on pulp bleaching since significant amounts of Cu(II) and Fe(II) are always found in pulp, and their presence is a major threat to the activity of the enzyme. Inhibitory effects of these ions (even 0.1 mM) on MnP from *P. chrysosporium* were previously reported (16). Our studies show that MnP from *T. versicolor* was also inhibited by these two ions.

We have thoroughly investigated the impact of various metal ions on MnP stability in the presence of high concentrations of H_2O_2 at this point. However, none of the metal ions studied stabilized MnP when pulp is present. We have, however, found that inactivation of MnP by H_2O_2 can be alleviated by addition of 1-HBT, TEMPO violuric acid, HBTO, and chlorpromazine. More importantly, in the presence of 1-HBT, MnP activity was stabilized even in the presence of pulp (Table 2). The prevention of the inactivation was through the conversion of the inactive form of MnP (Compound III) back to native enzyme, rather than through decomposing H_2O_2 (Fig. 1). Since Compound III can be formed from either Compound II or the native enzyme, use of 1-HBT should be an ideal method for the protection of MnP. From the UV-Vis spectra obtained, we could observe that 1-HBT functions as a substrate to MnP.

It has been demonstrated that the substrate binding site of the MnP Compound II from *P. chrysosporium* is so narrow that only small compounds such as Mn(II) and H_2O_2 can fit in. Therefore, a bulky peroxide might not be able to oxidize Compound II to Compound III. According to our results, sodium perborate inactivated the MnP from *T. versicolor* as much as did H_2O_2 probably because it is not large enough. However, mPCBA inactivated the enzyme to a lesser extent. This is in accordance with earlier results with MnP from *P. chrysosporium* (7).

In conclusion: 1-HBT is able to both protect and restore MnP activity by converting Compound III back to native enzyme both in the presence and in the absence of pulp. Mn(II) also stabilizes MnP activity by competing with H_2O_2 for Compound II in the absence of pulp. Hydroxyl free radicals and superoxide anion radicals are not involved in inactivation of the enzyme.

REFERENCES

1. Eriksson, K-E. L., R. A. Blanchette, and Ander, P. (1990) Microbial and enzymatic degradation of wood and wood components. Berlin: Springer-Verlag.
2. Boman, B., B. Frostell, M. Ek, and Eriksson, K-E. L. (1988) Nordic Pulp Paper Res. J. 1-1988-3, 13-18.
3. Reid, I.D. (1990) Tappi J. 73, 149-153.
4. Addleman, K., and Archibald, F. (1993) Appl. Environ. Microbiol. 59, 266-273.
5. Harazono, K., R. Kondo, and Sakai, K. (1996) Appl. Environ. Microbiol. 62, 913-917.
6. Kondo, R., K. Harazono, and Sakai, K. (1994) 60, 4359-4363.
7. Wariishi, H., L. Akileswaran and Gold, M. H. (1988) Biochemistry. 27, 5365-5370.
8. Paice, M. G., I. D. Reid, R. Bourbonnais, F. S. Archibald, and Jurasek, L. (1993) Appl. Environ. Microbiol. 59, 260-265.
9. Wariishi, H., K. Valli and Gold, M. H. (1992) J. Biol. Chem. 267, 23688-23695.
10. Pick, E. and, Keisari, Y. (1980) J. Immun. Methods. 38, 161-170.
11. Timofeevski, S. L. and Aust, S. D. (1998) Arch. Biochem. Biophys. 35, 287-295.
12. Wariishi, H., H. B. Dunford, I. D. MacDonald and Gold, M. H. (1989) J. Biol. Chem. 246, 3335-3340.
13. Cheton, P. L-B., and Archibald, F.S. (1988) J. Free Radicals Biol. Med. 5, 325-333.
14. Sundaramoorthy, M., K. Kishi, M. H. Gold, and Poulos, T. L. (1994) J. Biol. Chem. 269, 32759-32767.
15. Sutherland, G. R. J., and Aust, S. D. (1996) Arch. Biochem. Biophys. 332, 128-134.
16. Paszczynski, A., V-A. Huynh, and Crawford, R. (1986) Arch. Biochem. Biophys. 244, 750-765.

Biotechnology in the Pulp and Paper Industry
L. Viikari and R. Lantto (Editors)

Delving into the Fundamental LMS Delignification of High-kappa Kraft Pulps

R. P. Chandra[a], F. S. Chakar[a], L. Allison[a], D.H. Kim[a], A.J. Ragauskas[a], and T.J. Elder[b]

[a] Institute of Paper Science and Technology, Atlanta, Georgia.

[b] Auburn University, Auburn, Alabama.

The bleaching of high-kappa kraft pulps with a laccase mediator system (LMS) was found to provide 43-61% delignification following an (E+P) stage when violuric acid was used as the mediator. Pulp yields after the LMS(E+P) treatment were +99.9%. Molecular modeling of violuric acid (VA), N-hydroxybenzotriazole (HBT), and N-acetyl-N-phenylhydroxylamine (NHA) indicates an elevated unpaired electron density for violuric acid over the other mediators, which may account for VA's improved performance. Structural analysis of the residual lignin after LMS_{VA} treatments indicated that the LMS stage oxidizes primarily the phenolic structures of lignin. Full sequence ECF bleaching of high and low-kappa SW kraft pulps after a LMS_{VA}(E+P+O) or (LMS_{VA})(LMS_{VA})(E+P+O) indicated the pulps could be readily bleached to +85 TAPPI brightness. The physical strength properties of the fully bleached LMS-treated pulps exhibited tensile, tear, and burst indexes comparable to control ECF-bleached pulps.

1. INTRODUCTION

The bleaching of kraft pulp is an integral component in the production of high-value kraft pulps. This component of the industry has been historically tied to the use of bleaching chemicals such as chlorine, hypochlorous acid, or chlorine dioxide. In recent times, environmental concerns have resulted in an increased use of chlorine dioxide, hydrogen peroxide, oxygen, and ozone.[1] Although extensive studies have been performed to develop alternative biobleaching systems, few enzymatic treatments exhibit the delignification/brightening capabilities of modern chemical bleaching technologies. One of the few exceptions to this generalization has been the development of laccase mediator system (LMS) delignification technologies for kraft pulps. After the initial ABTS-laccase studies by Paice et al.,[2] several researchers have reported alternative mediators (see Figure 1) that extended the bleaching effects of LMS such that 30-65% delignification can be routinely achieved.[3]

Most of these studies have focused primarily on low lignin content pulps, typically of kappa 30 or less for SW and kappa 18 or less for HW kraft pulps. Recently, our research studies have focused on the use of LMS technologies on high-kappa kraft pulps. Our interest in this

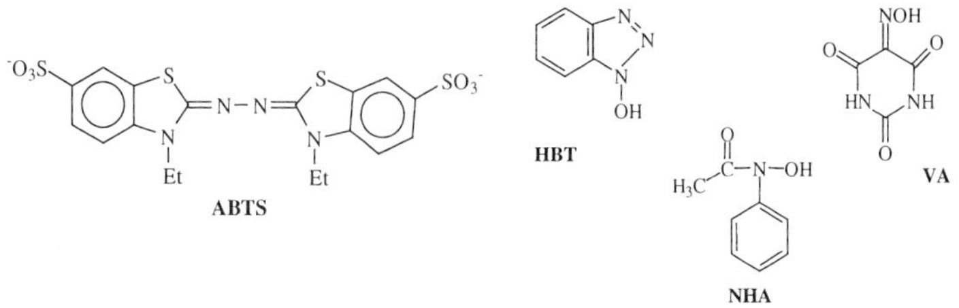

Figure 1. Commonly employed mediators for LMS biobleaching of kraft pulps.

application is based on the well-established phenomena that the selectivity of kraft pulping decreases as the pulping process completes the bulk phase and begins the residual phase. Several researchers[4] have noted that it is possible to improve pulp yields by halting the kraft cook prior to the residual phase and removing the additional lignin via oxygen delignification. In general, for SW kraft pulps, this requires employing an aggressive O or OO stage on a brownstock pulp having a kappa number > 40. The overall wood savings following this approach have been reported to be in the range of 2–6% for a modern pulp mill.

The application of LMS delignification as alternative to an extended O or OO stage appeared to be an attractive technology to delignifying high-kappa kraft pulps. Indeed, Haynes and Ragauskas[5] had reported that LMS treatment of cellulose led to no change in pulp viscosity, suggesting that laccase mediator delignification of kraft pulps is highly selective for lignin. Chakar et al. demonstrated the high selectivity of an LMS system with a kappa 70 SW kraft pulp yielding minimal losses in pulp viscosity.[6] Based on these observation, it appeared that an LMS stage could be used to process high-kappa kraft pulps with nominal loss in pulp carbohydrates providing improved pulp yields. LMS bleaching studies with a 70-kappa SW kraft pulp indicated that violuric acid was a superior mediator for the delignification of high-kappa kraft pulps compared with respect to either HBT or NHA.[6] The results of biobleaching a high and low lignin content SW kraft pulp with LMS(E+P), employing violuric acid as a mediator, are summarized in Table 1.

Table 1. LMS(E+P) Delignification Properties of SW Kraft Pulps.

Pulp	**SW-50.0 kappa**	**SW-27.5 kappa**
% Delignification	42.6	61.1
% Viscosity Retained	74.2	79.6
Yield	99.9	100.0

This paper further examines the nature of the changes in lignin during an LMS(E+P) stage and the integration of this biodelignification technology into modern ECF bleaching systems for the production of fully bleached kraft pulps.

2. RESULTS AND DISCUSSION

2.1. LMS Degradation of Lignin

As we have previously reported, the biobleaching chemistry of an LMS stage is influenced by the nature of the mediator and furnish employed.[7,8] For low-kappa pulps, it appears that the LMS_{HBT} system vigorously oxidizes C-5 condensed and noncondensed phenolics. For high-kappa pulps, this same delignification system removes primarily C-5 noncondensed phenolic structures. Furthermore, in comparing the reactivity of a LMS_{HBT} vs. LMS_{NHA} vs. LMS_{VA} system for high-kappa SW kraft pulps we observed that VA was approximately two fold more effective at removing lignin than HBT or NHA.[9]

To further explore the fundamental chemical basis by which the LMS_{VA} system delignified the SW kraft pulps presented in Table 1, the residual lignin from the kraft brownstocks and after an LMS_{VA}(E+P) were isolated and characterized by NMR. Residual lignin was recovered from the kraft pulps employing a standard[10] dioxane-water acidic hydrolysis procedure, which provided on average 47% lignin recovery. These lignin samples were then analyzed by quantitative ^{13}C and ^{31}P NMR techniques and these results are summarized in Figures 2–4. The ^{31}P NMR analysis technique is well suited for characterizing the hydroxyl functionality of lignin, whereas the ^{13}C NMR method provides a complete structural analysis of lignin.

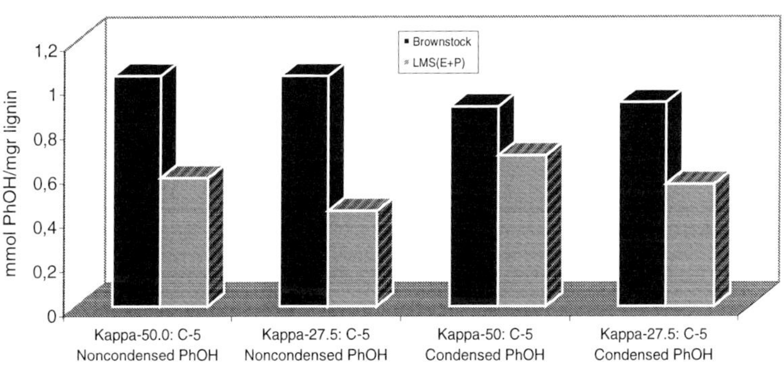

Figure 2. Changes in C-5 condensed and noncondensed phenolic guaiacyl content of residual lignin isolated from kappa 50.0 and kappa 27.5 brownstocks and after LMS_{VA}(E+P) treatments, as determined by phosphorylation and ^{31}P NMR analysis.

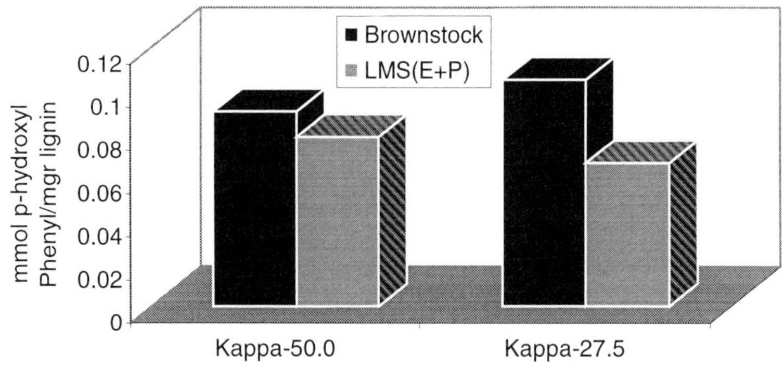

Figure 3. Changes in p-hydroxyl phenyl content of residual lignin isolated from kappa 50.0 and kappa 27.5 brownstocks and after LMS_{VA}(E+P) treatments, as determined by phosphorylation and ^{31}P NMR.

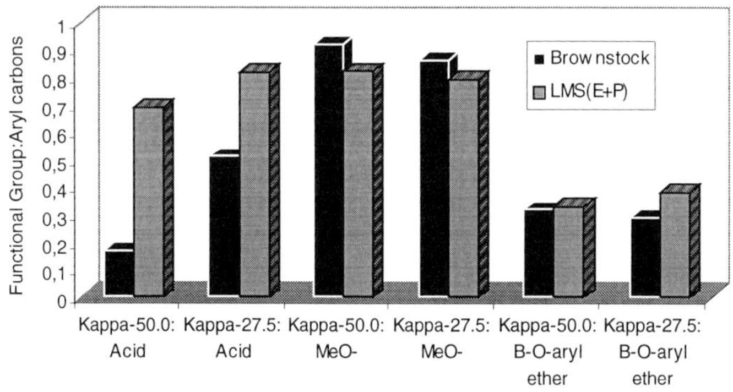

Figure 4. Changes in acid group, methoxyl, and β-O-aryl ether content of residual lignin isolated from kappa 50.0 and kappa 27.5 brownstocks and after LMS_{VA}(E+P) treatments, as determined by ^{13}C NMR.

The results of this analysis clearly indicated that the primary pathway of lignin degradation during an LMS_{VA}(E+P) treatment is the degradation of phenoxy groups with a preference for C-5 noncondensed phenoxy groups. This latter effect is attenuated for the low-kappa SW kraft pulp. It is interesting to note the slight decrease in p-hydroxyl phenyl groups. The loss in phenoxy groups is accompanied by a large increase in acid groups. It is also apparent that the β-O-aryl ether group is resistant to oxidation, and only slight demethoxylation was observed to occur under the biobleaching conditions employed.

The results of these studies are consistent overall with lignin model compounds studies with LMS but some differences do exist, such as the lack of reactivity with non phenolic and β-

O-aryl ether structures observed in this study versus LMS lignin model compound studies.[3c,f] This effect has been tentatively attributed to differences in LMS selectivity when reacting with one or two functional groups (i.e., lignin model compound) versus the lignin macropolymer containing an array of functional groups with which the oxidized mediator can react.

2.2. Mediator Modeling

It is now well established that the three most common N-hydroxyl mediators are unstable during an LMS stage and degrade.[11,12,13,14] The active species involved in lignin degradation is the RR´NO• radical. Computational results from PM3 indicate that the bond dissociation energy and electronic factors of the radical may contribute to the efficiency of the mediator for LMS delignification.[15] To further examine the fundamental principals governing the reactivity of the RR'NO• radical, we modeled its reactivity with a β-O-aryl ether structure, as shown in Figure 5.

Experimental LMS results have generally shown that laccase mediator systems employing violuric acid are superior to other first and second-generation mediators. The underlying factors controlling the behavior of laccase mediators has been the subject of some degree of speculation, since an understanding at this level would provide considerable insight into the development of mediators with improved efficacy. The current work addresses these questions by applying computational chemical methods in an examination of the energetics and electronics associated with the oxidation of a lignin model compound.

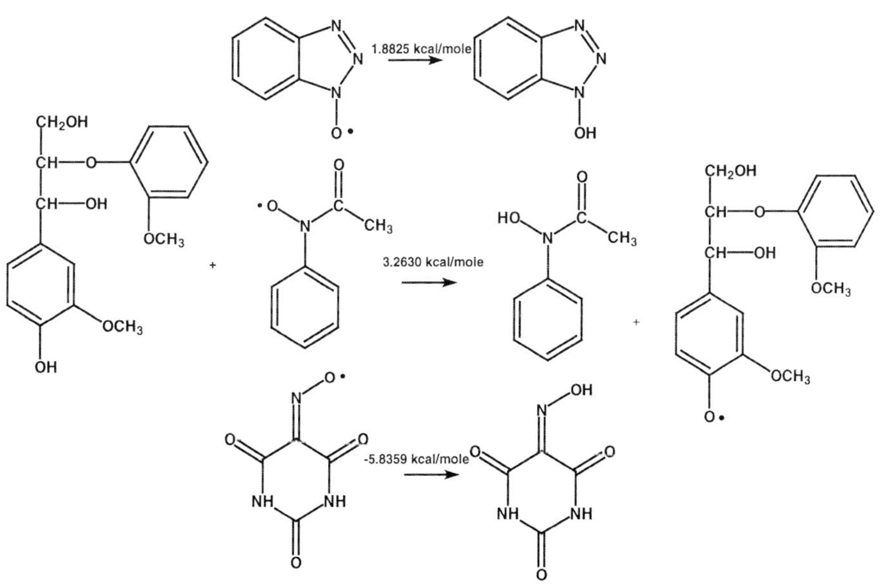

Figure 5. Computational results from modeling the oxidation of a model dilignol by the action of the RR'NO• radical from VA, HBT, and NHA.

From an energetic standpoint, it can be seen (see Figure 5) that the oxidation of the model dilignol by the action of the VA is an exothermic reaction, while HBT and NHA are both endothermic reactions. With respect to energetics, therefore, the reaction of VA should be preferred over the other mediators. Electronically, it was determined that the spin density (i.e., the degree of unpaired electron density in the free radicals) at the reactive oxygen atom is the largest in nitroxyl radical derived from VA, followed by NHA and HBT, which are quite similar. These data summarized in Table 2 are consistent with the literature,[3e] indicating the superior performance of violuric acid and a slight advantage of NHA over HBT.

Table 2. Calculated unpaired electron density and Mulliken charges for nitroxyl mediator.

Mediator	Unpaired electron density	Mulliken charge
VA	0.98	-0.097
NHA	0.92	-0.146
HBT	0.91	-0.158

It is also interesting to note the degree of delocalization of the unpaired electron as shown by its density among the other atoms in the various structures (Figure 6). The unpaired nitroxyl electron seems to be tightly constrained in VA, while HBT exhibits considerable delocalization, with NHA being somewhat intermediate in this regard. Finally, the Mulliken atomic charge at the reactive oxygen, although consistently negative, is less negative for violuric acid, such that its reduction may be favored over the other two mediators.

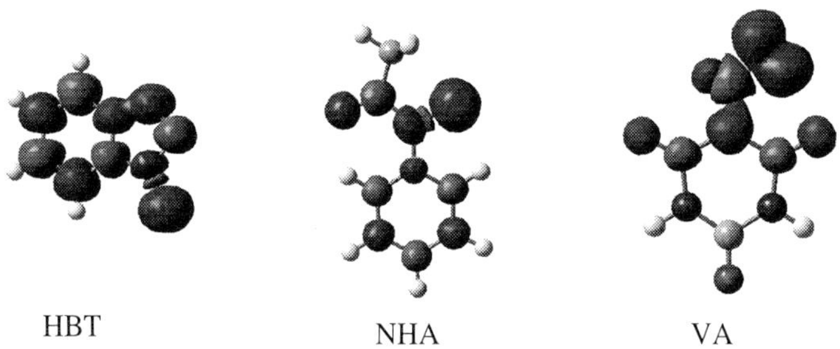

Figure 6. Spin density plots of the nitroxyl radical from VA, HBT, and NHA.

2.3. Full-Sequence Bleaching

Past studies by Poppius-Levlin et al.,[16] Sealey et al.,[3d] and others have all demonstrated that low-kappa pulps can be readily delignified with an LMS stage and then bleached to full brightness using either an ECF or TCF bleaching sequence. The nature of the physical pulp properties has not been as well established. This study examines the bleachability of high and low-kappa SW kraft pulps and defines their physical properties. The bleaching sequences examined are summarized in Table 3.

Table 3. Bleaching sequences applied to kappa 27.5 and 50.0 SW kraft pulp.

Kappa 27.5 SW kraft pulp	Kappa 50.0 SW kraft pulp
D(E+P+O)DED	OOD(E+P+O)D
LMS_{VA}(E+P+O)DED	$LMS_{VA}LMS_{VA}$(E+P+O)DED
OOD(E+P+O)D	LMS_{VA}OD(E+P+O)D
LMS(O)D(E+P+O)D	$OLMS_{VAD}$(E+P+O)D

The results of the bleaching sequence studies are summarized in Figures 7 to 10. In each case, it was possible to readily achieve a final brightness of ≈ 85.

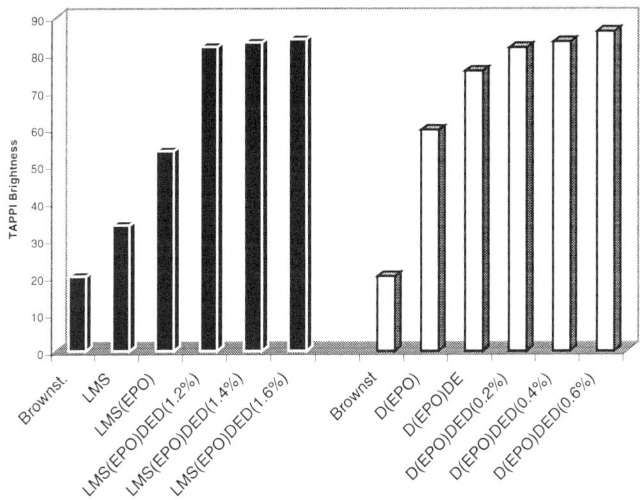

Figure 7. Brightness development for bleaching kappa 27.5 SW kraft pulp via LMS_{VA}(E+P+O)DED and D(E+P+O)DED.

Samples of the fully bleached pulp samples (TAPPI Brightness ≅85) were PFI refined to 2000 and 4000 revolutions and then analyzed for tensile, tear, and burst index values. These physical testing results are summarized in Table 4.

Table 4. Tensile, tear, and burst, index for fully bleached SW kraft pulp with a final TAPPI brightness of 84.5 ±0.50.

SW Kraft Pulp	Physical Properties Following 0–2000–4000 Revolutions		
	Tensile Index Nm/g	Tear Index mNm^2/g	Burst Index $kPam^2/g$
Kappa 27.5			
D(E+P+O)DED	15.7 – 69.4 – 75.4	12.4 – 18.4 – 15.8	0.8 – 6.0 – 6.7
LMS_{VA}(E+P+O)DED	22.8 – 77.6 – 87.6	13.9 – 14.5 – 11.5	1.4 – 6.2 – 6.8
OOD(E+P+O)D	23.9 – 80.3 – 87.9	15.2 - 12.2 – 11.0	1.6 – 6.3 – 6.8
LMS(O)D(E+P+O)D	24.4 – 80.4 – 87.2	16.6 – 12.8 – 10.5	1.5 – 6.2 – 6.8
Kappa 50.0			
OOD(E+P+O)D	41.1 – 98.5 – 104.4	19.1 – 13.4 – 11.4	3.3 – 8.5 – 9.4
LMS_{VA}OD(E+P+O)D	45.2 – 98.8 – 103.5	20.0 – 12.7 – 11.5	3.2 – 8.0 – 9.0
$OLMS_{VA}$D(E+P+O)D	40.8 – 101.6 – 102.5	22.7 – 13.0 – 11.4	2.8 – 8.5 – 9.6

The physical testing of the fully bleached pulps indicates that the LMS-delignified pulps have physical strength properties comparable to conventional ECF bleached pulps.

3. MATERIALS AND METHODS

3.1. Materials

Violuric acid and all other chemical reagents and solvents were purchased from Aldrich, Milwaukee, WI, and used as received except for *p*-dioxane, which was freshly distilled over $NaBH_4$ before use. Laccase from *Trametes villosa* was donated by Novo Nordisk Biochem. The kappa 27.5 SW kraft pulp was acquired from a southern USA pulp mill, whereas the kappa 50.0 SW pulp was laboratory-prepared from a conventional batch cook. The pulps were thoroughly washed, screened, centrifuged, fluffed, and stored at 4°C prior to use.

3.2. Pulp Characterization

All pulp testing was done according to TAPPI Test Methods.

3.3. Laccase Assay

Laccase activity was measured by monitoring the rate of oxidation of syringaldazine following literature methods.[8] In brief, the change in A_{530nm} of 0.001 per minute per mL of enzyme solution in a 100 mM potassium phosphate buffer (2.20 mL) and 0.216 mM syringaldazine in methanol (0.300 mL) was set to one unit (U) of activity. This test was done at 23°C and pH 4.5. The activity of the laccase was 1.87×10^6 U/mL of enzyme solution.

3.4. LMS Delignification

LMS_{VA} was accomplished as described in the literature[8] employing 1.87 x 10^7 U of laccase/gr o.d. pulp and a 4% charge of violuric acid. In brief, the LMS reaction was conducted with stirring at 9% csc, pH 4.5, 45.0°C, 120 psi oxygen pressure, and reaction time of 2 h.

3.5. Computational Methods

Ab initio calculations have been performed on the closed-shell and free radical forms of the HBT, NHA, VA, and a β-O-4 dilignol_using Gaussian 98W. These calculations were done with full geometry optimization at the HF/3-21G* and UHF/3-21G* levels for the closed-shell and free radical structures respectively.

3.6. Pulp Bleaching Conditions

Table 5 summarizes the bleaching conditions employed for the remaining bleaching stages described in Table 2.

Table 5. Summary of bleaching conditions employed.

Treatment	Bleaching Conditions
D_0	0.20 k.f., 10% csc, 50°C, 45 min.
E+P	2.5% NaOH, 0.05% $MgSO_4$, 10% csc, 70°C, 1 h
E+P+O	0.5% H_2O_2, 2.75% NaOH, 0.05% $MgSO_4$, 10% csc, 70°C, 1.0 h O_2 60 psi for 15 min. and then reduced 12 psi/5 min.
O	3.00% NaOH for low-kappa pulp, 4.00% for high-kappa pulp, 0.05% $MgSO_4$, 10% csc, 80 psi O_2, 95°C, 1.3 h
OO	3.00% NaOH for low-kappa pulp, 4.00% for high-kappa pulp, 0.05% $MgSO_4$, 10% csc, 80°C, 130 psi O_2 for 30 min., 95°C 60 psi for 60 min.
D_1	1% ClO_2, 10% csc, 70°C, 3 h
E	NaOH: 50% of equiv. Cl_2 charge, 10% csc, 70°C, 1 h
D_{final}	See figure for charge of ClO_2 employed, 10% csc, 70°C, 3 h

3.7. Isolation of Residual Lignins

The isolation of residual lignins was carried out following standard literature methods[7,8] employing an acidic dioxane:water solution. Lignin yields were in the range of 45-49% calculated as follows: % lignin yield = {mass of lignin isolated/ (initial kappa of brownstock) x 0.15)}x100.

3.8. NMR Characterization of Residual Lignins

The isolated residual lignin samples were analyzed using a 400 MHz Bruker DMX spectrometer. Quantitative ^{13}C NMR spectra were acquired and analyzed in accordance with established literature methods.[17] Lignin samples were also derivatized with 2-chloro-4,4,5,5-tetramethyl-1,3,2-dioxaphospholane and analyzed by ^{31}P NMR following literature methods.[18]

4. CONCLUSIONS

The use of LMS-delignification technologies can be applied to high-kappa kraft pulps that can be subsequently processed to yield high-quality fully bleached pulps. The key parameter involved in this biotechnology is the mediator used in conjunction with laccase. To date, we have observed that violuric acid is a superior agent for laccase. Nonetheless, the propensity of VA to degrade during the LMS stage significantly limits its practical applications to remove lignin. The challenge for commercial LMS systems is the development of catalytic mediators that can selectively remove > 60% of the lignin in pulp fibers in a cost-effective manner.

ACKNOWLEDGMENTS

The authors would like to thank the U.S. Department of Energy (DOE) and the member companies of the Institute of Paper Science and Technology for their support of these ongoing studies. This manuscript was prepared, in part, with the support of DOE Cooperative Agreement No. DE-FC36-99GO10374. However, any opinions, findings, conclusions, or recommendations expressed herein are those of the author(s) and do not necessarily reflect the views of the DOE. Also, we would like to thank Novo Nordisk Biochem for supplying the laccase. Portions of this work were used by F.S.C. and R.P.C. as partial fulfillment of the requirements for the Ph.D. degree at the Institute of Paper Science and Technology.

REFERENCES

1. McDonough, T.J., "Recent Advances in Bleached Chemical Pulp Manufacturing Technology; Extended Delignification, Oxygen Delignification, Enzyme Applications and ECF and TCF Bleaching", *TAPPI J.*, 78(3), 55(1995).
2. Bourbonnais, R.; Paice, M., "Oxidation of Non-Phenolic Substrates. An Expanded Role of Laccase in Lignin Biodegradation", *FEBS*, 267(1), 99(1990).

3a. Poppius-Levlin, K.; Wang, W.; Ranua, M.; Niku-Paavola, M.; Viikari, L., "Biobleaching of Chemical Pulps by Laccase/Mediator Systems", *1997 Biological Sciences Symposium: Proceedings, San Francisco*, October, 327.

3b. Nelson, P.; Chin, J.; Charles W.J.; Viikari, L.; Tenkanen, M., "Use of a Laccase Mediator Stage in Bleaching Eucalypt Kraft Pulps", *Appita J.*, 51(6): 451 (1998).

3c. Bourbonnais, R.; Paice, M.; Freiermuth, B.; Bodie, E.; Borneman, S., "Reactivities of Various Mediators and Laccases with Kraft Pulp and Lignin Model Compounds", *Applied Environmental Microbiology* 63(12): 4627(1997).

3d. Sealey, J.E.; Runge, T.M.; Ragauskas, A.J., "Laccase N-Hydroxybenzo-triazole Full-Sequence Bleaching with Hydrogen Peroxide and Chlorine Dioxide", *TAPPI J.*, 83(9), 66(2000).

3e. Amman, M., "The lignozym Process Coming Closer to the Mill", *9th International Symposium Wood Pulping Chemistry Conference Proceed.*, Montreal, Canada, pp. F4-1(1997).

3f. Bourbonnais, R.; Paice, M.G., "Oxidation of non-Phenolic Substrates. An Expanded Role for Laccases in Lignin Biodegradation", FEBS, 267(1), 99(1990).

4a. Parthasarathy, V., "Use of Digester Additives and Oxygen Delignification to Gain Bleached Pulp Yield and Achieve Environmental Objectives", 1998 *TAPPI Proceedings*: Breaking the Pulp Yield Barrier Symposium Proceedings, TAPPI PRESS, Atlanta, GA, 219.

4b. Magnotta, V.; Kirkman, A.; Jameel, H.; Gratzl, J., "High-Kappa-[No.] Pulping and Extended Oxygen Delignification to Increase Yield", 1998 *TAPPI Proceedings*: Breaking the Pulp Yield Barrier Symposium, TAPPI PRESS, Atlanta, GA, 165.

4c. Bokstrom, M.; Norden, S., "Extended Oxygen Delignification", 1998 *International Pulp Bleaching Conference Proceedings*, Helsinki, 1, 23.

4d. Chakar, F.S.; Lucia, L.; Ragauskas, A.J., "Extending the Limits of Oxygen Delignification", 2000 *International Pulp Bleaching Conference*, NS, Canada, 1, 123.

5. Haynes, K.; Ragauskas, A.J., "Effects of Laccase Mediator Delignification on Fiber Properties", *International Pulp Bleaching Conference*, Helsinki, Finland, 355(June 1998).

6. Laccase-Mediator Systems and Oxygen Delignification-A Comparative Study. Chakar, F.S.; Ragauskas, A.J.; McDonough, T.J., 2000 International Pulp Bleaching Conference, Halifax, NS, 59(June, 2000).

7. Chakar, F.S.; Ragauskas, A.J., "The Effects of Oxidative Alkaline Extraction Stages after $Laccase_{HBT}$ and $Laccase_{NHAA}$ Treatments – An NMR Study of Residual Lignin", *J. Wood Chem. Technol.*, 20(2), 169(2000).

8. Chakar, F.S.; Ragauskas, A.J., "The Kismet of Residual Lignins During LMS Delignification of High-Kappa Kraft Pulps", *Holzforschung*, 54, 647(2000).

9. Chakar, F.S.; Allison, L.; Kim, D.H.; Ragauskas, A.J., "The Path Forward to Practical Nascent Laccase Biobleaching Technologies", *TAPPI Pulping Conference Proceed.*, Boston, MA (Nov. 2000).

10. Froass, P.M.; Ragauskas, A.J.; Jian-er, J., "Chemical Structure of Residual Lignin from Kraft Pulp", *J. Wood Chem. Technol.*, 16(4), 347(1996).

11. Sealey, J.; Ragauskas, A.J., "Investigation of Laccase/N-Hydroxybenzotriazole Delignification of Kraft Pulp", *J. Wood Chem. Technol.*, 18(4), 403(1998).

12. Paice, M.G.; Bourbonnais, R.; Reid, I.; Archibald, F.S., "Kraft Pulp Bleaching by Redox Enzymes", *9th International Symposium on Wood and Pulping Chemistry Proceed.*, Montreal, Canada, PL1-1(1997).

13. Potthast, A.; Koch, H.; Fischer, K., "The Laccase-Mediator System Reaction with Model Compounds and Pulp", *9th International Symposium on Wood and Pulping Chemistry Proceed.*, Montreal, Canada, F2-1(1997).

14. Freudenreich, J.; Amann, M.; Fritz-Langhals, E.; Stohrer, J., "Understanding the Lignozym Process", *International Pulp Bleaching Conference Proceed.*, Helsinki, Finland, 71(1998).

15. Sealey, J.; Ragauskas, A.J.; Elder, T.J., "Investigations into Laccase-Mediator Delignification of Kraft Pulps", *Holzforschung*, 53, 498(1999).

16. Poppius-Levlin, K.; Wang, W.; Tamminen, T.; Hortling, B.; Viikari, L.; Niku-Paavola, M-L., "Effect of Laccase/HBT Treatment on Pulp and Lignin Structures", *J. Pulp Paper Science*, 25(3), 90(1999).

17. Robert, D., Carbon-13 Nuclear Magnetic Resonance Spectrometry. *In Methods in Lignin Chemistry*, Lin S; Dence, C., Eds., Springer Verlag, New York pp. 250-272 (1992).
18. Granata, A.; Argyropoulos, D.S., "2-Chloro-4,4,5,5-Tetramethyl-1,3,2-dioxaphospholane, a Reagent for the Accurate Determination of the Uncondensed and Condensed Phenolic Moieties in Lignins", *J. Agriculture Food Chem.*, 43, 1538(1995).

Biotechnology in the Pulp and Paper Industry
L. Viikari and R. Lantto (Editors)

Elucidating the Effects of Laccase on the Physical Properties of High-Kappa Kraft Pulps

Richard P. Chandra and Arthur J. Ragauskas

Institute of Paper Science and Technology, 500 10th St NW, Atlanta, Georgia, 30318, United States of America

Laccase was applied in combination with three phenolic acids to high-kappa kraft pulps to attach the phenolic acids to pulp fibers. Tensile index, burst index, apparent density, and z-directional tensile strength increased after pulp treatment with laccase and 4-hydroxybenzoic acid. Increases in strength properties were not accompanied by decreases in fiber strength or alterations of fiber morphology. It was hypothesized that the increase in paper strength properties was attributed to an increase in surface carboxylic acid groups that promote hydrogen bonding.

1. INTRODUCTION

Laccase is a robust oxidoreductase enzyme with a broad substrate range [1]. The predominant role played by laccase in pulp and paper research has been as an enzyme to be potentially employed for biologically bleaching kraft pulps. Laccase in combination with mediator compounds (LMS) has been shown to delignify kraft pulps [2]. The proposed role of the mediator compounds is to access the interior of the fiber and to enhance the oxidation capabilities of laccase [3]. Indeed, studies [2, 4] have shown that the LMS can delignify kraft pulps to the extent of commercial bleaching technologies. Studies conducted by Chakar and Ragauskas [5] have dealt with the application of new generation mediators such as NHA (N-acetyl-N-phenylhydroxylamine) and violuric acid. These authors applied the LMS to high-kappa kraft pulps to reveal the effect of LMS treatments on residual lignin structures in detail [5]. Even with the wealth of information gathered from studies on LMS treatments, several obstacles have hindered the commercial application of this technology, including the price and instability of mediators. With these factors in mind it is of great interest to explore alternative avenues to take advantage of the oxidative capabilities of laccase.

In nature, laccase is a versatile enzyme that has many roles including lignin biosynthesis, pigment production, plant pathogenesis, detoxification, and delignification [6]. Laccase functions both as a depolymerization and polymerization agent for compounds such as phenols, aminophenols, polyphenols, polyamines, certain inorganic ions, and aryl diamine compounds [1]. In the absence of mediator compounds, studies have shown that laccase polymerizes dissolved lignin model compounds and lignin preparations [7, 8]. Laccase has also been shown to possess the capability to couple soluble monomeric compounds to dissolved lignin preparations [9]. Remarkably, Lund and Ragauskas have recently shown that laccase could couple guaiacol sulfonate to lignin in dioxane:water solutions, rendering the resulting modified lignin macromolecule water-soluble [10]. Further studies involving

3. RESULTS AND DISCUSSION

Several studies have investigated the application of enzymes such as cellulases, hemicellulases, and pectinases for the modification of both kraft and mechanical pulps [18-24]. Some of the pulp properties affected by these enzymes include drainage, bleachability, paper strength, fiber strength, and retention of fillers and fines. Although they have been infrequently employed, oxidative enzymes such as laccases also possess attractive capabilities that may act to improve fiber properties in a unique manner compared to other types of enzymes.

The potential for laccase to modify paper physical properties has been overshadowed by its ability to degrade lignin. Felby [25, 26] has performed extensive work utilizing laccase enzymes to modify the bonding properties of medium density fiberboard made from beech-wood (*Fagus sylvatica*) thermomechanical pulp fibers (TMP). Hassingboe and Felby found that colloidal ligni acted as a natural laccase mediator that functioned to oxidize lignin resulting in fiber modification [27]. Kharazipoiur et al. [28] have also shown laccase pre-treatment of mechanical pulp furnish improves the properties of medium density fiber-boards. Furthermore, both Viikari et al. [13] and Haynes and Ragauskas [29] have also found that laccase enzymes can affect physical properties of paper sheets. Although laccase has been employed for fiber modification, the work mentioned thus far has focused on the ability of laccase to oxidize the fiber itself; however, laccase also possesses the ability to polymerize a variety of compounds.

In the work presented here, it was envisaged that laccase could polymerize water-soluble phenolic compounds with various functionalities to high-kappa pulp fibers. Laccase-facilitated attachment of compounds to high-kappa fibers presents many opportunities for custom modification of fibers to impart specific properties. Since previous work has shown that the introduction of carboxylic acid groups to pulp fibers results in paper strength improvements [30, 31], in this work three phenolic compounds with the carboxylic acid functionality were polymerized to fibers with laccase.

The first tests performed on the sheets from laccase treatments were those tests concerned with the strength of the fiber network. These tests included apparent density, burst, tensile, and z-directional tensile tests. The results of apparent density and burst testing are shown in Figures 2 & 3. Apparent density is an approximate measure of the tightness of sheet consolidation while the burst index tests the resistance of the sheet to rupture by a diaphragm. Laccase treatment with 4-hydroxybenzoic acid resulted in an increase in apparent density and burst index while there was little or no effect from treatment with syringic and vanillic acid. The ineffectiveness of vanillic acid and syringic acid was tentatively attributed to steric hindrance of the methoxy groups or the reactivity of the phenoxy radical. Ishihara et al. [32] and Leonowicz et al. [8] showed that vanillic and syringic acid undergo fewer polymerization reactions with laccase at a pH below 6.9. To gain a more definitive perspective of the paper strength increases that occur with laccase treatment with phenolic acids, further testing of the in-plane and transverse tensile strength were performed on treated and untreated sheets.

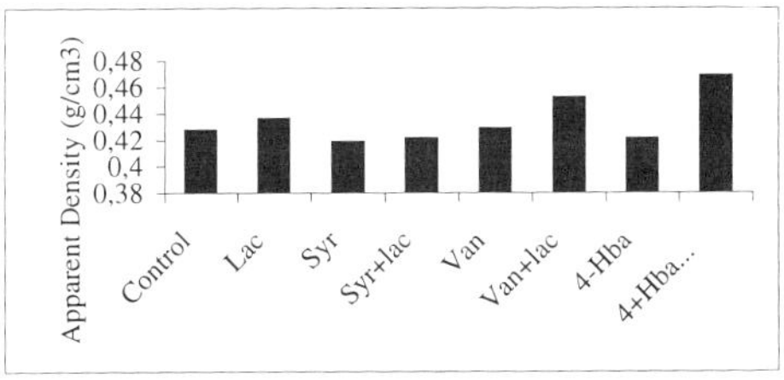

Figure 2: Apparent Density Measurements of High-Kappa Pulps Treated with Laccase and Phenolic Acids

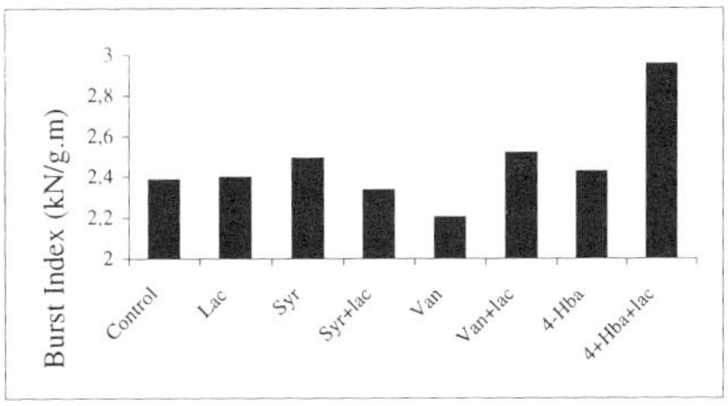

Figure 3: Burst Index Measurements of High-Kappa Pulps Treated with Laccase and Phenolic Acids

As shown in Figure 4, laccase treatment with 4-hydroxybenzoic acid resulted in tensile index increases of 15-20%. Felby et al. [25] treated a ferulated arabinoxylan adduct with laccase in the presence of Beech TMP fibers resulting in tensile strength increases. These authors rationalized their results by hypothesizing that laccase attached the ferulic acid portion of the adduct to the lignin on the fiber while the arabinoxylan served to fill gaps in the fiber network to increase tensile strength. It should also be noted that treatment with laccase in the absence of a phenolic acid resulted in increases in tensile strength. These results are similar to those obtained by Viikari et al. [13] with mechanical pulps.

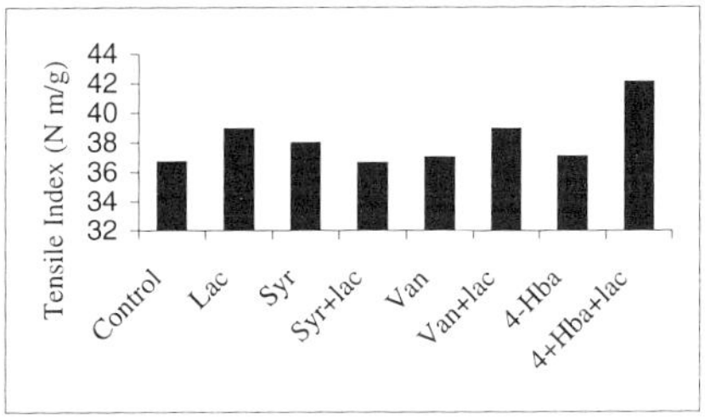

Figure 4. Tensile Index Measurements of High-Kappa Pulps Treated with Laccase and Phenolic Acids

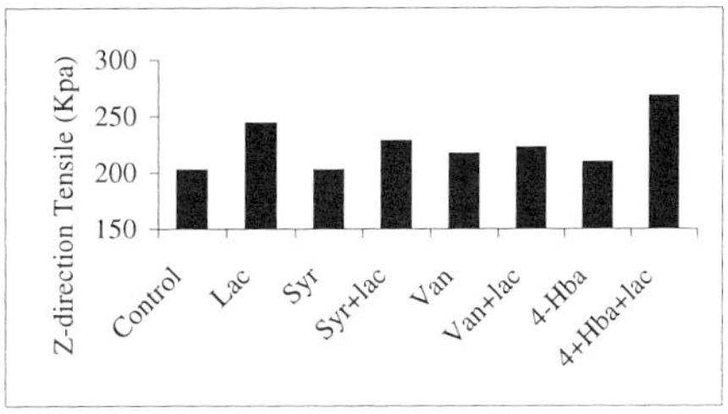

Figure 5. Z-direction Tensile Measurements of High-Kappa Pulps Treated with Laccase and Phenolic Acids

The z-direction tensile test has been shown to be an effective measure of internal bond strength for paper and paperboard [33]. This test measures paper strength in the transverse direction. The results of z-directional tensile strength are shown in Figure 5. Similar to the other tests, laccase treatment with 4-hydroxybenzoic acid resulted in increases in ZDT. However, unlike the other physical tests, laccase treatment with vanillic and syringic acids was also shown to increase the ZDT strength of paper sheets 10-20%.

Yamaguchi et al. [34, 35] utilized laccase to polymerize various phenolic compounds to form dehydrogenative polymers (DHP). The DHP's were subsequently coupled to thermomechanical pulp (TMP) in the presence of peroxidase for subsequent formation of paper by hot-pressing. Httermann [12] combined kraft lignin "activated" with laccase, with spruce sulfite pulp with subsequent sheet formation and press-drying. The resulting paper sheets in both the work of Yamaguchi et al [34, 35] and Huttermann [12] had enhanced paper strength properties. Unfortunately, it is difficult to compare the results obtained here to those of Yamaguchi et al. and Huttermann, since the sheets in the present study were not formed by heating. However, the results in the present study correlate with those obtained by Barzyk et al. [30]. Both in our previous work [15] and the work of Barzyk [30], fiber surfaces were enriched with carboxylic-acid groups. Furthermore, the data presented here strongly suggests that the increases in carboxylic acid groups on high-kappa fibers results in paper strength increases, similar to the results obtained by Barzyk [30]. Barzyk hypothesized that the increased paper strength was due to the specific placement of carboxylic acid groups at the fiber surface that enhanced the hydrogen bonding capability of fibers upon sheet formation [30]. In our previous work [15] we observed an eighty-percent increase in surface carboxylic acid groups. Therefore, it was hypothesized that the increase in the paper strength in the work described here is due to a similar mechanism as theorized by Barzyk [30]. Although laccase treatment increased the tensile index of sheets, there was no indication of any damage or physical alteration of pulp fibers occurring during laccase treatments. To determine whether laccase treatment had an effect on fiber strength or curl, zero-span tensile testing and fiber quality analysis were performed.

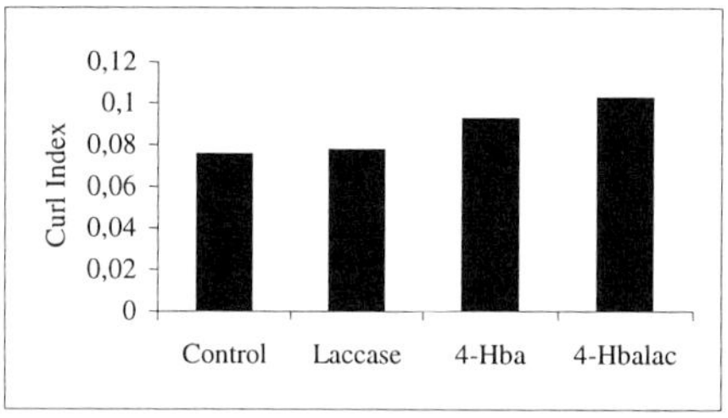

Figure 6. Curl Index Measurement of High-Kappa Pulps Treated with Laccase and 4-Hydroxybenzoic Acid Performed on a Fiber Quality Analyzer

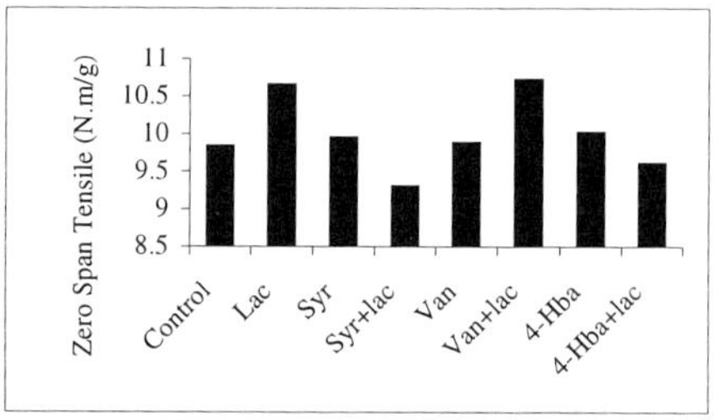

Figure 7. Zero Span Measurements of High-Kappa Pulps Treated with Laccase and Phenolic Acids

The results of curl testing on the fiber quality analyzer are shown in Figure 6. The degree of curl of pulp fibers has been shown to have adverse effects on the tensile strength of pulps [36]. In the work performed here, handling of pulp at high consistency was expected to impart curl on the fibers that would be subsequently relinquished upon re-suspension of fibers at low consistency [36]. Tests were performed on laccase treatments with 4-hydroxybenzoic acid since these treatments had the most influence on the physical pulp properties. All of the samples tested essentially contained straight fibers. Therefore, pulp treatments were shown to have minimal effects on fiber curl which corresponds to the increase in tensile strength observed with samples treated with laccase and with laccase + 4-hydroxybenzoic acid. As shown in Figure 7, laccase treatments had only minimal effects on the results of the zero-span tensile test. The zero-span tensile test has been established as a strong indicator of the strength of fibers in the paper sheet [37]. Therefore, laccase-facilitated grafting of phenolic acids to pulp fibers achieved increases in paper strength without compromising fiber strength or altering fiber morphological characteristics.

4. CONCLUSIONS

Laccase-facilitated coupling reactions resulted in increases in paper strength properties. The strength increases were obtained without significant decreases in fiber strength. This technology is beneficial and can be exploited for the improvement of existing pulp products or the creation of new product platforms.

ACKNOWLEDGEMENTS

The authors would like to thank the Institute of Paper Science and Technology and its member companies for their support for these ongoing studies. The gift of laccase by Novo Laboratories Inc. and pulp by Temple Inland is gratefully appreciated. Portions of this work were used by R.P.C. as partial fulfillment of the requirements for the Ph.D. degree at the Institute of Paper Science and Technology.

REFERENCES

1. Yaropolov, A.I., Skorobogatko, O.V., Vartanov, S.S., Varfolomeyev, S.D..Applied Biochemistry and Biotechnology, 49, (1994), 257-280.
2. Paice, M.G., Bourbonnais, R., and Reid, I.D..Tappi Journal, 78, No.9 (1995), 161-168.
3. Paice, M.G., Bourbonnais, R., Reid, I.D., Archibald, F.S., and Jurasek, L., Journal of Pulp and Paper Science, 21, No.8 (1995), 280-284.
4. Chakar, F.S., Ragauskas, A.J., TAPPI 1998 Pulping Conference, Montreal, Canada,(1998),109-117.
5. Chakar, F.S., and Ragauskas, A.J., Sixth European Workshop on Lignocellulosics and Pulp, Bordeaux, France, (2000), 53-56.
6. Call, H.P., Mucke, I., Journal of Biotechnology, 53, (1997), 163-202.
7. Leonowicz, A., Szklarz,G., and Wojta-Wasilewska, M.., Phytochemistry, 24, No.3 (1985), 393-396.
8. Leonowicz, A., Edgehill, R.U., and Bollag, J-M., Archives of Microbiology, 137, (1984), 89-96.
9. Lund, M., Felby, C., Bjerrum, M., 7th International Conference on Biotechnology in the Pulp and Paper Industry, Vancouver, Canada, (1998), C129-C142.
10. Lund, M. and Ragauskas, A.J., Applied Microbiology and Biotechnology, Paper Accepted for Publication, (2001).
11. Mai, C., Milstein, O., Huttermann, A., Applied Microbiology and Biotechnology, 51, (1999), 527-531.
12. Huttermann, A., Kharazipour, A., Schindel, K., Fastenrath, M., Noetzold, S., Schroeter, M., 7th International Conference on Biotechnology in the Pulp and Paper Industry, Vancouver, Canada, (1998), A207-A209.

13. Viikari, L., Harkki, T., Niku-Paavlova, M-L., Buchert, J., and Popplus-Levlin, K., 7th International Conference on Biotechnology in the Pulp and Paper Industry, Vancouver, B.C., (1998), A121-A124.
14. Laine, J., Stenus, P., Carlsson, G., and Stroh, G..Cellulose, No.1 (1994), 145-160.
15. Chandra, R.P. and Ragauskas,A.J., Submitted Paper, (2001).
16. Sealey, J., Ragauskas, A.J., Enzyme and Microbial Technology, 23, (1998), 422-426.
17. Scallan, A.M., and Grignon, J., Svensk Paperstidning, No.2 (1979), 41.
18. Mansfield, S. and Saddler, J.N., Journal of Pulp and Paper Science, 25, No.3 (1999), 84-89.
19. Oksanen, T., Pere, J., Buchert, J., and Viikari, L., Cellulose, 4, (1997), 329-339.
20. Reid, I., Ricard, M.. Enzyme and Microbial Technology, 26, No.2 (2000), 115-123.
21. Wong, K.K.Y., Kibblewhite, P.R., and Signal, F.A., Journal of Wood Chemistry and Technology, 19, No.3 (1999), 203-212.
22. Farrel, R., Senior, D., and Viikari, L., Pulp Bleaching: Principles and Practice, Reeve, D. and Dence, C., Editors. 1996, Tappi Press, 363-367.
23. Kantelinen, A., Niku-Paavlova, M.-L., Ranua, M., and Suurnakki, A., International Conference on Cellulose and Cellulose Derivatives:Physico Chemical Aspects and Industrial Applications,(1998),43-48.
24. Buchert, J., Suurnakki, A., and Viikari, L., 211th National Meeting of the American Chemical Society, New Orleans, LA, (1996),15-24.
25. Felby, C., Olesen, P.O., Hansen, T.T., 6th International Conference on Biotechnology in the pulp and paper industry, Vienna, (1996),
26. Felby, C., Pedersen, L.S., and Nielsen, B.R.. Holzforchung, 51, No.3 (1997), 281-286.
27. Hassingboe, J., Lawther, J.M., Felby, C., 7th International Conference on Biotechnology in the Pulp and Paper Industry, Vancouver, British Columbia, (1998), A125-A128.
28. Kharazipour, A., Huttermann, A., Luedmann, H.D., Journal of Adhesion Science and Technology, 11, No.3 (1997), 419-427.
29. Haynes, K. and Ragauskas, A.J., International Pulp Bleaching Conference, Helsinki, Finland, (1998), 355-359.
30. Barzyk, D., Page, D.H., and Ragauskas, A.J., Journal of Pulp and Paper Science, 23, No.2 (1997), 59-61.
31. Engstrand, P., Htun, M., Olander, K., and Sjogren, B., Appita 6th International Symposium on Wood and Pulping Chemistry, Mebourne, Australia, (1991),75-79.
32. Ishihara, T..Mokuzai Gakkaishi, 29, No.11 (1983), 801-805.
33. Koubbaa, A., and Koran, Z..Tappi, 103-111, No.3 (1995),
34. Yamaguchi, H., Maeda, Y., Sakata, I.,.Mokuzai Gakkaishi, 38, No.10 (1992), 931-937.
35. Yamaguchi, H., Maeda, Y., Sakata, I.,.Mokuzai Gakkaishi, 40, No.2 (1994), 185-190.
36. Barbe, M.C., Jordan, B.D., and Page, D.H..Journal of Pulp and Paper Science, 10, No.3 (1984), J74-79.
37. Page, D.H., Seth, R.S., and El-Hosseiny, F., 8th Fundamental Research Symposium Held at Oxford, London,(1985),77-91.

Biotechnology in the Pulp and Paper Industry
L. Viikari and R. Lantto (Editors)

New developments in enzyme-assisted delignification and bleaching

H.P. Call

Bioscreen e.K., 52531 Uebach-Palenberg, Heinsberger Strasse 15, Germany

The paper focuses on new applications of certain hydrolases, mainly lipases, in an enzyme-assisted oxidizing system useful *e.g.* for delignification/bleaching of pulps, the so-called hydrolase mediated oxidation system (HOS). This new system has as its main components: lipases, special fatty acid compounds, special ketone compounds and peroxides. The paper also focuses on new approaches to delignify or to bleach pulp with a new class of oxidation systems of enzymatically-generated redox agents *in situ*: reactive oxygen species (ROS) or reactive nitrogen species (RNS). These reactive ROS or RNS species (radicals, ions or reactive compounds) are produced in combination with H_2O_2 or superoxide from precursors which are able to release NO, NO^+ and/or NO^-, dicyclopentadienyl transition metal complexes (mainly iron complexes) or special sulfur compounds or activated sulfite which can react with ketones to form dioxirane or similar types of compounds.

1. INTRODUCTION

After the discovery of lignolytic enzymes (especially lignin peroxidase [LiP] and manganese peroxidase [MnP]), the pulp industry in particular had great expectations with regard to implementing cell-free enzymatic delignification and/or a bleaching system in its industrial processes in the foreseeable future. In spite of the enormous efforts undertaken, it had to be conceded that the use of lignolytic enzymes alone would not succeed, the molecular weight of the enzymes applied being too large to be able to penetrate the fibre walls. Therefore the idea arose very quickly of using these enzymes together with chemical compounds which can be oxidized by the corresponding lignolytic enzymes. These enzymatically-generated oxidants (also called mediators) can diffuse into the fibre walls and oxidize the residual lignin structures.

The most promising mediated systems so far are the manganese peroxidase system, i.e. MnP in combination with Mn^{2+} and H_2O_2 (1) and particularly the laccase mediator system (LMS-system), *i.e.* the combination of laccase and special mediator compounds like NOH-compounds, cation radicals or chelated transition metals (2-14). Nowadays – besides the above-mentioned oxidoreductase systems, which are not yet commercialized - only the xylanases are used in mill scale, but more as indirect lignin removal agents *i.e.* as bleach boosters (15, 16).

A possible alternative to the existing enzymatic and chemical delignification/bleaching concepts can be the recently published hydrolase mediated oxidation system (HOS) (17). The main components of this system are a hydrolase, especially a lipase, special environmentally-safe and cheap ketone and fatty acid compounds, which are also used e.g. in the food and

Some examples of treated softwood and hardwood pulps and the corresponding results with regard to kappa reduction (delignification), brightness and viscosity are shown in Table 2.

Table 2. Treatment of pulp with the aid of the hydolase mediated oxidation system [HOS]

pulp treatment	**delignification**	**viscosity** (ml/g)	**ISO-brightness gain***
A) kraft/ **hardwood**			
untreated	-	900	-
treated + extraction	33%	840	13,7%
B) kraft/**softwood**			
untreated	-	850	-
treated + extraction	34,5%	790	15,5%
C) sulfite/**softwood**			
untreated	-	840	-
treated + extraction	42,5	780	10,5%
D) sulfate/**softwood** (O_2-delignified)			
untreated	-	880	-
treated + extraction	49,5%	820	11,5

Treatment conditions: pH 4; 45°C; 4h retention time; 12,5% consistency; other conditions HOS (see under materials and methods)
* calculated as % brightness gain based on the brightness of the untreated pulp

The results presented clearly indicate that one can obtain good performance, selectivity (less viscosity loss) and brightness in all cases. Figure 1 demonstrates a possible mechanism of lipase action under these special conditions.

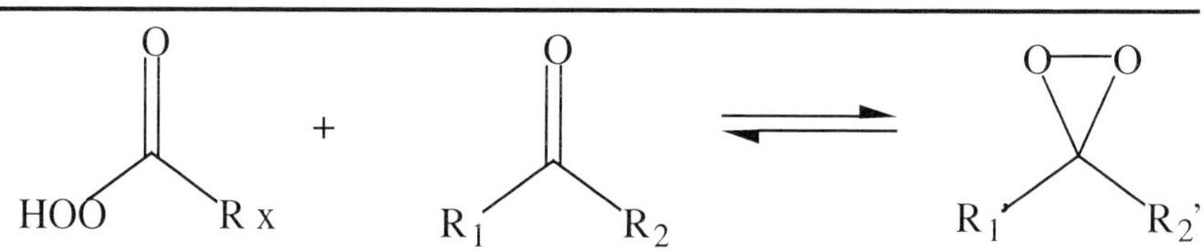

(B) + **(C)** = **(D)**

(A) R – COOH → **(B)** R – COOOH
(H_2O_2) → **(H_2O)**

lipase (other hydrolases)

A = fatty acid; B = peracid; C = ketone; D = activated oxygen species e.g. dioxirane; co-oxidant = H_2O_2; R, Rx = fatty acid rests; R_1, R_2, R_1', R_2' = ketone compound rests

Figure 1. Possible mechanism of the lipase mediated generation of activated oxygen species.

It is also important to consider whether the delignification performance is caused by the oxidation power of the system or whether it is also influenced by possible side-activities of the fungal enzyme used (*e.g.* xylanase activity). Table 3 shows the respective results. There is a clear indication that the oxidation abilities of the system are specifically responsible for the system performance, due to the important fact that the purified enzyme (without any traces of endoglucanases and xylanases) is also able to delignify strongly.

Table 3. Influence of different system components on the system performance, softwood pulp (O_2-delignified).

treatment conditions	delignification [%]
pulp untreated (K 16,6)	0
pulp + ketone; fatty acid; H_2O_2; without enzyme	7,5
only Lipase (crude) (1x)	12
only Lipase (crude) (10x)	19,2
lipase (crude) (1x) + ketone + H_2O_2 without fatty acid	12,6
lipase (crude) (1x) + fatty acid + H_2O_2 without ketone	14,2
lipase (crude) (1x) + ketone + fatty acid + H_2O_2 = **complete system**	41,3
only Lipase (purified)	8,1
lipase (purified) (1x) + ketone + fatty acid + H_2O_2 = **complete system**	30,7

Treatment conditions: pH 4; 45°C; 4h retention time; 12,5% consistency; lipase: (see materials and methods): 1x = 20g pure enzyme/to pulp; 10x = 200g pure enzyme/to pulp; other components of HOS: (see materials and methods)

Some new results obtained with several proteinases instead of lipases are illustrated in Figure 2. The results demonstrate that other hydrolases (related to lipases) which are commercially available and mostly very cheap also have the general ability to perform reactions such as these.

3.3. The use of special new peroxide-activating red/ox compounds, such as ferrocene

It was discovered that ferrocene or special derivatives of ferrocene in combination with mainly enzymatically generated peroxide could delignify and bleach pulps very efficiently under special conditions without showing significant loss in viscosity due to possible Fenton-type behaviour. In recent years many approaches have been undertaken to use chelated transition metal complexes as delignification agents *per se* or as activating systems for O_2-delignification or peroxide bleaching. The favoured complexes contain Fe, Cu, Mn, Mo, Va, W etc. or, in the case of peroxometalates, a combination of several metals. One driving force for the continuously growing research in this field has been the increasing mechanistic insight and knowledge regarding the active centre of the metallo-enzymes (oxidoreductases) and the challenge to mimic their properties (28, 29).

The new system presented contains only complexed iron 2^+ with cyclopentadienyl ligands and GOD as the enzyme for the slow release of H_2O_2 from glucose in the presence of oxygen, an important factor, as mentioned above. It is shown that a possible Fenton reaction was avoided under optimized pH, temperature, consistency and retention time conditions and at an appropriate peroxide concentration (in the range of 0.5 to 1 %). Moreover, it seems very likely that radicals with lower oxidation potential could be generated by adding special compounds e.g. sodium salts of sulfate, sulfite or peroxydisulfate ($S_2O_8^{2-}$). It is also very important to note that with the aid of this new system, a non-enzymatic generation of NO-radicals from the corresponding NOH, NO-compound (*e.g.* violuric acid etc.) and other LMS mediator types becomes possible, thus mimicking the action of laccases. The main benefit of these kinds of combined red/ox systems is the better strength properties of the pulp treated, the lower chemical demand and the higher performance in comparison to the peroxide/ferrocene system. An overview of the results obtained is presented in Table 5.

Table 5. Delignification performance of dicyclopentadienyl-iron complexes (ferrocenes) and some in situ generated radical species.

compound	**radical precursor**	**delignification**	**viscosity loss**
ferrocene + H_2O_2	-	32,4%	14,5%
ferrocene + H_2O_2	violuric acid (2 kg/ton pulp)	40,5%	10,9%
ferrocene + H_2O_2	sodium sulfite (2 kg/ton pulp)	38,4%	9,5%
ferrocene + H_2O_2	sodium peroxodisulfate (2 kg/ton pulp)	39,4%	11,5%

Treatment conditions: 4h; 60°C; pH 4; 12,5% consistency; concentrations: ferrocene: 0,125 kg/ton pulp; H_2O_2: 5 kg/ton pulp

3.4. Use of dioxirane compounds generated from special sulfur compounds or activated sulfite

Dioxiranes are well-established highly selective and powerful oxidants and are useful for many oxidation reactions in organic chemistry but also for delignification reactions in pulp bleaching. It has recently been discovered that some special compounds containing sulfur, *e.g.* organosulfonic peracid or sulfonimidic peracids produced from sulfonyl azoles (30) or similar compounds, can generate in the presence of ketones and H_2O_2 at slightly alkaline pH range dioxiranes. The main drawback to this chemical generation in respect of pulp delignification or bleaching is the relatively high cost of these substances, but particularly the fact that the sulfonic acid which remains after the reaction has no further oxidizing activity and cannot be recycled. Therefore, the slow generation of the active sulfonic peracid by the simultaneous enzymatic production of peroxide by *e.g.* glucose and GOD in the presence of oxygen is a very important prerequisite for the economic feasibility and performance of the new delignification and/or bleaching method presented. It is obvious that the slow liberation of the active component is superior to the complete addition of the peroxide (Table 6).

Table 6. Use of dioxirane compounds generated from organosulfonic acids precursors.

precursor P + H_2O_2 (completly added)	**delignification [%] 1kgP/to pulp**	**5kg P/topulp**	**ketone (2kg/to pulp)**
A	20,7	27,0	+
B	25,3	29,3	+
C	24,5	27,3	+
A	17,7	20,7	-
precursor P + GOD + glucose (H_2O_2 generation)			
A	25,7	30,9	+
B	29,5	33,0	+

Treatment conditions: 4h; 50-60°C; pH 6-7,5; 12,5% consistency; GOD concentrations (crude): ~ 5g per ton pulp; glucose concentration: 5kg/ton pulp; precursor P: organosulfonic acids (A,B,C/ see materials and methods); ketone: acetone; H_2O_2 concentration: 5kg/ton pulp

Another recent approach (31, 32, 33) was to use sulfite at alkaline pH range in the presence of sufficient oxygen and under the catalytic effect of Cu^{2+} ions for the generation of the peroxymonosulfuric acid anion (SO_5^{2-}) as a bleaching agent produced on site. The disadvantage of this method is the difficulty of maintaining the balance between SO_5^{2-} and SO_3^{2-}, as otherwise a reaction to the inactive SO_4^{2-} ion occurs. The slow addition of sulfite under intensive stirring is one unsatisfactory method of reducing these difficulties. We assumed that the generation of active SO_3^{2-} radicals and the simultaneous production of superoxide probably lead to the SO_5^{2-} ion. In addition, we found that the addition of special ketone compounds could further enhance the delignification reaction, probably through the generation of dioxirane or similar activated oxygen species (data not shown).

4. SUMMARY

The best results of the new delignification and/or bleaching methods presented here, including cost estimation are summarized in Table 7. Besides the recently published hydrolase-mediated oxidation system (HOS), some other new methods for delignification/ bleaching of pulp have been presented. The cost structure and the performance of these new systems are comparable. The final brightness level after conventional ECF bleaching conditions (including e.g. a HOS-based delignification step) can be ca. 90% ISO brightness or more. A possible role of peroxynitrite/ peroxynitrous acid *in vivo* delignification by fungi is thought to be likely due to its additional involvement in lipid peroxidation.

Table 7. Examples of some highlighted results.

system	**delignification**	**cost estimation**
A) HOS	41%	~ 10 US$/ton pulp
B) GOD glucose/nitrite-system	40%	~ 10 US$/ton pulp
C) ferrocene/GOD/ glucose system	46%	~ 12 US$/ton pulp

It can be summarized that:

1. Multistage bleaching with different enzymatic oxidizing systems based on different mechanisms of lignin attack is feasible.
2. All chemicals used (with the exception of ferrocene) are known in the pulp industry and are cheap and environmentally safe.
3. The enzymes used (catalase, peroxidase, glucose oxidase, lipase) are bulk enzymes, commercially available and cheap.
4. The delignification rate with all new systems is high with good strength properties.
5. The conditions of treatment need no special equipment.
6. Not only subsequent bleaching steps, but also a combination of different systems in one stage is possible.

REFERENCES

1) Paice, M.G., Manganese peroxidase catalyzed bleaching of kraft pulp, In: Tappi Proc. Biological Science Symp., 1997, 343-345
2) Harvey, P.J., Schoemaker, H.E. and Palmer, J.M., Veratryl alcohol as a mediator and the role of radical cations in lignin biodegradation by Phanerochaete chrysosporium, FEBS Lett. 195, 1986, 242-246
3) Call, H.P., Process for producing cellulose from lignin containing raw materials using an enzyme or microorganism while monitoring and maintaining the redox potential, WO 88/ 03190, 1988
4) Bourbonnais, R., Paice, M.G., Oxidation of non-phenolic substrates, an expanded role for laccase in lignin biodegradation FEBS, 267, 1990, 99-102
5) Call, H.P., Process for the delignification of material containing lignocellulose, bleaching, and treatment of waste water by means of laccases with extended efficiency. WO 92/ 20857, 1992
6) Bourbonnais, R., Paice, M.G., Demethylation and delignification of kraft pulp by Trametes versicolor laccase in the presence of 2,2'-azinobis-(3-ethylbenzthiazoline-6-sulphonate), Appl. Microbiol. Biotechnol., 36, 1992, 823-827.
7) Call, H.P., Process for changing, breaking down or bleaching lignin, material containing lignin or similar substances, WO 94/ 29510, 1994a

8) Call, H.P., Mücke, I., State of the art of enzyme bleaching and disclosure of a breakthrough process, In: Proc. Int. Non-Chlorine Bleaching Conf., Amelia Island, Fl, USA, 1994b
9) Call, H.P., Multicomponent system for modifying, decomposing or bleaching lignin, lignin containing materials or similar substances and method of using this system, WO 96/ 18770, 1996
10) Call, H.P., Mücke, I., History, overview and application of mediated lignolytic systems, especially laccase-mediator-systems (Lignozym^R-process), Journal of Biotech-nology, 53, 1997a, 163-202
11) Call, H.P., Multicomponent system for changing, reducing or bleaching lignin, lignin-containing materials or similar substances as well as processes for its application, WO 97/ 60041, 1997b
12) Call, H.P., A Multicomponent system for changing, degrading or bleaching lignin, lignin containing materials or similar substances and processes for its use, WO 97/ 60039, 1997c
13) Call, H.P., Multi-component system for modifying, decomposing or bleaching lignin, materials containing lignin or similar substances and methods for producing the same, WO 98/ 55684, 1998
14) Bourbonnais, R., Rochefort, D., Paice, M.G., Renaud, S. and Leech, D., Transition metal complexes : a new class of laccase mediators for pulp bleaching, TAPPI Journal, Vol 83, No. 10, 2000, 68
15) Viikari, L., Ranua, M., Kantelinen, A., Linko, M. and Sundquist, J. Bleaching with enzymes, Biotechnology in the Pulp and Paper Industry, Proc. 3rd Int. Conf., Stockholm, 1986, 67-69
16) Viikari, L. Kantelinen, A. Sundquist, J. and Linko, M. Xylanases in bleaching: From an idea to the industry, FEMS Microbiology Reviews, 13, 1994, 335-350
17) Call, H.P., Oxidation and bleaching system with enzymatically produced oxidizing agents, WO/ 98/ 59108
18) Call, H.P., Combination of hydrolytic enzyme mediated delignification and bleaching systems with enhanced oxidative enzyme systems, In: Proc./Abstracts 219th ACS National Meeting, San Francisco, USA, 2000, Cell 0153
19) Call, H.P., New developments and further improvements of hydrolase mediated oxidative delignification and bleaching of pulp, In: Proc. Tappi Pulping/ Process & Product Quality Conf., 2000, Boston, USA
20) Kazlauskas, R. and Bornscheuer, U.T., Biotransformations with lipases, In: Biotechnology Vol. 8a, Wiley-VCH, 1998, 37-191
21) Rüsch, M. gen. Klaas and Warwel, S., Lipase-catalysed peroxy fatty acid generation and lipid oxidation, In: Enzymes in Lipid Modification, Bornscheuer, U.T. ed. Wiley-VCH, 2000, 116-127
22) Suchy, M. Argyropoulos, D.S., Catalysis and activation of oxygen and peroxide delignification of chemical pulps; a review, In: Proc. Tappi Pulping/Process & Product Quality Conf., 2000, Boston, USA
23) Denmark, S.E., Forbes, D.C., Hays, D.S., DePue, J.S. and Wilde, R.G., Catalytic epoxidation of alkenes with Oxone, J. Org. Chem. 60, 1995, 1391-1407
24) Hollocher, Th., Enzymes of bacteria, plants and fungi that produce free nitrogen oxides In: Nitric Oxide Research, John Wiles & Sons, 1998, 119-146

25) Ku, P.W.C., Hsieh, J.S., Jayawant, M.D. and Larry, L., Chlorination stage elimination by using nitrosation pretreatment before oxygen delignification/ oxygen reinforced extraction, TAPPI J., 75 (10), 1992, 146-151
26) Ohi, H., McDonough, T.J., Mechanism of nitrogen dioxide pretreatment for oxygen bleaching of kraft pulp, Mokuzai Gakkaishi, 38, 1992, 707-712
27) Filizzola, M.J., Vidal, T. and Colom, J.F., ECF bleaching sequences with nitrogen oxide pretreatments, Invest. Tec. Pap. 32 (132), 1995, 54-74
28) Call, H.P., Process of enzymatic bleaching of cellulose, WO 91/14823, 1991
29) Messner, K., Fackler, K., Koller, K., Lampaipis, P., Srebotnik,E. Gindl,W. and Watanabe,T., Depolymerization of lignin by coordinated copper and peroxide and its effect on wood, In: Proc. Tappi Pulping/Process & Product Quality Conf., 2000, Boston, USA
30) Schulz, M. Kluge, R., Organosulfonic and sulfonimidic peracids – new oxidants for diastereoselective and enantioselective oxidations of organic substrates, In: Peroxide Chemistry, Wiley-VCH, 2000, 113-127
31) Devuyst, E.A.P., Ettel, V.A., Mosoiu, M.A., An oxidant of unexpected utility, Chemtech 9 (7), 1979, 426-427
32) Shaharuzzaman, M., Bennington, C.P.J., The effect of mixing on the generation of alkaline peroxymonosulfate, In: Proc. 2000 International Pulp Bleaching Conf., Halifax, Canada, 2000, 267- 273
33) Bouchard, J., Magnotta, V. and Berry, R., Improving oxygen delignification with peroxymonosulphate: the(Opx) process, In: Proc. 2000 International Pulp Bleaching Conf., Halifax, Canada, 2000, 97-102

Biotechnology in the Pulp and Paper Industry
L. Viikari and R. Lantto (Editors)

In vivo and *in vitro* biobleaching of unbleached hardwood kraft pulp by a marine fungus, *Phlebia* sp. MG-60

X. Li, R. Kondo and K. Sakai

Graduate School of Bioresource and Bioenvironmental Sciences, Kyushu University, Hakozaki 6-10-1, Higashi-ku, Fukuoka, 812-8581, Japan

A marine fungus, *Phlebia* sp. MG-60, was employed for the first time to bleach the unbleached hardwood kraft pulp (UKP) *in vivo* and the pulp brightness was improved from 32.1 % to 62.1% (ISO) and kappa number decreased by 11 points after 8 days of incubation. The pulp brightness bleached by *Phlebia* sp. MG-60 was much higher than that by *Phanerochaete chrysosporium* at the same incubation conditions. Activities of ligninolytic enzymes from the fungal cultures were determined, and MnP activity was much higher than LiP and laccase activities.

UKP was also bleached by crude enzymes produced by *Phlebia* sp. MG-60 *in vitro*. With only 0.5 mM H_2O_2 continual addition, the pulp was brightened by about 10 points in 6 hours of incubation when the crude enzymes with 20U of MnP were added. The impact of only initial addition of H_2O_2 on the bleaching of UKP was also investigated.

1. INTRODUCTION

Biological treatments of kraft pulp provide alternative approaches to improve the environmental and cost performance properties. After Kirk and Young first reported that *Phanerochaete chrysosporium* could partially delignify softwood unbleached pulp (1), white rot fungi and their extracellular ligninolytic enzymes, lignin peroxidase (LiP), manganese peroxidase (MnP) and laccase were also employed to bleach unbleached hardwood kraft pulp (UKP) (2-7). Research results have demonstrated that white rot fungi and their enzymes have shown significant advancement as replacements for chemical bleaching agents over the past decade

P. chrysosporium, an outstanding lignin degradation fungus, and its MnP and LiP have been widely studied and used to bleach UKP, decolorize synthetic dyes and treat pulp and paper mill effluent (8-11). Other continental white rot fungi have shown a good application prospect on biodegradation and bioremediation (12-15). Marine fungi play an important role as primary decomposer in marine substrate, such as mangrove and sea grasses. The ability of marine fungi to degrade organic materials has been demonstrated (16-19). In our laboratory, a marine fungus, *Phlebia* sp. MG-60, was isolated from mangrove stands based on its delignification and decolorization ability (20). In the present paper, we report that a marine fungus was able to degrade lignin efficiently.

2. MATERIALS AND METHODS

2.1. Pulp

UKP was obtained from a pulp and paper mill in Japan. Its brightness and kappa number is 32.1% (ISO) and 16.2, respectively. It was once-air-dried.

2.2. Fungi

The strain, MG-60, and *P. chrysosporium* ME-466 were maintained in potato dextrose agar (Difco Laboratories) slants at 4°C before used.

2.3. Enzyme Preparation

A liquid culture medium was monitored for enzyme production, which contained 10 g/l of glucose, 1.2 mM of ammonium tartrate, 20 mM of sodium acetate, 1.0 g/l of Tween 80 and Kirk's trace elements (21) and its pH was adjusted to 4.5. MG-60 was incubated on potato dextrose agar in petri plates (diameter 9 cm) at 30°C for 7 days. One quarter of each plate was homogenized in the liquid medium with a Waring blender and then inoculated into a total volume of 200 ml of the liquid medium in 500 ml Erlenmeyer flasks. After 10 days of incubation at 30 °C and 150 rpm, the supernatant was separated from the mycelium by filtration with glass fiber and 0.45 μm membrane filters (Toyo Roshi Kaisha, Ltd., Japan) and then used as crude enzyme to bleach UKP.

2.4. Enzyme assays

MnP activity was determined by using 2,6-dimethoxyphenol as a substrate at 470 nm and one unit of enzyme activity is defined as 1 μmol products produced in one minute (22). Enzyme was added to a solution containing 1.0 mM 2,6-dimethoxyphenol and 1.0 mM $MnSO_4$ in 50 mM malonate buffer (pH 4.5) and the reaction was initiated with hydrogen peroxide (0.2 mM) in a final volume of 1 ml. As control, hydrogen peroxide (0.2 mM) was replaced by same volume of water. Lignin peroxidase activity was estimated by measuring the amount of veratryl aldehyde formed from 0.1 mM veratryl alcohol (VA) in 50 mM succinate buffer (pH 3.0) containing 0.2 mM hydrogen peroxide (4). Laccase activity was measured by monitoring the oxidation of 2,6-dimethoxyphenol at 470 nm in 0.1 M phosphate buffer (pH 6.0) (4). One unit of enzyme activity is defined as 1 μmol products produced in one minute.

2.5. Pulp treatments

For biobleaching *in vivo*, *Phlebia* sp. MG-60 and *P. chrysosporium* ME-466 were incubated for 7 days in potato dextrose broth (Difco Laboratories) liquid culture in petri dish (d = 15cm) at 30°C. After separation from the liquid culture, the mycelia were washed with distilled and autoclaved water and homogenized with autoclaved water or sea salt solution. After sterilization with 6ml of distilled water or sea salt solution, UKP of 4.0 g was mixed with 8 ml of mycelium solution and cultivated stationarily at 30°C. For enzyme activity assays, enzymes were extracted from the treated pulp by 300 ml of 50 mM malonate buffer (pH 4.5) with 0.5% Tween 80 at 4 °C for 2 hours.

For biobleaching *in vitro*, UKP was suspended at the consistency of 1% in 50 mM malonate buffer (pH 4.5), which contained 0.1 mM $MnSO_4$, 0.05% Tween 80, and proper MnP solution. The enzymatic reaction was initiated with hydrogen peroxide and continued at 37 °C. As a control, UKP was treated in the same way without MnP addition.

2.6. Effect of H_2O_2 on biobleaching *in vitro*

Hydrogen peroxide at proper concentration was continuously added at the flow rate of 3 ml/h with a peristaltic pump or only initially added to the enzyme reaction system at 37 °C. The properties of the bleached pulp were tested to evaluate the effect of H_2O_2 on biobleaching *in vitro*.

2.7. Pulp properties

Handsheets were prepared with a Buchner funnel from the treated pulp that had been washed with distilled water and homogenized for 30 s in a Waring blender. After air-drying on blotter, each handsheet was tested for brightness and kappa number. Brightness was determined using a colorimeter (CR-200, Minolta, Japan). The kappa number was determined using a standard method (ISO302-1981).

3. RESULTS

3.1. Effect of fungal treatment on bleaching

The unbleached hardwood kraft pulp was inoculated with MG-60 and *P. chrysosporium* and incubated for 14 days and 12 days, respectively. After treatment with MG-60, marked brightness increase was observed after day 6 of incubation (Fig.1). The pulp was brightened 30.0 points by MG-60, while the brightness was increased 23.2 points by *P. chrysosporium* on day 8. The pulp kappa numbers decreased along with the brightness increase as shown in Fig.2.

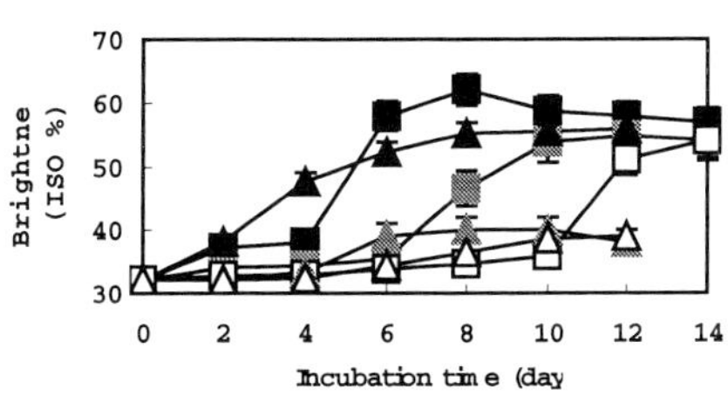

Fig.1 Brightness increase of UKP during biobleaching with the fungus MG-60 at 0% (■), 3% (▩) and 5% (□) sea salt concentrations and *P. chrysosporium* at 0% (▲), 3% (▲) and 5% (Δ) sea salt concentrations

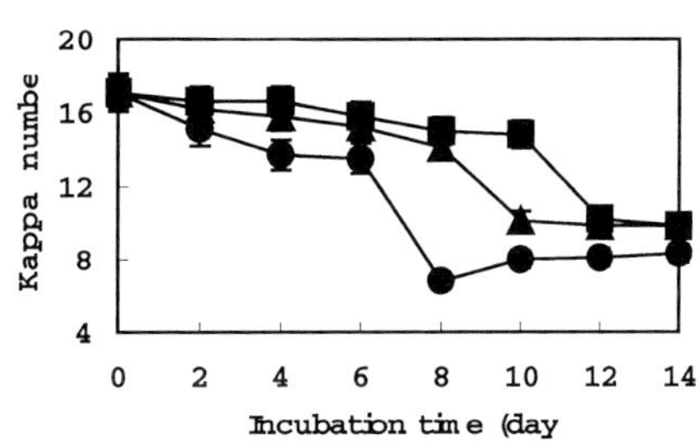

Fig. 2 Decrease of kappa number of UKP bleached with the fungus MG-60 at 0% (•), 3% (▲) and 5% (▪) sea salt concentrations

3.2. Enzyme production during fungal bleaching

MnP, LiP and laccase activities in the fungal culture were determined when UKP was bleached with MG-60. Peak production of MnP occurred after 8 days of fungal inoculation (Fig. 3). Laccase activity was low and only trace amount of LiP activity was detected. The results in Fig.1 and Fig.3 indicate that the pulp brightness increase was coincident with the production rate of MnP and / or laccase during the biobleaching process.

3.3. Effect of crude enzyme on bleaching

Crude enzyme secreted by MG-60 at day 10 of incubation was prepared and MnP, LiP and laccase activities were examined. As similar to the results obtained with the fungal treatment of pulp, only trace amount of LiP was secreted and laccase activity was much lower than MnP activity. To initiate the bleaching reaction, 0.5 mM H_2O_2 was continuously added into the pulp suspension containing 0.1 mM $MnSO_4$ at the rate of 3 ml/h for 24 h. The pulp brightness increased by about 10 points, compared with the control when the crude enzymes containing 20U of MnP activity was added (Fig.4). Decrease of kappa number along with the brightness increase was also observed from Fig. 4.

Biobleaching *in vivo* requires longer time than that *in vitro*. As usually reported, treatment time of pulp with MnP was as long as 24 hours. The present results in Fig. 5 illustrated that the pulp brightness increased sharply about 6 points during the first 3 hours and then there was slight change in the following period. This implies that pulp treatment with the crude MnP could be accomplished in shorter than 6 hours.

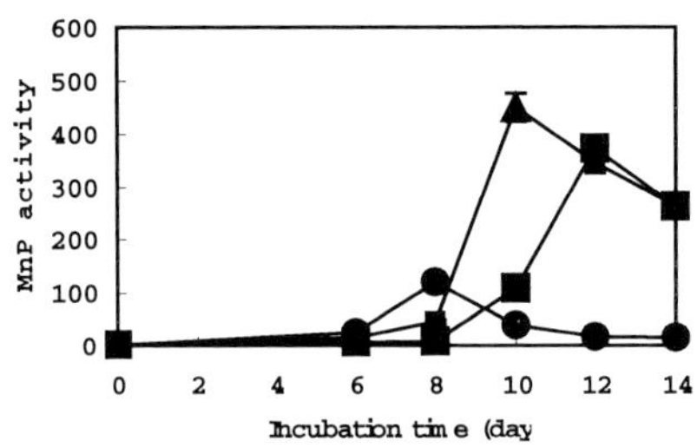

Fig. 3A Time course of MnP activity during UKP bleached with the fungus MG-60 at 0% (●), 3% (▲) and 5% (■) sea salt concentrations

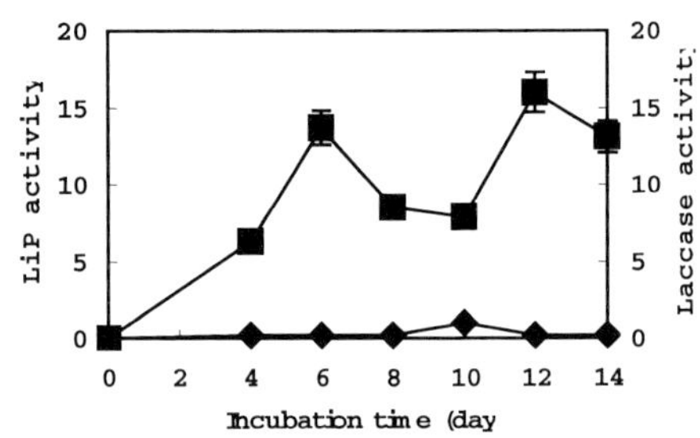

Fig. 3B Time course of LiP (♦) and laccase (■) activities during UKP bleached with the fungus MG-60 at 0% sea salt concentration

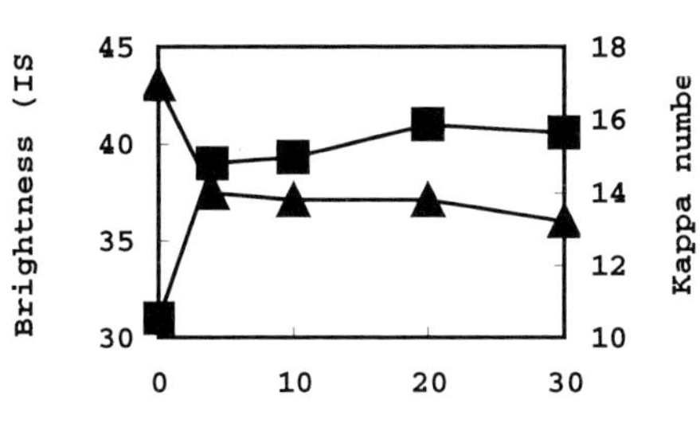

Fig. 4 Effect of MnP activity on bleaching of UKP at 37°C for 24 h. (brightness (■) and kappa number (▲))

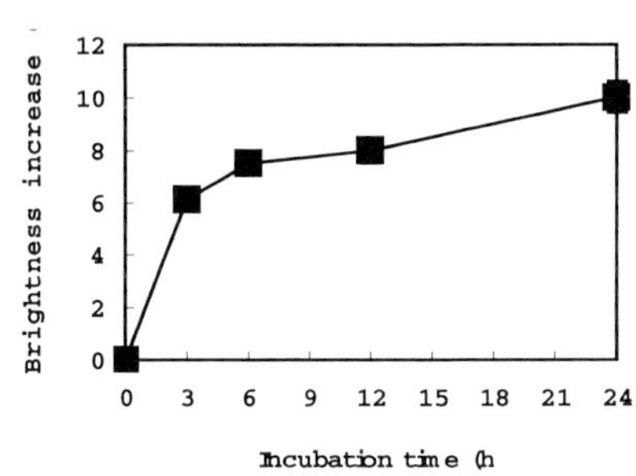

Fig. 5 Time course of pulp brightness increase during a treatment with the crude enzyme solution secreted by MG-60

3.4. Effect of H_2O_2 on biobleaching *in vitro*

MnP requires a source of hydrogen peroxide, but it is inhibited by high concentrations. Effect of H_2O_2 on UKP treatment with the crude MnP was determined and the results are shown in Fig. 6. The pulp brightness began to decrease when the concentrations of continuously added H_2O_2 were higher than 0.5 mM. The pulp was brightened about 10 points when 0.5 mM of H_2O_2 was continuously added to the reaction system.

When hydrogen peroxide was only initially added, 20 U of MnP dosage could increase the pulp brightness about 6 points during the first 3 hours. In the first 3 hours, no obvious difference was observed between the results when hydrogen peroxide was continuously or initially added at the same dosage of MnP and H_2O_2 concentration (Fig. 7).

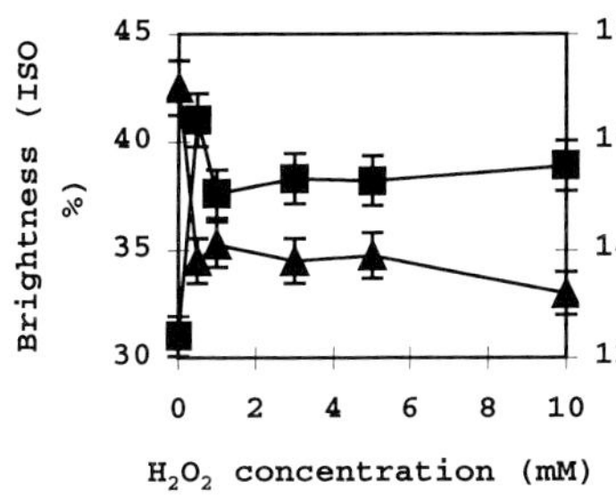

Fig. 6 Effect of hydrogen peroxide concentration on bleaching of UKP the crude enzyme solution containing 20 U MnP of MG-60 at 37°C for 24 h. (brightness (■) and kappa number (▲))

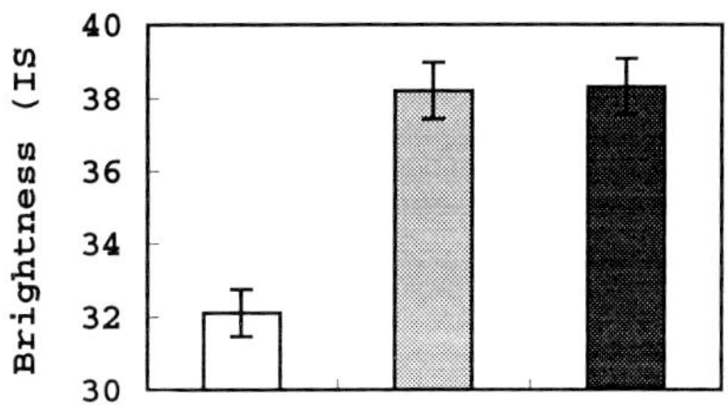

Fig. 7 Comparison of brightness increase after 3 hours enzyme bleaching with continuous (▩) or initial (■) addition of hydrogen peroxide. (control: □)

4. DISCUSSIONS

P. chrysosporium is regarded as an outstanding fungus to delignify, whereas the brightness increase of the pulp treated with *Phlebia* sp. MG-60 was even higher than that with *P. chrysosporium* under the same incubation conditions. Therefore, *Phlebia* sp. MG-60 could be an ideal microorganism for study of fungal degradation of lignin.

Kondo *et al.* reported that MnP plays an important role in the bleaching of UKP by white rot fungi (22). Paice *et al.* and Katherine *et al.* also reported that there was a strong positive correlation between the level of MnP and brightness increase (7,23). However, Katagiri *et al.* reported that there was no correlation between the brightness increase and the MnP production during biobleaching *in vivo* (24). Moreover, it has been reported that MnP and/or laccase from white rot fungi might have properties enabling them to oxidize nonphenolic substrates directly (25). To understand the role of these enzymes on delignification during fungal treatment of UKP with MG-60, their activities during incubation were determined. Trace amount of LiP activity was detected and laccase activity was much lower than MnP activity. It seems that MnP played a more important role in delignification during biobleaching *in vivo*. We used MG-60 to bleach UKP under sea salt conditions. The brightness was less increased along with the increase of sea salt concentration even though its MnP production was strongly enhanced and laccase activity was inhibited by 3% sea salts (data not shown). These results pointed out that only high MnP activity was not enough to delignify, and MnP and/or laccase group from *Phlebia* sp. MG-60 might play an important role on biobleaching *in vivo.*

MnP requires H_2O_2 as a cosubstrate and catalyst of the oxidation of Mn^{2+} to Mn^{3+}, but the sensitivity of the enzyme to H_2O_2 has been a problem. Hydrogen peroxide can be generated from glucose and glucose oxidation in fungal bleaching process. It has to be explored how and how much H_2O_2 is added to the enzyme bleaching system. With addition of 10 mM H_2O_2, 20 U MnP from *Phanerochaete sordida* YK-624 was reported to brighten UKP 10 points and decrease kappa number about 6 points (22). At the same MnP dosage, however, the *Phlebia* sp. MG-60 crude enzyme brightened UKP over 10 points with only 0.5 mM of H_2O_2 addition. It is suggested that as low as 0.5 mM of H_2O_2 was enough for the crude enzyme of *Phlebia* sp. MG-60 to bleach UKP.

The operation of enzymatic bleaching process would become convenient and simple if H_2O_2 can be once added to the bleaching system. Some effort has been done on this aspect. It was once reported that MnP activity was inactivated by concentrated H_2O_2 above 0.1 mM (26). Surprisingly, when 0.5mM H_2O_2 was once added at initial time of biobleaching system of MG-60 crude enzyme, the brightness increased over 6 points in the first 3 hours which could be compared with the result of continual addition of H_2O_2. We therefore inferred that the *Phlebia* sp. MG-60 MnP had fine stability to hydrogen peroxide. Although the pulp was brightened about 6 points in the first 3 hours after addition of H_2O_2, the brightness kept at the same level in the following period. The brightness at only initial addition of H_2O_2 condition did not reach the level at continuous addition of same amount of H_2O_2. This implies that the MnP activity maybe partly inhibited by the once-added H_2O_2. Another possibility was that MnP had reacted with the high concentration of H_2O_2 before it penetrated into the pulp fiber.

Hydrogen peroxide of 0.5 mM was either once added or continuously added to the system, the pulp was brightened over 6 points during the first 3 hours by *Phlebia* sp. MG-60 crude MnP. This several hour treatment and its corresponding achievement could be possibly accepted by the pulping and papermaking industry.

REFERENCES

1. T. K. Kirk and H. H. Young, Biotechnol. Lett., 1 (1979) 347-352.
2. M. T. Moreira, G. Feijoo, S. A. Reyes, J. Lema and Jim A. Field, J. Biotechnol., 53 (1997) 237-251.
3. M. Tuomela, M. Vikman, A. Hatakka and M. Itävaara, Bioresource Technology., 72 (2000) 169-183.
4. H. Hirai, R. Kondo and K. Sakai, Mokuzai Gakkaishi., 40(1994) 980-986.
5. K. Harazono, R. Kondo and K. Sakai, Journal of the Japanese Technical Association of the Pulp and Paper Industry, 50(1996) 59-68.
6. J. E. Sealey, T. M. Runge, and A. J. Ragauskas, Tappi Journal, 83 (2000) 66.
7. M. G. Paice, I. D. Reid, R. Bourbonnais, F. S. Archibald, and L. Jurasek, Appl. Environ. Microbiol., 59 (1993) 260-265.
8. M. H. Gold, H. L. Youngs and M. D. S. Gelpke, METAL IONS IN BIOLOGICAL SYSTEMS, VOL 37: MANAGANESE AND ITS ROLE IN BIOLOGICAL PROCESSES. 37 PG: 559-586, 2000.

9. T. Iimori, S. Miyawaki, M. Machida and K. Murakami, J. Wood Sci., 44 (1998) 451-456.
10. J. K. Glenn, M.H. Gold, Appl. Environ. Microbiol., 45 (1983) 1741-1747.
11. H. H Yusef, K. M. Ghanem, H. Y. El-Kassas, Fresenius Environmental Bulletin, 8(1999) 668-677.
12. Y. Kashino, T. Nishida, Y. Takahara, K. Fujita, R. Kondo and K. Sakai, Tappi Journal, 76 (1993) 167-171.
13. I. D. Reidand, M. G. Paice, FEMS Microbiology Reviews, 13 (1994) 369-376.
14. A. Sethuraman, D. E. Akin and E. E. L. Eriksson, Appl. Microbiol. Biotechnol., 52 (1999) 689-697.
15. J. Jellison, C. A. Jasalavich, International Biodeterioration & Biodegradation, 46(2000, Special Issue SI) 241-244.
16. A. B. Cribb and J. W. Cribb. University of Queensland, Department of Botany, 3 (1955) 78-81.
17. J. Kohlmeyer and E. Kohlmeyer, Mycologal., 63: 831-861.
18. J. H. Block, P. Catalfomo, G. H. Constantine and P. W. Kirk, Mycologal., 65 (1973) 488-491.
19. E. A. Cronin and F. J. Post, Mycologal., 69 (1977) 846-847.
20. X. Li, R. Kondo, K. Sakai, Accepted by J. Wood Sci., (2001).
21. M. Tien and T. Kirk, Methods Enzymol., 161 (1988) 238-249.
22. R. Kondo, K. Harazono and K. Sakai, Appl. Environ. Microbiol., 60 (1994) 4359-4363.
23. K. Addleman and F. Archibald, Appl. Environ. Microbiol., 59(1993) 266-273.
24. N. Katagiri, T. Yuji and T. Nishida, Appl. Environ. Microbiol., 61(1995) 617-622.
25. F. H. Férié and M. H. Gold, Appl. Enviorn. Microbiol., 57 (1991) 2240-2245.
26. P. Bajpai and P. Bajpai. Biotechnology in the pulp and paper industry. Printed by Pira International, Leatherhead, Uk, 1998. Pp57-58.

Biotechnology in the Pulp and Paper Industry
L. Viikari and R. Lantto (Editors)

Flax pulp bleaching and residual lignin modification by laccase-mediator systems*

S. Camarero,[1] O. García,[2] T. Vidal,[2] J. Colom,[2] J.C. del Río,[3] A. Gutiérrez,[3] M.J. Martínez,[1] and A.T. Martínez[1]

[1]CIB, CSIC, Velázquez 144, E-28006 Madrid, Spain
[2]ETSII, UPC, Colom 11, E-08222 Terrassa, Spain
[3]IRNAS, CSIC, PO Box 1052, E-41080, Sevilla, Spain

Enzymatic bleaching of flax alkaline pulp is intended here by using laccase-mediator systems (LMS) based on *Pleurotus eryngii, Trametes versicolor* and *Pycnoporus cinnabarinus* laccases and two mediators. The efficiency of the different LMS treatments was compared in terms of brightness, kappa number and viscosity modifications. Furthermore, changes in the molecular structure of residual lignin and lignin/carbohydrate ratio were analyzed by pyrolysis-gas chromatography-mass spectrometry of the treated pulps. *Pycnoporus cinnabarinus* and *T. versicolor* laccases in the presence of 1-hydroxybenzotriazole (HBT), were the most effective for flax pulp delignification. Up to 20% ISO brightness increase with respect to the initial pulp and a decrease of kappa number from 9 to 3 were attained after alkaline extraction of the pulps treated with those LMS. *Pycnoporus cinnabarinus* laccase plus HBT gave also the best selectivity in lignin removal. Analytical pyrolysis of the LMS-treated pulps showed a selective removal of lignin against cellulose and a strong alteration of the residual lignin with preferential removal of the syringyl units. *Pycnoporus cinnabarinus* LMS_{HBT} gave the best results in terms of decrease of lignin/cellulose and syringyl/guaiacyl ratios, in agreement with data from pulp parameters. For the above reasons laccase from *P. cinnabarinus* and HBT were selected for further assays including a peroxide step. Up to 89% relative delignification rate (attaining 1.3 kappa number) and an increase of 44% ISO brightness were attained after LMS_{HBT}-peroxide bleaching. Ongoing studies focus on improvement of a LMS_{HBT}-based totally chlorine free sequence for flax pulp bleaching.

1. INTRODUCTION

Pulp and paper manufacture from non-woody fibers is a widespread reality in Asia, South America and eastern Europe. Beside cereal straw, the leading non-woody fiber and one of the oldest sources of paper pulp, several crops are grown up for their content in long fibers. These textile fibers are mainly used for high added-value products in developed countries. In fact, high-quality pulps for specialty papers (tea bags, filters, cigarettes, bibles, condensers, etc) are manufactured from textile fibers such as flax, hemp, abaca, kenaf, jute or sisal.

---

* This work has been funded by the Spanish project 2FD97-0896-C02-02 and the EU project QLK3-99-590.

During last years new pulping concepts have arisen to overcome the limitations (i.e. chemical recovery and scaling up) from the application of conventional processes to these type of raw materials [1]. At the same time improvement of pulp quality and environmentally-sound bleaching sequences are searched. In this sense, the use of fungal laccases in the presence of mediators offers the possibility to develop new totally-chlorine free (TCF) sequences to bleach different types of pulps. The potential of these laccase-mediator systems (LMS) must be carefully evaluated since differences in the structure and redox potential of the enzyme as well as reactivity of the oxidized mediator can affect delignification rate and consequently pulp bleaching results [2-5]. Great advances have been achieved during the last decade in the understanding of LMS chemical and enzymatic mechanisms. However, most LMS studies have been carried out onto softwood and hardwood kraft pulps [6;7] and few enzymatic bleaching trials have been performed on non-woody materials.

The present study focuses on bleaching and delignification of a non-woody pulp, namely flax alkaline pulp. We first compared the efficiency of laccases from three white-rot fungi, *Pleurotus eryngii, Trametes versicolor* and *Pycnoporus cinnabarinus*, and two mediators, 2,2'-azinobis(3-ethylbenzothiazoline-6-sulfonic acid) (ABTS) and 1-hydroxybenzotriazole (HBT), for lignin removal from flax pulp. In this way, brightness, kappa number and viscosity of the LMS-treated pulps were determined. Moreover, the structure of the residual lignin in pulps was analyzed by pyrolysis-gas chromatography-mass spectrometry (Py-GC-MS) [8], a powerful analytical tool which provides information in terms of lignin/carbohydrate and *p*-hydroxyphenyl:guaiacyl:syrinyl (H:G:S) ratios, even when the kappa number is low. Both, pulp properties and pyrolysis evaluation were used for selection of the best LMS to be applied in a TCF sequence for flax pulp bleaching.

2. MATERIAL AND METHODS

2.1. Materials

The initial flax pulp with 36% ISO brightness, 11 kappa number and 900 mL/g viscosity was provided by CELESA mill (Spain). The initial raw material, flax (*Linum usitatissimum*), contained 15% of core fibers (xylem), and the pulps were obtained by soda anthraquinone cooking. ABTS and HBT were purchased from Boehringer and Aldrich respectively.

2.2. Enzyme production and activity determination

Laccases were produced by *P. eryngii* ATCC 90787 (= IJFM A169) grown in glucose-peptone medium [9] with 100 μM $MnSO_4$, and *T. versicolor* IJFM A136 and *P. cinnabarinus* IJFM A720 grown in glucose-ammonia medium [10]. In all cases laccase production was induced by addition of 150 μM $CuSO_4$. Cultures were harvested at the point of maximal laccase activity for the enzymatic bleaching assays. Laccase activity was determined by monitoring the OD_{436} change of 5 mM ABTS oxidation to its cation radical (extinction coefficient at 436 nm 29300 mM^{-1} cm^{-1}) in 100 mM acetate buffer, pH 5. One unit of enzyme activity was defined as the amount of enzyme that transforms 1 μmol of substrate per minute.

2.3. Laccase-mediator treatments

The laccase-mediator treatments (L-stage) were carried out in duplicate with 10 g of flax pulp at 2% consistency in 50 mM tartrate buffer, pH 4. Enzyme and mediators dosages were 10-20 U of laccase (the whole liquid cultures being used) and 20 mg of ABTS or HBT per gram of pulp. Tween 80 (0.05% w/v) was also added as surfactant. Flasks were kept under O_2

atmosphere for 24 h, at 160 rev/min and 30°C. In subsequent experiments, L-stage was performed with 20 U/g of *P. cinnabarinus* partially purified laccase (ultrafiltered by 3 kd cut off) and 4% (w/w) of HBT relative to pulp. As controls, pulps were treated under identical conditions but without enzyme and mediator.

2.4. Post LMS stages

LMS-treatments were followed by an alkaline extraction of the pulps (E-stage consisting of 1.5% NaOH treatment for 1 h at 60°C). Subsequent bleaching with hydrogen peroxide (P-stage) consisted of 3% H_2O_2 in 1.5% NaOH, for 2 h at 90°C. Pressurized peroxide bleaching (P_O-stage) was carried out with 3% H_2O_2 in 1.5% NaOH, for 2 h at 90°C in the presence of 5 bar O_2). A reductive step (R-stage) was applied under the following conditions: 2% $NaBH_4$, 30 min at 20°C. All post-LMS treatments were carried out at 5% pulp consistency. Brightness, kappa number and viscosity were measured at the different stages according to ISO 302, ISO 5351/1 and ISO 3688 standards, respectively.

2.5. Analytical Pyrolysis

Pyrolysis was performed in duplicate with a Curie-point pyrolyzer coupled to a Varian Saturn 2000 GC/MS equipment, using a 30 m x 0.25 mm DB-5 column (film thickness 0.25 μm). Approximately 1 mg of sample was deposited on a ferromagnetic wire, then inserted into the glass liner and immediately placed in the pyrolyzer. The pyrolysis was carried out at 610°C for 3.5 seconds. The chromatograph was programmed from 40°C (1 min) to 300°C at a rate of 6°C/min. The final temperature was held for 20 min. The injector, equipped with a liquid carbon dioxide cryogenic unit, was programmed from -30°C (1 min) to 300°C at 200°C/min, while the gas chromatography-mass spectrometry (GC-MS) interface was kept at 300°C. Pyrolysis products were identified by comparison with those reported in the literature and in the Wiley and Nist computer libraries.

Since some of the lignin-derived compounds are minor peaks in most pyrograms, their areas were integrated in single-ion chromatographic traces corresponding to their molecular ions. Furthermore, for relative estimation of polysaccharide and lignin removal, selected compounds were quantified as representative markers for cellulose and the different lignin units. Then, the decrease of lignin/cellulose ratio and the changes in lignin S/G ratio were calculated from the peak areas of 4-hydroxy-5,6-dihydro(2*H*)-pyran-2-one (*m/z* 114) as cellulose marker, compared with the following lignin markers: 4-methylguaiacol (*m/z* 138), 4-ethylguaiacol (*m/z* 152), 4-vinylguaiacol (*m/z* 150) and *trans*-4-propenylguaicol (*m/z* 164) as guaiacyl markers; and 4-methylsyringol (*m/z* 168), 4-ethylsyringol (*m/z* 182), 4-vinylsyringol (*m/z* 180) and *trans*-4-propenylsyringol (*m/z* 194) as syringyl markers.

3. RESULTS AND DISCUSSION

3.1. Selection of the best LMS for flax pulp bleaching

Table 1 shows that the laccases from *P. cinnabarinus* and *T. versicolor* were the most effective for flax pulp delignification, in all cases the best results being obtained with HBT as mediator. An increase of 10-15% ISO brightness with respect to the control pulp, which represents 20% ISO brightness increase with respect to the initial pulp, was attained after alkaline extraction of the pulps treated with *P. cinnabarinus* and *T. versicolor* laccases in the presence of HBT. At the same time a significant decrease of kappa number occurred (from 8.3 of the control to values around 3). *Pycnoporus cinnabarinus* laccase plus HBT gave also the best selectivity in lignin removal measured as the ratio between delignification efficiency and reduction of pulp viscosity (3.2 compared with 2.2 of *T. versicolor* LMS_{HBT}).

Table 1
Pulp properties and Py-GC-MS analysis after flax pulp treatment with different LMS followed by alkaline extraction (treatments were carried out by adding the pulp to fungal cultures with high laccase activity, 10-20 U/g pulp)

	Pulp properties			Pyrolysis analysis	
	Brightness (%)	Kappa number	Viscosity (mL/g)	Lignin/Cellulose decrease (%)	S/G ratio
Initial pulp	40.0	9.0	750	0	0.23
Control pulp	45.3	8.3	710	0	0.17
T. versicolor laccase + ABTS	48.4	6.3	434	15	0.07
T. versicolor laccase + HBT	60.2	3.3	600	56	0.03
P. cinnabarinus laccase + ABTS	34.2	8.5	565	0	0.12
P. cinnabarinus laccase + HBT	55.3	2.9	678	69	0.02
P. eryngii laccase + ABTS	33.4	9.8	586	18	0.11
P. eryngii laccase + HBT	35.8	8.0	607	30	0.11

Lignin removal as well as changes produced in the residual lignin of alkaline pulps after the LMS treatments were examined by Py-GC-MS. This degradative technique permits the identification of a series of products as derived from either cellulose or lignin polymers. In all cases, the main peaks corresponded to carbohydrate-derived compounds whereas the lignin-derived peaks were much lower. However, by monitoring individual ions corresponding to the M_w of selected lignin markers (representative compounds derived from the three types of lignin units) more information on lignin content and composition could be obtained (Fig. 1). This fact is specially interesting for pulps with very low kappa number (e.g. those enzymatically treated where kappa number has strongly decreased). By using this methodology, the presence of a G-rich GS-type residual lignin in flax pulps was revealed. The analytical pyrolysis of the pulps treated with the different LMS showed a selective removal of lignin against cellulose (Table 1). Moreover, a strong alteration of the residual lignin with preferential removal of S-units was observed. The use of laccases from *P. cinnabarinus* and *T. versicolor* and HBT as mediator resulted in higher delignification rates compared with ABTS utilization. *Pycnoporus cinnabarinus* LMS_{HBT} gave the best results in terms of lignin removal and modification of S/G ratio. Figure 2 illustrates the preferential removal of two selected lignin markers against cellulose marker with *P. cinnabarinus* LMS_{HBT} compared with LMS_{ABTS} and control. The lowest lignin/cellulose ratio given by *P. cinnabarinus* LMS_{HBT} after Py-GC-MS analyses coincides with the strong brightness increase and the highest (and most selective) lignin removal showed by kappa number and viscosity data.

In the light of these results, it appears that the capability of the above enzymatic systems to degrade lignin in flax pulp fibers and to modify the composition of the residual polymer is proved. Moreover, a correlation between some pulp properties and pyrolysis data could be established, showing the potential of this technique for lignin analysis when small pulp samples are used in bleaching optimization studies. Furthermore, by sample permethylation before Py-GC-MS analyses [11;12] more information on the structure of residual lignin could be obtained, similar to that provided by other techniques used for pulp analysis [13;14].

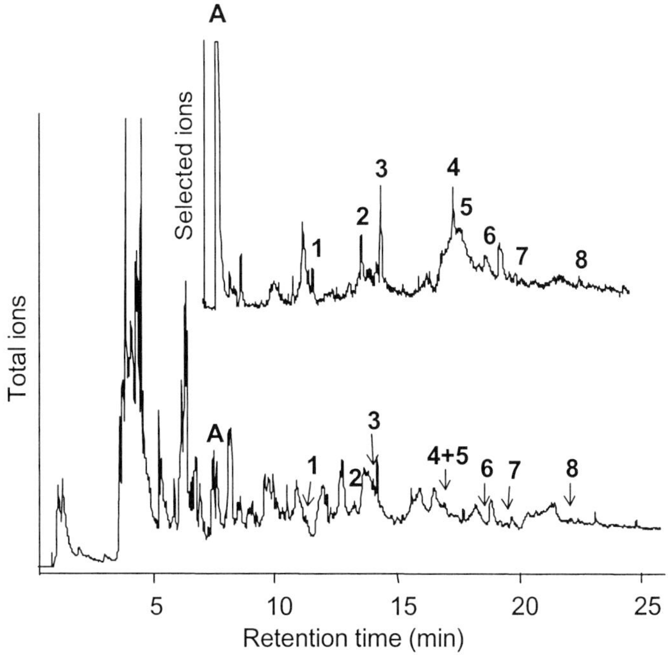

Figure 1. Total-ion Py-GC-MS chromatogram of initial flax pulp (11 kappa number) showing all carbohydrate and lignin-derived compounds released upon pyrolysis. Inset is the reconstructed ion-chromatogram for the selected marker for cellulose (A, 4-hydroxy-5,6-dihydro(2*H*)-pyran-2-one; *m/z* 114) compared with the following lignin markers: 1, 4-methylguaiacol (*m/z* 138); 2, 4-ethylguaiacol (*m/z* 152); 3, 4-vinylguaiacol (*m/z* 150); 4, *trans*-4-propenylguaicol (*m/z* 164); 5, 4-methylsyringol (*m/z* 168); 6, 4-ethylsyringol (*m/z* 182); 7, 4-vinylsyringol (*m/z* 180); and 8, *trans*-4-propenylsyringol (*m/z* 194) (1-4 as guaiacyl markers, and 5-8 as syringyl markers).

3.2. LEP bleaching sequence using laccase from *P. cinnabarinus* and HBT

Because of the results described above, *P. cinnabarinus* laccase was selected for further assays using HBT as mediator. Then, the LMS_{HBT} performance on flax pulp was followed by alkaline extraction and peroxide bleaching, and final values of 80% ISO brightness and kappa number around 1 were reached after the LEP sequence (Table 2). It is worth saying that the flax pulp used in this study is very resistant against bleaching in part because of the presence of core fibers in the raw material, and 80% ISO brightness is a value difficult to be reached in chemical TCF sequences. For this reason, the brightness obtained does not correspond with that expected from the very low kappa value attained (using other

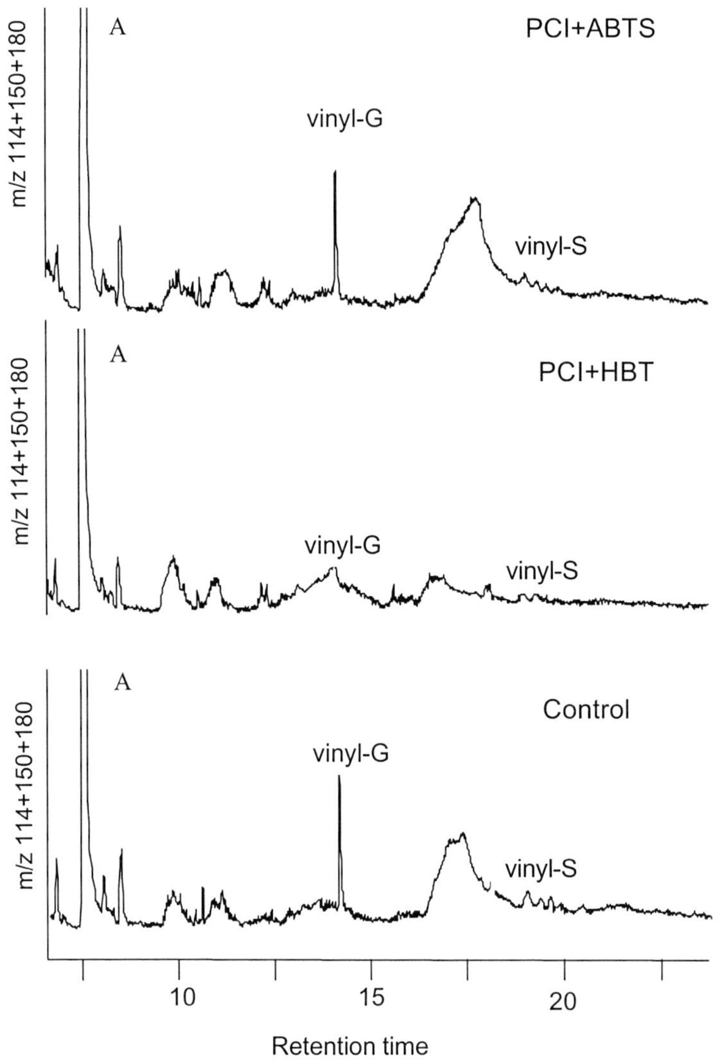

Figure 2. Py-GC-MS chromatograms of pulps treated with *Pycnoporus cinnabarinus* laccase in the presence of ABTS and HBT compared with the control. The sum of three single ions corresponding to the molecular ions of the cellulose marker (A, 4-hydroxy-5,6-dihydro(2*H*)-pyran-2-one, 114 mz), and two main lignin markers, 4-vinylguaiacol (*m/z* 150) and 4-vinylsyringol (*m/z* 180), is represented.

raw materials, e.g. hardwoods, up to 90%. ISO brightness would be attained for this kappa value [15]). The high delignification rate (89%) and the strong increase of brightness (from 36% to 81% ISO) obtained after LEP, showed that L-stage improves the poor capacity of hydrogen peroxide as delignifying agent, increasing also its bleaching properties. At the same time the P-stage significantly enhances bleaching results obtained after LE stages.

Table 2
Brightness, kappa number and viscosity of flax pulp treated with *P. cinnabarinus* laccase (20 U/g) in the presence of HBT followed by alkaline extraction (LE) and a peroxide bleaching stage (LEP)

	LE			LEP		
	Brightness (%)	Kappa number	Viscosity (mL/g)	Brightness (%)	Kappa number	Viscosity (mL/g)
Initial pulp	36.7	10.2	855	57.1	6.9	630
Control pulp	41.8	8.9	920	61.3	5.5	683
Laccase + HBT	59.4	2.8	613	80.6	1.3	470

Pyrolysis of the LMS_{HBT}-treated pulp before and after P-stage showed that pulp delignification was mainly due to laccase-mediator action since only a small increase of lignin removal was produced when a P-stage was added to the LE sequence (Table 3). The same could be said for modification of the lignin remaining in pulp. Py-GC-MS results evidenced that the main modification of residual lignin composition, as shown by S/G values, was due to the enzymatic treatment (Table 3). The S/G ratio decreased from 0.35 (control pulp) to 0.03 after LMS-treatment followed by alkaline extraction. Peroxide bleaching of the LE sample resulted in the completely disappearance of S units.

Table 3
Relative removal of lignin from Py-GC-MS analyses (with respect to lignin/cellulose ratio in the control pulp before peroxide) and changes of S/G ratio of flax pulp treated with *P. cinnabarinus* laccase (20 U/g) in the presence of HBT followed by alkaline extraction (LE) and a peroxide bleaching stage (LEP)

	LE		LEP	
	Lignin/Cellulose decrease (%)	S/G ratio	Lignin/Cellulose decrease (%)	S/G ratio
Control pulp	0	0.35	16	0.29
Laccase + HBT	85	0.03	92	0

The rate of lignin removal obtained by pyrolysis after LE sequence was very similar to delignification rate given by kappa number determination. Moreover, correlation between kappa values and pyrolysis data of the LMS_{HBT}-treated pulps before and after P bleaching was established with regression coefficient R^2= 0.90 (Fig 3).

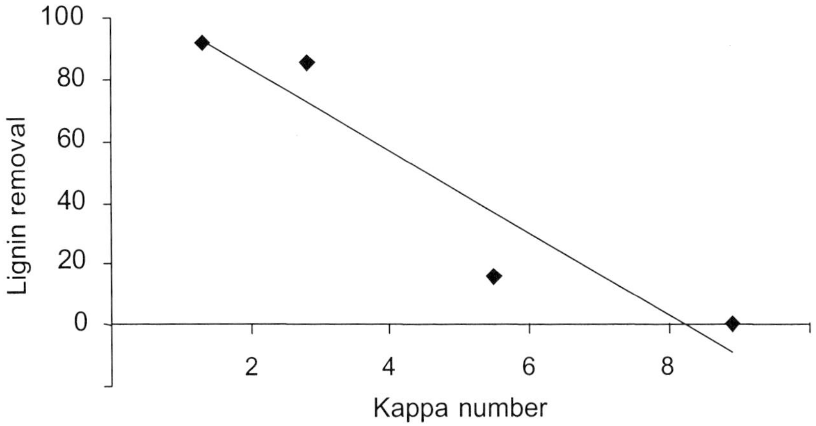

Figure 3. Correlation between kappa number and Py-GC-MS estimation of lignin removal from flax pulp after LMS_{HBT} treatment followed by E and EP stages.

3.2. Optimization of the LEP bleaching sequence

Further studies for optimization of the bleaching sequence after LMS_{HBT} application showed that the LEP sequence can be easily reduced to a two-step LP sequence since similar results in terms of brightness and kappa number were obtained when suppressing the alkaline extraction before the P-stage (Table 4). It seems that the colored lignin-degradation products (including quinones) formed during the enzymatic treatment, in part responsible for pulp color and kappa number, can be removed by the peroxide stage (under alkaline conditions) without a previous alkaline extraction. Moreover, viscosity was improved when suppressing the E-stage.

Concerning viscosity data, it is known the decrease of pulp viscosity in alkaline medium after oxidative bleaching (e.g. with ozone). This rapid degradation of cellulose is caused by carbonyl groups formed in the course of oxidative bleaching, which cause the cleavage of cellulose chains by β-elimination in alkaline medium (as used for viscosity estimation). This undesirable effect can be avoided by reducing the carbonyl groups to alcohols by $NaBH_4$ previously to alkaline stage. When applying the latter treatment to the LMS_{HBT}-treated pulp, viscosity values were much higher than those from standard viscosity determination.

The studies for optimization of a LMS_{HBT}-based TCF sequence for flax pulp bleaching continued with the pressurization of the P-stage (Po) (Table 4). No significant improvements were achieved in kappa number and viscosity values with this new LPo sequence compared to previous LP. However, near 82% ISO brightness was attained. Oxygen pressurized reactor was also used to perform the LMS_{HBT} treatment with different reaction times. In spite of brightness and kappa number improvements after the enzymatic treatment, no progress in pulp parameters were found after the P-stage (with respect to those under similar conditions but without pressurized oxygen). However, shorter reaction times showed reduction of viscosity losses. Finally, studies for the definition of a statistical model that provides the optimal conditions (in terms of enzyme and mediator doses and reaction time) for the LMS_{HBT}

treatment of flax pulp, are currently being developed. Therefore, pulp properties would be predicted as a function of the three main variables of the LMS treatment.

Table 4
Comparison of brightness, kappa number and viscosity of flax pulps treated with LMS_{HBT} after different post-LMS treatments: alkaline extraction (E), alkaline extraction followed by peroxide stage (EP), peroxide stage (P), pressurized peroxide stage (Po) and reductive treatment with $NaBH_4$ (R)

	Brightness (% ISO)	Kappa number	Viscosity (mL/g)
Initial pulp	37.3	10.11	1025
Initial control	38.0	13.7	1065
Initial LMS_{HBT}	41.6	9.5	835
Control E	39.1	9.4	1000
LMS_{HBT} E	49.3	5.0	780
Control EP	62.0	5.5	530
LMS_{HBT} EP	76.4	1.4	465
Control P	63.2	5.4	685
LMS_{HBT} P	79.8	1.6	725
Initial control R	42.5	9.1	1070
Initial LMS_{HBT} R	50.7	6.5	1010
Control Po	65.4	4.6	610
LMS_{HBT} Po	81.5	1.3	640

REFERENCES

1. G. Moore, Nonwood fibre applications in papermaking, Pira International, Leatherhead, Surrey, UK 1996.
2. F. Xu, W. S. Shin, S. H. Brown, J. A. Wahleithner, U. M. Sundaram, E. I. Solomon, BBA Protein.Struct.Mol.Enzym. 1996, 1292 303-311.
3. R. Bourbonnais, M. G. Paice, B. Freiermuth, E. Bodie, S. Borneman, Appl.Environ.Microbiol. 1997, 63 4627-4632.
4. J. Freudenreich, M. Amann, E. Fritz-Langhals, J. Stohrer, Proc.Intern.Pulp Bleaching Conf., Helsinki, 1-5 June 1998, 1 71-76.
5. K. C. Li, R. F. Helm, K. E. L. Eriksson, Biotechnol.Appl.Biochem. 1998, 27:239-243 239-243.
6. F. S. Archibald, R. Bourbonnais, L. Jurasek, M. G. Paice, I. D. Reid, J.Biotechnol. 1997, 53 215-236.
7. P. Bajpai, P. K. Bajpai, Biotechnology in the Pulp and Paper Industry, Pira International, Leatherhead, Sussex, UK 1998.

8. S. Camarero, P. Bocchini, G. C. Galletti, A. T. Martínez, Rapid Commun.Mass Spectrom. 1999, 13 630-636.
9. M. J. Martínez, F. J. Ruiz-Dueñas, F. Guillén, A. T. Martínez, Eur.J.Biochem. 1996, 237 424-432.
10. F. Guillén, A. T. Martínez, M. J. Martínez, Appl.Microbiol.Biotechnol. 1990, 32 465-469.
11. S. Camarero, G. C. Galletti, A. T. Martínez, Rapid Commun.Mass Spectrom. 1997, 11 331-334.
12. S. Camarero, P. Bocchini, G. C. Galletti, M. J. Martínez, A. T. Martínez, J.Anal.Appl.Pyrolysis 2001, 58/59 413-423.
13. K. Poppius-Levlin, W. Wang, T. Tamminen, B. Hortling, L. Viikari, M.-L. Niku-Paavola, J.Pulp Paper Sci. 1999, 25 90-94.
14. F. S. Chakar, A. J. Ragauskas, J.Wood Chem.Technol. 2000, 20 169-184.
15. P. J. Nelson, C. W. J. Chin, L. Viikari, M. Tenkanen, Proc.Intern.Pulp Bleaching Conf., Helsinki, 1-5 June 1998, 2 349-354.

Biotechnology in the Pulp and Paper Industry
L. Viikari and R. Lantto (Editors)

Biomimetic Pulp Bleaching with Copper Complexes and Hydroperoxides

Karin Fackler[a], Ewald Srebotnik[a], Takashi Watanabe[b], Pongsak Lamaipis[a], Miha Humar[c], Crtomir Tavzes[c], Marjeta Sentjurc[d], Franc Pohleven[c], and Kurt Messner[a]

[a]Institute for Biochemical Technology and Microbiology, University of Technology, Vienna, Austria.
[b]Wood Research Institute, Kyoto University, Uji, Japan
[c]University of Ljubljana, Biotechnical Faculty, Department of Wood Science and Technology, Slovenia
[d]Jozef Stefan Institute, Ljubljana, Slovenia

The efficiency of a low molecular weight delignification system consisting of copper (II) and hydroperoxides in the presence of the copper coordinating compound 4-aminopyridine (4-AP) was evaluated using oxygen bleached kraft pulp (OKP). A kappa number reduction of 30% at pH 11.3 could be achieved by one single copper/ 4-AP/ H_2O_2 stage. Consecutive treatment with alkaline peroxide resulted in a cumulative kappa number reduction of 52%. When H_2O_2 was replaced by the lipid hydroperoxide model compound cumene hydroperoxide (CHP), the kappa number reduction by the single copper stage increased to 52%. However, the selectivity of delignification was higher when H_2O_2 rather than organic hydroperoxides were used. The pH optimum for the selectivity of the process at pH 10 to 11 correlated well with that for the stability of the copper/ 4-AP complex. Studies that were carried out on ^{14}C-labeled lignin and cellulose model compounds further demonstrated the selectivity of the coordinated copper system for lignin structures. Electron spin resonance (ESR) studies suggested that the presence of the copper coordinating compound suppressed the production of free OH-radicals.

1. INTRODUCTION

Due to environmental problems caused by chlorine, and chlorine dioxide used for pulp bleaching, totally chlorine free (TCF) bleaching sequences based on oxygen and reactive oxygen species such as ozone and alkaline peroxide have been developed. Alternative selective bleaching methods applying ligninolytic enzymes of white rot fungi, particularly laccase in the presence of a mediator (LMS®)[1] are under investigation.

A number of white rot fungi are capable of selectively removing lignin from sites where enzymes cannot penetrate. Therefore, various low molecular weight compounds have been proposed to be involved in the ligninolytic mechanism of these organisms [3-5]. Recently, it has been shown that *Ceriporiopsis subvermispora* produces unsaturated lipids during cultivation on wood meal [6]. Peroxidation of these lipids can be catalysed by chelated manganese (III) generated by manganese peroxidase [7-9].

Transition metal complexes other than chelated manganese (III), particularly iron (II) and copper (I) may also play an important role in lignin biodegradation. Besides their ability to react with oxygen species such as hydrogen peroxide and lipid hydroperoxides they form the active centers of lignin peroxidase, manganese peroxidase, and laccase. Kawai and Ohashi [12] reported that copper-ethylenediamine is able to oxidise phenolic lignin in the presence of molecular oxygen. Recently, we have shown that pyridine coordinated copper in the presence of hydroperoxides is able to oxidise various synthetic dyes, and delignify ^{14}C labeled kraft pulp [2,13]. Even thermomechanical pulp (TMP) was delignified with Cu/ 4-AP in the presence of organic hydroperoxides [14].

Catalytic systems consisting of copper, hydroperoxides, and copper coordinating compounds containing aromatic nitrogen atoms have a great potential for application in TCF as well as ECF bleaching of pulp. Although the reduced forms of iron and copper are commonly regarded as Fenton reagents that produce hydroxyl radicals in the presence of hydrogen peroxide - and copper exhibits an even higher reaction rate with H_2O_2 than iron [10] - according to our findings, the free hydroxyl radical does not seem to be the exclusive reactive intermediate of the copper/ 4-AP/ H_2O_2 reaction. With the lipid hydroperoxide model compound t-butylhydroperoxide, alkoxyl, peroxyl and carbon centered radicals were formed by copper when appropriate coordinating compounds were present [11].

In this work, experiments were performed to demonstrate the selectivity of the process for lignin structures. A new promising copper complex, that is stable also at alkaline bleaching conditions, is successfully tested for the bleaching of kraft pulp.

2. MATERIALS AND METHODS

2.1. Materials

Chemicals were purchased in p.a. grade from Merck and Sigma and used without further purification. Side-chain β-^{14}C-labeled synthetic guaiacyl lignin (^{14}C-DHP, 0.01mCi/mmol) and PEG-4000, ^{14}C-labeled at the terminal hydroxyethyl group (^{14}C-PEG-4000), 19.2 mCi/g, Amersham) were kindly provided by Kenneth E. Hammel (FPL Madison, Wisconsin, USA). The ESR spintrap 5-(diethoxyphosphoryl)-5-methyl-1-pyrroline *N*-oxide (DEPMPO) was provided by the Jozef Stefan Institute, Ljubljana, Slovenia. Oxygen bleached kraft pulp (OKP: kappa number = 16.0, viscosity = 990 ml/g) was from the Austrian paper company Pöls.

2.2. Depolymerisation of ^{14}C- DHP

Reactions contained 18000 dpm of ^{14}C-DHP, 0.25% Tween 20, 125μM $CuSO_4$, and 29mM pyridine in 30 ml of water. H_2O_2 (0.9 M stock solution) was added continously during 72 hours with a rate of 1.5 mM/h. Excess peroxide was destroyed after the reaction by addition of solid catalase after which 5 ml DMF were added. The clear mixture was concentrated to a small volume by rotary evaporation and transferred into a centrifuge tube. The flask was purged a few times with DMF and the collected DMF-phases were centrifuged in a Sorvall RC5C centrifuge equipped with a F21S rotor at 15000 rpm for 15 minutes. An aliquot of the sample (5 ml) was fractionated by gel permeation chromatography (GPC) using a 2 x 35 cm Sephadex LH 60 column in DMF that contained 0.08 M LiCl. Fractions of 1.5 ml were collected and mixed with 1.5 ml Optiphase HiSafe 3 scintillation cocktail (HVD Life Sciences) and assayed for radioactive products by liquid scintillation counting.

2.3. Oxidation of ^{14}C- PEG-4000

The substrate (9000 dpm of ^{14}C-PEG-4000) was reacted at room temperature in an aqueous solution containing 125 μM $CuSO_4$ and 29 mM pyridine. H_2O_2 (100 mM stock solution) was added continuously during 24 hours with a rate of 2.5 mM/h. The initial volume of the assay was 300 μl. Excess peroxide was destroyed after the reaction by addition of solid catalase. The whole aqueous sample was subjected to GPC using the same column as for the fractionation of DHP but without LiCl. Fractions of 1.5 ml were collected mixed with 1.5 ml of scintillation cocktail and assayed for radioactive products by liquid scintillation counting.

2.4. Bleaching of OKP with copper(II)/ 4-aminopyridine and hydroperoxides

Bleaching sequence Q-Cu-Q-P: the Q-stage was carried out at 10 % pulp consistency with 0.5 % disodium-EDTA on pulp for one hour at 90 °C. After washing the pulp fibers, an aqueous solution containing $CuSO_4$ and 4-AP was added and the consistency was brought to 10 %. Prior to addition of the peroxide (H_2O_2 or CHP), the pH was adjusted to 11.3 with 4 N NaOH. The mixture was incubated for 4 hours at 90 °C (Cu-stage). The concentrations of reagents are indicated in the text. Another Q-stage was carried out as described above. Finally, the pulp was treated with alkaline peroxide (P-stage) at 10 % consistency (2 % NaOH, 1 % H_2O_2, 0.2% DTPA, and 0.05 % $MgSO_4$ x 7 H_2O on pulp) for three hours at 90 °C

Bleaching sequence Cu-E: The pulp was treated with copper, 4-AP and H_2O_2 without the pre-chelating stage. The Cu-stage was carried out as described above in the presence of 10μM $CuSO_4$, 1 mM 4-AP, and 1 % H_2O_2 on pulp. The pH was adjusted with 4 N NaOH or 4 N H_2SO_4. After washing the pulp fibers, an alkaline extraction with 2 % NaOH on pulp was performed for three hours at 60°C at 10 % pulp consistency.

Pulp analyses: Kappa numbers were determined according to DIN 54357, but in a smaller scale. Viscosities were measured in copper ethylendiamine using the SCAN-C 15:62 method that was slightly modified regarding the cellulose concentration in the assay.

2.6. Free radical production by the copper/ 4-aminopyridine/ H_2O_2-system

The concentrations in the assay were 50 μM Cu(II), 5 mM 4-aminopyridine, 25 mM DEPMPO, and 20 mM H_2O_2. $CuSO_4$, 4-AP, and the spintrap DEPMPO were mixed at room temperature before addition of H_2O_2. ESR-spectra were recorded with a Bruker ESP-300 X-band ESR spectrometer 2 min after mixing. The parameters applied were: Microwave frequency 9.62 GHz, microwave power 20 mW, modulation frequency 100 kHz, center field 341.5 mT, field width 14 mT, and modulation amplitude 0.1 mT.

2.7. pH dependence of the stability of the copper/ 4-aminopyridine complex

An aqueous solution containing 10 mM $CuSO_4$ and 500 mM 4-AP was prepared. The pH was adjusted with 4 N H_2SO_4, or 4 N NaOH. The relative concentrations of the blue copper/ 4-AP complex was determined spectrophotometrically at 590 nm.

3. RESULTS AND DISCUSSION

3.1. Depolymerisation of ^{14}C- DHP

In the presence of pyridine as the coordinating compound, the copper/ H_2O_2 system extensively depolymerized DHP to material with a molecular weight in the range of lignin dimers or monomers (Figure 1), which were eluted in fractions 31 and 34, respectively (not shown). In the absence of pyridine, the depolymerisation of the synthetic lignin was much less

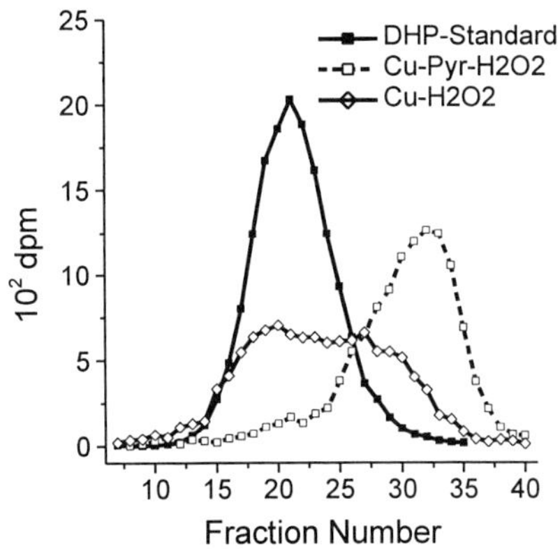

Figure 1. Depolymerisation of ^{14}C-labeled synthetic lignin by copper/ H_2O_2 in the presence and absence of pyridine.

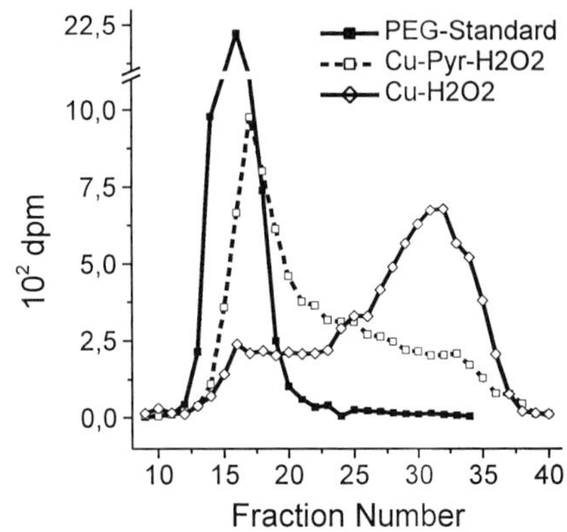

Figure 2. Depolymerisation of ^{14}C-labeled polyethylene glycol by copper/ H_2O_2 in the presence and absence of pyridine.

effective. Some low molecular weight material was formed but most of it remained polymeric. However, it appears more likely that repolymerisation reactions predominated this system. The total recoveries of radiolabel compared to the DHP standard were similar in both reactions, i. e. 75% in the presence and 70% in the absence of pyridine.

3.2. Oxidation of ^{14}C- PEG-4000

PEG-4000 was used as a model substrate for cellulose because it has successfully been used to study the mechanism of cellulose depolymerisation caused by brown rot fungi [17]. Furthermore, the reaction of free hydroxyl radicals with both substrates, PEG and cellulose, leads to hydrogen abstraction, and consequently to fragmentation of the polymer by β-elimination [16, 17]. In contrast to the results obtained with DHP, the copper/ H_2O_2 system was not sufficiently reactive to efficiently depolymerise PEG-4000 when pyridine was present (Figure 2). In the absence of pyridine, however, most of the initially added PEG-4000 was extensively depolymerised to low molecular weight products that were eluted at the positions of lignin dimers and monomers (see above). The results obtained with DHP and PEG-4000 collectively suggest that the presence of a metal ligand such as pyridine not only suppresses the attack of copper/ H_2O_2 on cellulose but also drives the ligninolytic reactions towards depolymerisation.

3.3. Delignification of OKP with the sequence Q-Cu-Q-P

Table 1 shows kappa numbers and viscosities of OKP after treatment with the coordinated copper system in the presence of H_2O_2 or CHP carried out after a chelating stage with EDTA.With H_2O_2, up to 30 % kappa number reduction was obtained after Q-Cu-Q, which was less effective than an alkaline peroxide treatment with 1 % H_2O_2 (36% kappa reduction).

Table 1. Kappa numbers and viscosity after the bleaching sequence Q-Cu-Q-P

		Kappa Number	Kappa-Reduction	Viscosity [ml/g]	Viscosity Loss
OKP		16.0	---	990	---
25µM Cu; 2.5mM 4-AP	after Q-Cu-Q	11.2	30 %	829	16 %
0.5% H_2O_2 on pulp, pH_i=11.3	after Q-Cu-Q-P	8.0	50 %	720	27 %
50µM Cu; 5mM 4-AP	after Q-Cu-Q	11.5	28 %	783	21 %
0.5% H_2O_2 on pulp, pH_i=11.3	after Q-Cu-Q-P	7.7	52 %	670	32 %
50µM Cu; 5mM 4-AP	after Q-Cu-Q	7.7	52%	368	63%
4% CHP on pulp, pH_i=11.3	after Q-Cu-Q-P	5.3	67%	343	65%
Alkaline Peroxide	Q-P	10.2	36 %	n.d.	---
	Q-P-P	9.3	42 %	828	16 %

However, after a consecutive alkaline peroxide stage, the kappa number was 7.7 after Q-Cu- Q-P and 9.3 after Q-P-P. The viscosity values, however were also lower, when OKP was treated with the Cu/ 4-AP/ H_2O_2-system. With CHP, the kappa number was reduced to 7.7 (52% kappa reduction) demonstrating the much stronger oxidative reaction that is initiated by this system. The selectivity of the process, however, was much lower than in the coordinated copper system with H_2O_2.

3.4. Influence of pH on the selectivity of a Cu/ 4-AP/ H_2O_2-stage

The lowest viscosity losses of the pulp were found when the copper/ coordinator treatment was carried out at an initial pH of 11.0 (Figure 3). Under these conditions the viscosity after Cu-E was 843 (15% loss) compared to 597 (40% loss) at pH 7.0. The kappa numbers were also lower at higher pH, probably at least partially due to the higher reactivity of the pulp and synergistic alkaline peroxide effects. The viscosity maximum was highly correlated to the stability region of the Cu(II)/4-AP-complex (Figure 4). When the Cu-stage was initiated at pH between 11 and 11.3., most of the copper remained coordinated to 4-AP in the whole process, during which the pH dropped to 9.2. The difference in reactivity and oxidative species produced at different pH values were demonstrated in the ESR studies described below.

3.5. Delignification of OKP with coordinated iron and manganese

When the 4-AP complexes of iron and manganese were applied at the same reaction conditions, the kappa number reductions were significantly lower than achieved with copper (Table 2). The kappa reductions of 21% and 44% that were obtained with Fe(III) after the sequences Q-Me-Q, and Q-Me-Q-P, respectively, did not significantly differ from those obtained in the control reaction without metal ions and without 4-AP (21% and 43%). With the Fenton active metal ions Fe(II) and Mn(II), however, the kappa numbers even increased compared to the control. The lowest kappa numbers after Q-Me(Cu)-Q and Q-Me(Cu)-Q-P sequences were observed with coordinated copper: 28% and 52%, respectively.

3.6. Influence of pH on free hydroxyl radical production by Cu/ 4-AP/ H_2O_2

The signal intensity of the free hydroxyl radical spin adduct of the spintrap DEPMPO was monitored by ESR (Figure 5). The published hyperfine splitting parameters of DEPMPO-OH are 14.0 G for A_N, 13.0 G for $A_{H\beta}$, 0.27 G (3 H) for $A_{H\gamma}$, and 47.5 G for A_P [15] and correlated

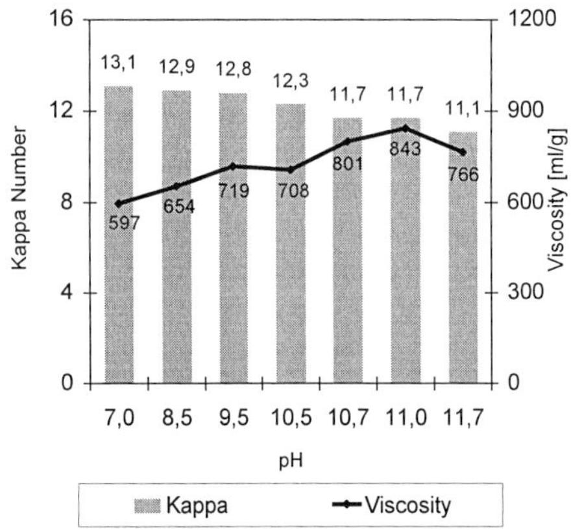

Figure 3. pH dependence of kappa numbers and viscosities after the bleaching sequence Cu-E.

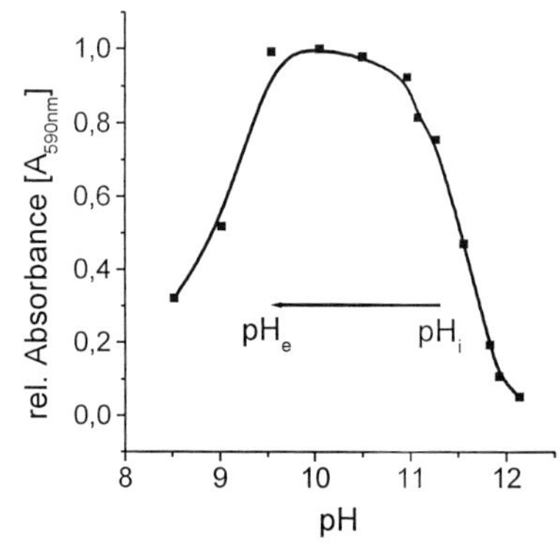

Figure 4. pH course of relative concentration of the copper coordinated to 4-AP.The arrow indicates the change of pH during a typical Cu-stage.

well with the values observed for the copper/ 4-AP/ H_2O_2 system. Free hydroxyl radicals were produced by copper in the presence of 4-AP at all investigated pH values, but exhibited a strong pH dependence. At pH 9.8, their production was strongly suppressed compared to more acidic or more alkaline conditions. In Figure 6, the spectra at pH 6.1 and 9.8 are compared. The pH value for minimum free hydroxyl radical production highly correlated with the results of the pulp bleaching experiments (Figure 3) and the stability region of the complex (Figure 4). Thus, only coordinated copper produced oxidative species different from free hydroxyl radicals, that allow for the selective oxidation of lignin structures. Hydroxyl radicals and other radical species initiate radical chain reactions during the bleaching stage, and are regarded to be responsible for unselective bleaching results due to carbohydrate degradation [16].

3. SUMMARY

The reactions of copper complexes with hydroperoxides are highly oxidative and open new strategies for pulp bleaching. In the presence of pyridine, the lignin model DHP was extensively depolymerised, whereas the depolymerisation of the cellulose model polyethylene

Table 2. Bleaching of OKP with various transition metal complexes

Bleaching Sequence	50 μM Cu(II), 5 mM 4-AP 0.5% H_2O_2 o.p.	50 μM Fe(II), 5 mM 4-AP 0.5% H_2O_2 o.p.	50 μM Fe(III), 5 mM 4-AP 0.5% H_2O_2 o.p.	50 μM Mn(II), 5 mM 4-AP 0.5% H_2O_2 o.p.	Control (no metal, no coordinator)
Q-Me-Q	11,5	13,2	12,7	13,8	12,6
Q-Me-Q-P	7,7	9,9	8,9	9,6	9,1

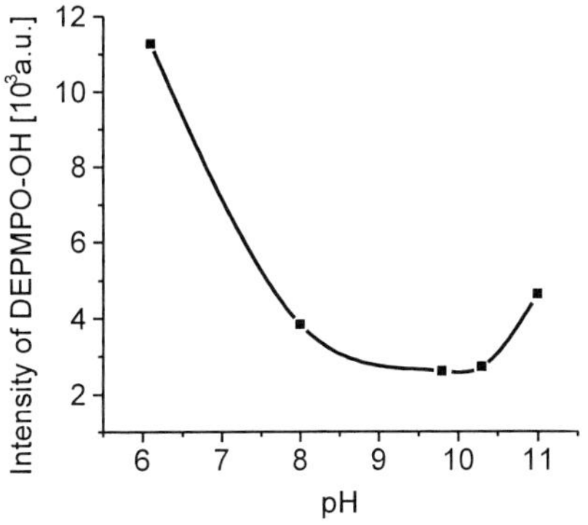

Figure 5. pH course of the relative intensity of the OH-radical spin adduct obtained wit copper/ 4-AP/ H_2O_2.

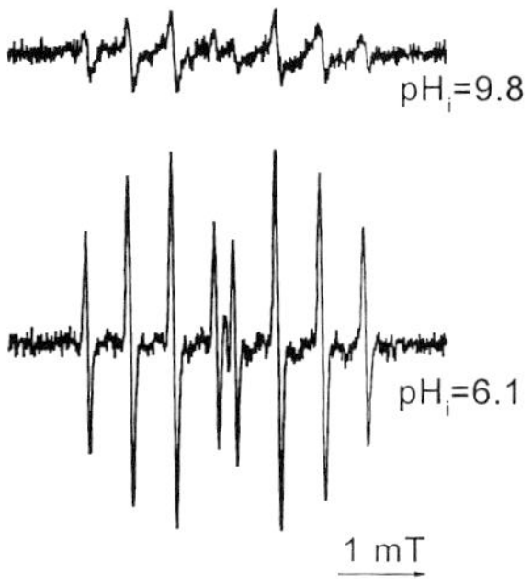

Figure 6. ESR spectra obtained with copper/ 4-AP/ H_2O_2 with the spin trap DEPMPO at pH 9.8 and 6.1.

glycol was suppressed, demonstrating the increased selectivity of the process for lignin structures compared to reactions with copper/H_2O_2 alone. When oxygen prebleached kraft pulp (OKP) was delignified with the coordinated copper system, the amount of delignification by a Cu-stage followed by a P stage, was 10% (20% relative) better than with two alkaline peroxide treatments of the same pulp. The viscosity losses, however, were also higher.

ESR studies revealed that the formation of free hydroxyl radicals was strongly suppressed at pH values around 10, at which the coordination of 4-AP to copper is optimal. However, best delignification results as well as highest selectivity was obtained at an initial pH of 11.3, which can be explained by the fact that the pH dropped to 9.2 in the course of the reaction. These results suggest that the oxidative reactions responsible for pulp delignification proceed via a pathway different from the Fenton reaction, comprising oxidative intermediates other than the free hydroxyl radicals.

When the organic hydroperoxide CHP was applied for pulp bleaching, we observed very strong delignification of OKP, which was, however, accompanied by severe viscosity losses. Lamaipis et al. [14] could show by ESR that in the case of organic hydroperoxides, free radical production is much stronger in the presence of the coordination compound than in its absence.

The high pH optimum of the 4-AP complex compared to previously described copper/coordinator complexes makes this compound a promising candidate for pulp bleaching at alkaline conditions. However, further optimisation, particularly of the organic hydroperoxide component and intensive screening for novel coordination compounds will be necessary to improve the selectivity of pulp bleaching with copper.

ACKNOWLEDGEMENTS

We greatfully acknowledge the following support: Österreichische Nationalbank Jubiläumsfondsprojekt Nr. 8086 and Kompetenzzentrum Holz Project 1.1. Basic Research and Methodology.

REFERENCES

1. Call, H.-P., Mücke, I. (1997) History, overview and applications of mediated ligninolytic systems, especially laccase-mediator-systems (Lignozym-process). J. Biotechnol. 53, 163-202.
2. Watanabe, T., Koller, K., Messner, K. (1998) Copper-dependent depolymerization of lignin in the presence of fungal metabolite, pyridine. J. Biotechnol. 62, 221-230.
3. Harvey, P. J., Shoemaker, H. E., Palmer, J. M. (1986) Veratrylalcohol as a mediator and the role of radical cations in lignin degradation by *Phanerochaete chrysosporium.* FEBS Lett. 195, 242-246.
4. Wariishi, H., Valli, K., Gold, M. H. (1992) Manganese(II) Oxidation by Manganese Peroxidase from the Basidiomycete *Phanerochaete chrysosporium*. Kinetic Mechanism and Role of Chelators. J. Biol. Chem. 267(33), 23688-23695.
5. Bourbonnais, R., Paice, M. G. (1990) Oxidation of non-phenolic substrates. An expanded role of laccase in biodegradation. FEBS Lett. 267, 99-102.
6. Enoki , M., Watanabe, T., Nakagame, S., Koller, K., Messner K., Honda, Y., Kuwahara, M. (1999) Extracellular lipid peroxidation of selective white-rot fungus, *Ceriporiopsis subvermispora*. FEMS Microbiol. Lett. 180, 205-211.
7. Watanabe, T., Katayama, S., Enoki, M., Honda, Y., Kuwahara, M. (2000) Formation of acyl radical in lipid peroxidation of linoleic acid by manganese-dependent peroxidase. Eur. J. Biochem. 267 (13), 4222-4231.
8. Bao, W., Fukushima, Y., Jensen, K. A., Moen, M. A., Hammel, K. E.(1994). Oxidative degradation of non-phenolic lignin during lipid peroxidation by fungal manganese peroxidase. FEBS Lett. 354, 297-300.
9. Kapich, A. N., Hammel, K. E. (1999) Peroxyl radicals are potential agents of lignin biodegradation, FEBS Lett. 461, 115-119.
10. Halliwell, B., Gutteridge, J. M. C. (1989) Free Radicals in Biology and Medicine. 2nd Ed. Oxford Univ. Press (Clarendon), Oxford.
11. Ueda, J., Saito, N., Ozawa, T. (1996) Detection of free radicals produced from reactions of lipid hydroperoxide model compounds with Cu(II) complexes by ESR spectroscopy. Arch. Biochem. Biophys. 325, 65-76.
12. Kawai, S., Ohashi, H. (1993) Degradation of lignin substructure model compound by copper(II)-amine complex as a model catalyst of fungal laccase. Holzforschung 47, 97-102.
13. Messner, K., Koller, K., Fackler, K., Srebotnik, E., Watanabe, T. (1999) Pulp bleaching by a copper based lignin depolymerisation system. 10th ISWPC, Yokohama, June.
14. Lamaipis, P., Watanabe., T., Messner, K., unpublished data.
15. Frejaville, C., Karoui, H., Tuccio, B., Le Moigne, F., Culcasi, M., Pietri, R. L., Tordo, P. (1995) 5-(Diethoxyphosphoryl)-5-methyl-1-pyrroline-*N*-Oxide: A New Efficient Phosphorylated Nitrone for the in Vitro and in Vivo Spin Trapping of Oxygen-centered Radicals. J. Med. Chem., 38, 258-265.
16. Gierer, J.(1997) Formation and involvement of superoxide ($O_2^{\cdot-}/HO_2\cdot$) and hydroxyl ($HO\cdot$) radicals in TCF bleaching processes: A review. Holzforschung, 51, 34-46.
17. Kerem, Z., Bao, W., Hammel, K. E.(1998) Rapid polyether cleavage via extracellular one-electron oxidation by a brown-rot basidiomycete. Proc. Natl. Acad. Sci. U.S.A., 95, 10373-10377

Biotechnology in the Pulp and Paper Industry
L. Viikari and R. Lantto (Editors)

Enzymatic degradation of oxalic acid for prevention of scaling

N.-O. Nilvebrant[a], A. Reimann[a], F. de Sousa[a], P. Cassland[b], S. Larsson[a], F. Hong[b] and L.J. Jönsson[b,c]

[a]STFI, Swedish Pulp and Paper Research Institute, P.O. Box 5604, SE-114 86 Stockholm, Sweden
[b]Applied Microbiology, Lund University/Lund Institute of Technology, P.O. Box 124, SE-221 00 Lund, Sweden
[c]Division for Chemistry, Karlstad University, SE-651 88 Karlstad, Sweden

Formation of calcium oxalate incrusts, scaling, may cause severe problems in the pulp and paper industry. Enzymatic degradation of oxalic acid provides a novel approach to eliminate the problems with calcium oxalate precipitation. The performance of two oxalate-degrading enzymes, oxalate oxidase from barley and oxalate decarboxylase from *Aspergillus*, was tested in model experiments with respect to catalytic efficiency under different conditions, including pH, temperature and enzyme concentration. Oxalate decarboxylase was found to be more sensitive to temperature variations than oxalate oxidase, which was selected for further experiments. Authentic samples from pulp bleach plants were used to test the performance of oxalate oxidase. The results showed that oxalic acid could be degraded enzymatically also in the industrial bleaching filtrates, which were obtained from D-, E-, O-, OP-, PO-, Q-, QP-, and Z-stages. The bleaching filtrates contained compounds that inhibited the action of oxalate oxidase. The degree of inhibition was strongly dependent on the filtrate and could be alleviated by dilution.

1. INTRODUCTION

One of the goals of the pulp and paper industry is a sustainable production of pulp and paper in ecological balance with nature. Recycling of the water in the bleach plant is one of the most important steps towards the closed mill. However, a high degree of system closure leads to high levels of dissolved organic and inorganic substances in process liquors. This can lead to problems with precipitation of *e.g.* oxalates. The formation of oxalate incrusts has frequently led to problems in production plants, such as clogging of pipework and filters.

Oxalic acid is found naturally in wood, but is also formed during the pulping process. Bleaching with strong oxidising agents like ozone, oxygen and hydrogen peroxide leads to the oxidation of dissolved lignocellulosic materials and the formation of various acids like oxalic, acetic and formic acid. Among wood constituents, lignin is the most important source of oxalic acid (Krasowski and Marton 1983; Nilvebrant and Reimann 1996; Elsander *et al.* 2000).

In wood and consequently in pulp mills, calcium ions are present in high concentrations and can form insoluble compounds like calcium oxalate and calcium carbonate (Ulmgren 1997). This has resulted in increased process problems as extensive scaling and formation of deposits. Presently, there is no satisfactory method for prevention of the formation of calcium oxalate incrusts in the production of pulp and paper. However, a method has been proposed for reducing the risk of calcium oxalate scaling in bleach plants by adding magnesium ions, which forms a soluble complex with oxalic acid (Jönsson *et al.* 1999). The present paper describes a new approach to deal with the oxalate problem in the pulp and paper industry, namely treatment of process waters with oxalate-degrading enzymes. This approach has the advantage that it can actually eliminate the oxalic acid, which otherwise will accumulate in a closed process.

Oxalate oxidase (oxalate: oxygen oxidoreductase, EC 1.2.3.4, OXO) catalyses the conversion of oxalate and molecular oxygen to carbon dioxide and hydrogen peroxide (Figure 1). The best-known oxalate oxidases have been isolated from plants such as wheat and barley (reviewed by Dunwell *et al.* 2000). The cereal oxalate oxidases belong to a group of proteins known as germins (Lane 2000). Members of this family are oligomeric, refractory to hydrolysis by certain proteases (such as pepsin and trypsin), SDS-tolerant and stable at relatively high temperatures (Lane 2000; Kotsira and Clonis 1998). Oxalate decarboxylase (EC 4.1.1.2) catalyses the conversion of oxalate to formate and carbon dioxide (Figure 1). Many filamentous fungi secrete oxalate decarboxylases (Dutton *et al.* 1994; Dunwell *et al.* 2000). Both oxalate oxidase and oxalate decarboxylase are members of the diverse cupin superfamily, which is defined by having a β-barrel core structure and certain conserved motifs (Dunwell *et al.* 2000).

$$HOOC\text{-}COO^- + H^+ + O_2 \xrightarrow{\textit{Oxalate oxidase}} 2\,CO_2 + H_2O_2$$

$$HOOC\text{-}COO^- \xrightarrow{\textit{Oxalate decarboxylase}} HCOO^- + CO_2$$

Figure 1. Reactions catalysed by oxalate oxidase and oxalate decarboxylase.

Oxalate-degrading enzymes have been used for medical applications, mainly in assays for determination of oxalate in blood and urine. Several other potential applications have been suggested for oxalate oxidase and oxalate decarboxylase. In the present study, we have investigated the performance of oxalate-degrading enzymes in model experiments and in authentic samples from the pulp and paper industry.

2. EXPERIMENTAL

2.1. Enzyme preparations

The enzymes used were obtained commercially. Oxalate oxidase from barley seedlings (0.71 units/mg solid) was obtained from Sigma-Aldrich (St. Louis, MO, USA). Oxalate decarboxylase from *Aspergillus* sp. (20 units/0.5 ml) was obtained from Boehringer-Mannheim (Mannheim, Germany).

2.2. Determination of oxalic acid

The chromatographic equipment consisted of a Dionex DX 500-series ion chromatograph equipped with a conductivity detector (ED-40). An IonPac AS4A anion-exchange column (250 mm X 4 mm I.D.) and an AG4A guard column were used. A membrane suppressor

(ASRS-I) was used to convert the eluent and the separated anions to their respective acid forms and to lower the conductivity of the eluent relative to that of the ions of interest. All instrumentation, including columns and suppressor, was supplied by Dionex (Sunnyvale, CA, USA). The eluent used consisted of a mixture of 1.7 mM $NaHCO_3$ and 1.8 mM Na_2CO_3. For all the analyses, a flow rate of 2.0 ml/min was used and 20 μl of sample was injected. Ions were identified by their retention times compared to those of standards. Quantification of oxalic acid was accomplished by measuring the peak area and comparing it to a calibration curve. A detailed description of the methods used for determination of oxalic acid and calcium oxalate was presented and discussed previously (Reimann *et al.* 2000).

2.3. Enzymatic treatment of oxalic acid solutions and bleaching filtrates

The enzymes were compared with respect to catalytic efficiency under different conditions, such as pH, temperature, reaction time and enzyme concentrations. The concentration of oxalic acid in the model experiments was 40 mg/l, the reaction time was 30 min (if not stated otherwise) and phosphate was used as buffer. The optimal conditions for the decomposition of pure oxalic acid solutions and an authentic industrial bleach plant filtrate were investigated by the use of fractional factorial experimental design done with the MODDE 4.0 software from Umetri (Umeå, Sweden).

3. RESULTS AND DISCUSSION

The catalytic performance of oxalate oxidase and oxalate decarboxylase was first examined using model experiments. To investigate the influence of pH, temperature and enzyme concentration, an experimental design was made using MODDE. The purpose of the experimental factorial design was to conduct and plan the experiments in order to extract as much information as possible from the data collected in the smallest number of runs and reveal possible interaction effects between the experimental variables (Lundstedt *et al.* 1998). In order to cover the entire area of interest with as few experiments as possible, all variables were changed simultaneously and the influences of the factors on the response were investigated. A reduced factorial design (15 instead of 27 experiments) was performed with three replicated centre points to estimate the experimental errors (Figure 2-4), except for the results in Figure 5 that was a full factorial. Values for the fit of the models were given as the explained variation (R^2) and the predicted variation (Q^2).

It should be kept in mind that the response surfaces obtained are models designed by the program and that minor variations may be below a reasonable confidence interval. The response surfaces will, however, provide a convenient overview of the reaction based on the range of the chosen variables. The variables were given maximum and minimum values as indicated in Table 1.

Table 1. Maximum and minimum values of the variables used in the experimental design. The time of the reaction was set to 30 min.

Variable	Oxalate decarboxylase		Oxalate oxidase	
	Low level	High level	Low level	High level
pH	3	8	3	8
Temperature (°C)	26	48	26	48
Enzyme concentration (mU/μmole oxalic acid)	180	990	63	347

Table 3. Oxalic acid degradation by oxalate oxidase in kraft mill bleaching stage filtrates. The reaction time was 30 min, the temperature was 37 °C and the pH was 4. Oxalate oxidase was added to a concentration of 600 mU per μmole oxalic acid.

Filtrate	Initial concentration of oxalic acid (mg/l)	Degradation of oxalic acid (μg oxalic acid per h and μg oxalate oxidase)
D	19	121
O	720	15
OP	200	26
PO	70	30
Z	160	36

Table 4. Degradation of oxalic acid in authentic bleaching filtrates. The reaction time was 2 h, the temperature was 22 °C and the pH was 3.8.

Filtrate	Initial concentration of oxalic acid (mg/l)	Degradation of oxalic acid (μg oxalic acid per h and μg oxalate oxidase)
D_0	19	15
E	153	100
OP	48	3
Q	56	18
QP	65	29
Z	39	6

Since the rate of oxalic acid degradation was much lower in process waters than in model solutions containing oxalic acid, it was concluded that compounds that inhibit the action of oxalate oxidase were present in bleaching filtrates. The results also indicated more efficient degradation of oxalic acid in the D- and E-stage filtrates than in the other ones tested. Probably, a large number of filtrates from different types of bleaching stages need to be investigated to provide reliable predictions regarding the performance of oxalate-degrading enzymes in authentic process waters.

Since certain metal ions possibly inhibit oxalate oxidase (Kotsira and Clonis 1998), chelation of metal ions in the filtrates before enzymatic treatment could tentatively result in improved degradation of oxalic acid. It was therefore decided to carry out a third experimental design with different amounts of a complexing agent (DTPA) added. Also, the sample concentration was used as a variable in this experimental design. The temperature, pH and enzyme concentration were fixed. The sample chosen for this study was a Z-stage bleach plant filtrate used previously (Table 3). Table 5 shows the upper and lower values used for the factors that were varied in this third experimental design.

Figure 5 shows the obtained response surface plot of the degradation of oxalic acid in the Z-stage filtrate at a given reaction time, with varying dilution and concentration of DTPA. Dilution of the filtrate with water and, as a consequence, lower concentration of inhibitors resulted in a higher degree of degradation of oxalic acid. By diluting the filtrate four times,

Table 5. Maximum and minimum values of the variables used in the third experimental design. The high level of sample concentration represents the undiluted Z-stage filtrate. The time of the reaction was 30 min, the temperature was 37 °C, the pH was 4 and the concentration of enzyme was 600 mU/μmole oxalic acid.

Variable	Oxalate oxidase	
	Low level	High level
DTPA conc. (mg/l)	0	300
Sample concentration (mg/l oxalic acid)	40	160

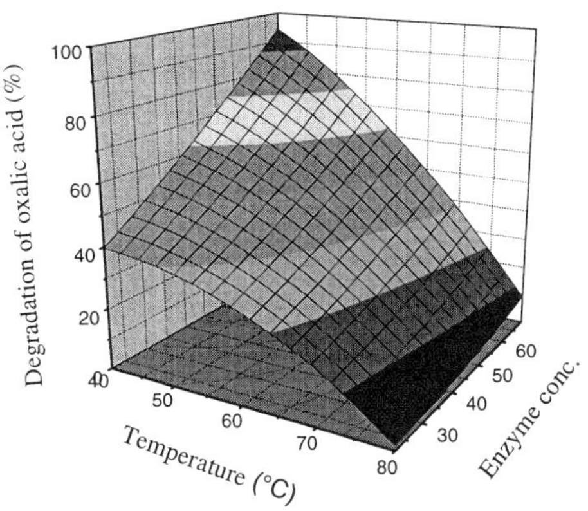

Figure 4. Model of a 15 min degradation of oxalic acid by oxalate oxidase showing the effect of temperature and enzyme concentration (mU/μmole).
R^2=0.95, Q^2=0.52, RSD=9.9.

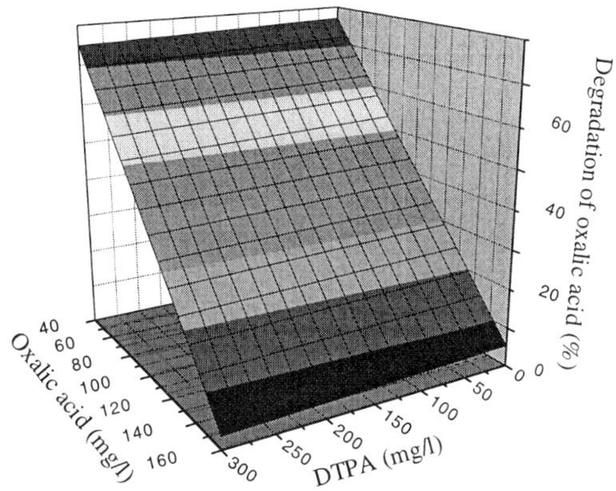

Figure 5. Response surface showing the effect of dilution and DTPA addition on the degradation of oxalic acid in a Z-stage filtrate by oxalate oxidase.
R^2=0.99, Q^2=0.98, RSD=1.6.

while holding the enzyme concentration constant, the degradation of oxalic acid was increased from 17 % to 75 %. The highest level of oxalic acid in the experiment, 160 mg/l or 1.8 mM, was far below the level reported for substrate inhibition, 4 mM (Kotsira and Clonis 1997). Substrate inhibition is therefore not likely to have contributed to the positive effect observed for the dilution. The addition of DTPA, however, did not result in improved oxalate degradation (Figure 5). This could be due to that either metal ions were not responsible for the inhibition in this filtrate or that the concentration of DTPA was insufficient to alleviate the inhibition. Nevertheless, the effect of dilution of the filtrate showed the presence of inhibiting compounds and was a possible approach to obtain a higher degree of degradation of oxalic acid, although non-practical.

4. CONCLUSIONS

The performance of two oxalate-degrading enzymes, oxalate oxidase and oxalate decarboxylase, were tested in a novel approach to deal with problems caused by calcium oxalate precipitations in the pulp and paper industry.

Oxalate oxidase showed less dependence on variations in temperature than oxalate decarboxylase in the experimental region studied, and was selected for treatment of industrial bleach plant filtrates.

The use of oxalate-degrading enzymes proved to be a possible approach to eliminate oxalic acid in a variety of authentic bleaching filtrates.

Bleach plant filtrates contain compounds that are inhibitory to the action of oxalate oxidase. The degree of inhibition was strongly dependent on the filtrate and could be alleviated by dilution.

ACKNOWLEDGEMENTS

The financial support of the Swedish National Board for Industrial and Technical Development (NUTEK) is gratefully acknowledged.

REFERENCES

Chiriboga J., Biochem. Biophys. Res. Commun. 11 (1963) 272.

Dumas B., Sailland A., Cheviet J.-P., Freyssinet G. and Pallett K., C. R. Acad. Sci. Paris, Science de la vie/Life science 316 (1993) 793.

Dunwell J.M., Khuri S. and Gane P.J., Microbiol. Mol. Biol. Rev. 64 (2000) 153.

Dutton M.V., Kathiara M., Gallagher I.M. and Evans C.S., FEMS Microbiol. Lett. 116 (1994) 321.

Elsander A., Ek M. and Gellerstedt G., Tappi J. vol. 83 no. 2 (2000) 73.

Emiliani E. and Bekes P., Arch. Biochem. Biophys. 105 (1964) 488.

Jönsson T., Rådeström R., Tomani P. and Ulmgren P., Proc. SPCI 6th International Conference on Available Techniques, Stockholm (1999) 194.

Kotsira V.P. and Clonis Y.D., Arch. Biochem. Biophys. 340 (1997) 239.

Kotsira V.P. and Clonis Y.D., Arch. Biochem. Biophys. 356 (1998) 117.

Lane B.G., Biochem. J. 349 (2000) 309.

Krasowski J.A. and Marton J., J. Wood Chem. Technol. 3 (1983) 445.

Lundstedt T., Seifert E., Abramo L., Thelin B., Nyström Å., Pettersen J. and Bergman R., Chemometrics and Intelligent Laboratory Systems 42 (1998) 3.

Nilvebrant N.-O. and Reimann A., 4th European Workshop on Lignocellulosics and Pulp (EWLP), Stresa, Italy (1996) 485.

Reimann A., de Sousa F. and Jansson M.-B., 6th European Workshop on Lignocellulosics and Pulp (EWLP), Bordeaux, France (2000) 543.

Requena L. and Bornemann S., Biochem. J. 343 (1999) 185.

Sugiura M., Yamamura H., Hirano K., Sasaki., Morikawa M. and Tsuboi M., Chem. Pharm. Bull. 27 (1979) 2003.

Ulmgren P., Nordic Pulp Paper Res. J. 12:1 (1997) 32.

Biotechnology in the Pulp and Paper Industry
L. Viikari and R. Lantto (Editors)

Expression of thermophilic xylanases in fungal hosts

P. L. Bergquist[a,b], V. S. J. Te'o[a], M. D. Gibbs[a], A. C. E. Cziferszky[c], F. P. de Faria[d], M. O. Azevedo[e], and K. M. H. Nevalainen[a]

[a]Department of Biological Sciences, Macquarie University, Sydney, Australia

[b]Division of Molecular Medicine, University of Auckland Medical School, Auckland, New Zealand

[c]Institute of Biochemical Technology and Microbiology, Section of Microbial Biochemistry, Technical University of Vienna, Vienna, Austria

[d]Departamento de Biotecnologia, FAENQUIL, Lorena-SP, Brazil

[e]Departamento de Biologia Celular, IB, UnB, Brasilia-DF, Brazil

Fungal expression systems provide a promising, industrially-relevant alternative to bacteria for heterologous enzyme production. We are developing the yeast *Kluyveromyces lactis* and the filamentous fungus *Trichoderma reesei* for the extracellular production of thermophilic enzymes for the pulp and paper industry. Gram amounts of largely pure xylanase A from *Dictyoglomus thermophilum* have been produced in *K. lactis* chemostat culture. *D. thermophilum xynB* gene has been reconstructed according to *T. reesei* codon preferences and substantial amounts of enzyme produced in *T. reesei*. A heterologous fungal gene, *Humicola grisea xyn2* was expressed in *T. reesei* without codon modification. Initial amounts of the XYN2 protein were of a gram per liter range in shake flask cultivations and the gene product was processed by the heterologous host in a similar manner to its endogenous xylanases.

1. INTRODUCTION

Effective production of thermophilic hydrolases is a prerequisite for their industrial application. For industrial processes of large volume, such as wood pulp bleaching, enzymes must be produced at less than US$2.25/g, and the final cost of enzyme treatment should not exceed US$4.50 per tonne of pulp to provide a competitive alternative/additive to eliminate or decrease the amount of chlorine dioxide used in pulp delignification. Therefore, cost-effective production of enzymes for the processing industries makes the appropriate selection of the host-vector expression system critical.

Commercially-manufactured xylanases are typically produced by mesophilic filamentous fungi such as *Trichoderma reesei* and *Aspergillus niger*, who are excellent protein secretors by nature. However, mesophilic fungal xylanases may not be sufficiently thermostable for processes where enzymes active at high temperatures (up to 85°C and pH 9 or more) would have a competitive advantage. Therefore, enzymes produced by thermophilic microorganisms have gained attention and a number of genes encoding hemicellulases have been isolated from various thermophilic bacteria [1] and fungi [2]. Relatively low productivity, technical problems in the fermentation process and unfavorable production economics associated with the cultivation of thermophilic microorganisms on a large scale [3,4] has led to a search for other, industrially-relevant microorganisms as production hosts for enzymes from thermophilic microorganisms.

Industrial strains of filamentous fungi are capable of secreting tens of grams of homologous extracellular protein into their growth medium [5,6]. In contrast, many bacterial enzymes produced in fungal hosts are not effectively secreted but seem to be trapped in the mycelium [7,8]. In general, the yields for bacterial enzymes secreted from fungi have been on the order of 10-20 mg/l [9]. This indicates that while fungal systems have great potential for extracellular protein production, more research is needed into the molecular and physiological adjustments that would lead to successful expression of heterologous gene products in fungi. We have chosen to explore the unicellular yeast *K. lactis* and the filamentous fungus *T. reesei* as hosts for heterologous protein production to allow significant scale-up and to simplify down-stream processing of thermophilic enzymes.

2. XYLANASE GENES FROM THERMOPHILIC BACTERIA EXPRESSED IN *KLUYVEROMYCES LACTIS*: EFFECT OF PLASMID ARCHITECTURE AND GROWTH MEDIUM

Standard *Saccharomyces cerevisiae* genes such as *TRP1*, *URA3* and *LEU2* complement auxotrophic mutations in *K. lactis* and have thus been used as transformation markers in *K. lactis* plasmids [10]. As there are significant similarities between genes sequenced from the two yeasts [11], it is anticipated that development of *K. lactis* as an expression host will benefit from the current large-scale effort to determine gene functions in *S. cerevisiae*.

Various heterologous genes have been expressed in *K. lactis* with protein yields from 20mg/l (amylase) to a few grams per liter (bovine prochymosin) from both plasmid-borne and integrated genes [10,16].

The pKD1-based [12] expression vector pCWK1 (Figure 1) has been used for the insertion of the *Dictyoglomus thermophilum* and *Thermotoga maritima* FjSS3B.1 *xynA* genes [13-15]. *K. lactis* transformant strains examined were able to promote secretion of the *T. maritima* FjSS3B.1 xylanase XynA with the same electrophoretic mobility as the enzyme produced in *E. coli*. Recombinant *T. maritima* XynA comprised over 95% of the total protein secreted from the transformed *K. lactis* strain CBS1065 (Centraalbureau voor Schimmelculturen, Utrecht, The Netherlands). The *T. maritima* XynA enzyme produced in *K. lactis* has optimal activity at 90°C and a half-life comparable to the results reported for this enzyme produced in *E. coli* [18]. Similarly, the *D. thermophilum* XynA protein produced in *K. lactis* was also shown to have identical biochemical characteristics to its counterpart produced in *E. coli* [13]. The two recombinant xylanases produced in *K. lactis* are biologically active [13,14] and are not hyperglycosylated, which was the case with the thermophilic *Caldicellulosiruptor saccharolyticus* xylanase, XynA expressed in *S. cerevisiae* [19].

Parameters that influence xylanase secretion levels in *K. lactis* include plasmid architecture and composition of the growth medium. The use of full pKD1 sequence vectors resulted in significantly higher levels of extracellular *D. thermophilum* XynA through *LAC4*-directed expression as compared to expression with other pKD1 *ori* plasmid derivatives that we have constructed [14]. The carbon source influenced secretion levels by modulating promoter strength, which in turn affects the mitotic stability of plasmids [13]. Expression of *xynA* in CBS1065 under the direction of *LAC4* was incompletely repressed in non-inducing glucose medium, even though plasmid stability remained high. However, full induction of the *LAC4* promoter on galactose (without glucose) resulted in drastic decrease in plasmid stability, leading to lower levels of secreted xylanase. The *K. lactis* killer toxin signal sequence [17] was processed by the *K. lactis* signal peptidase after Gln-Gly to release a correctly processed XynA.

Using the strain CBS1065 as the expression host combines reasonable secretion levels on non-selective medium without optimization of induction (120 mg/l in shake-flask culture) with high plasmid stability on YEP plus glucose. Thus, the recombinant *K. lactis* CBS1065 producing *T. maritima* xylanase A is suitable for scaled-up fermentation. The approach holds promise, since about 20-30-fold improvements in a chemostat culture have been recorded in preliminary experiments with *K. lactis* transformant strains harboring the recombinant *D. thermophilum xynA* gene (Harris and Bergquist, unpublished). Further improvement of the yields of recombinant xylanases in *K. lactis* involves optimization of fermenter cultivations for the manufacture of substantial amounts of enzyme for other basic biochemical studies and for larger-scale bleaching trials. In order to reach this goal, it will be necessary to develop an appropriate induction strategy to minimize the effects of plasmid instability with pCWK1 seen in the absence of geneticin selection.

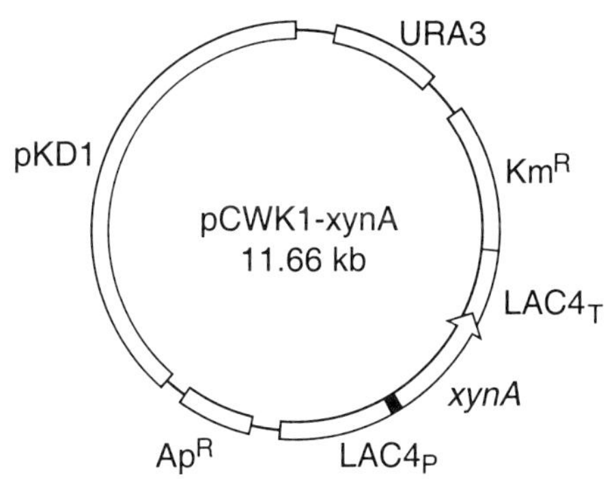

Figure 1. *K. lactis* expression vector pCWK1-*xynA*. The 11.66 kb vector consists of the pKD1 plasmid of *K. drosophilarum*, the pUC19 plasmid and the *LAC4* promoter and terminator region from *K. lactis*. The *D. thermophilum* Rt46B.1 *xynA* gene has been inserted in the single *Mlu*I after the killer toxin signal sequence (shown in black). Transformants can be selected based on geneticin resistance. The *URA3* gene of *S. cerevisiae* allows integration of the construct into the yeast genome or for selection in antibiotic-free medium when used in suitable *K. lactis* strains.

In summary, a prerequisite for the expression of the AT rich *D. thermophilum xynB* gene in *T. reesei* was the construction of a synthetic gene according to the codon usage of *T. reesei* genes. Production of XYNB was greater from a vector where the gene was fused to the *cbh1* signal sequence. We believe that fusion of the synthetic gene using the codons preferred by the expression host to a mature endogenous carrier protein [21] is less crucial in order to facilitate effective transcription and secretion. Moreover, optimizing the cleavage of secretion signal and facilitating enzyme processing have improved the yield further. In an earlier work by White and Hinde [24], production of *B. circulans* xylanase in *T. longibrachiatum* was hampered with inefficient cleavage of the *cbh1* secretion signal in the expression host and presumably also by the presence of three codons in the foreign gene that are never used by *T. reesei xyn1, 2* or *cbh1* genes [31].

3.2. Expression of a xylanase from thermophilic fungus *Humicola grisea* in *T. reesei*

Thermophilic fungi can grow at temperatures exceeding 45°C and have a strong capacity to hydrolyse plant biomass. Their hydrolases are in general more heat stable than the corresponding enzymes produced by mesophilic fungi [2].

A cDNA gene encoding a family 11 xylanase (*xyn2*) has been isolated recently from *Humicola grisea* var *thermoidea* [32] cultivated from Brazilian soil [33]. The *H. grisea* XYN2 gene product is highly active at 70°C, pH 6.5 (Faria et al. unpublished). The *xyn2* gene was cloned into *T. reesei* as a fusion to the *cbh1* signal sequence and the CBHI core-linker [34]. Culture medium from a transformant harboring the plasmid with *xyn2* fused to the *cbh1* secretion signal showed a prominent band at the correct molecular mass that also exhibited xylanase activity on a zymogram assay. This band was isolated and the N-terminal sequence determined. The sequence obtained, QVTPNAE, was shown to coincide with a Kex2-like cleavage site in alignments against *T. reesei* xylanases XYNI and II showing that it was processed in a similar manner to give mature enzyme secreted into the supernatant [34]. The recombinant XYN2 yields were of the order of 0.5-1g per liter in non-optimized shake flask cultivations and activities up to about 50 000 nkat/ml have been obtained by small scale fermentation.

4. CONCLUSIONS

Low cost, simple cultivation and minimal amounts of downstream processing can be achieved with a production system where the enzyme is effectively secreted into the cultivation medium. In this regard, fungi provide an attractive option to industrial bacterial systems. Modification of the incoming gene and the expression vector at a molecular level are examples of key operations required for better secretion, stability and processing of the gene product. Basic information can be extracted from public databases to facilitate the construction of an expression plasmid "interpretable" to the production host. Each gene can be modified, for example, in terms of codon usage and by introducing signals for extracellular transport and processing. However, it is evident that more studies are needed into the expression biology of the chosen host organism, especially regarding conditions for effective product fermentation. It is reasonable to assume that the yields of recombinant xylanases produced in fungal hosts will be considerably improved by following the strategies discussed.

REFERENCES

1. N. Kulharni, A. Shendye and M. Rao, FEMS Microbiol. Revs., 23 (1999) 411.
2. R. Maheswari, G. Bharadwaj and M.K. Bhat, Microbiol. Mol. Biol. Revs., 64 (2000) 461.
3. F.T. Robb and A.R. Place AR. (eds.), Archaea A laboratory Manual Cold Spring Harbor Press, Cold Spring Harbor NY, 1995.
4. M.W.W. Adams, Adv. Prot. Chem., 48 (1996) 1.
5. H. Nevalainen, P. Suominen and K. Taimisto, J. Biotechnol., 37(1994) 193.
6. R.M. Berka, F.T. Bayliss, P. Bloebaum, D. Cullen, N.S. Dunn-Coleman, K. Kodama, K. Hayenga, R. Hitzeman, M. Lamsa, M.M. Przetak, M. Rey, L. Wilson and M. Ward, In J.W. Keely JW and T.O. Baldwin (eds.), Applications of Enzyme Biotechnology, Plenum Press, New York (1991) 273.
7. C.A.M.J.J. Van den Hondel, P.J. Punt and R.F.M. van Gorcom, In J.W. Bennet and L.L. (eds.), More Gene Manipulation in Fungi. Academic Press, San Diego (1991) 396.
8. D.J. Jeenes, D.A. MacKenzie, I.N. Roberts and D.B. Archer (1991), Biotechnology and Genetic Engineering Reviews, Intercept Ltd Andover, UK, Vol. 9 (1991) 327.
9. D.I. Gwynne, F.P. Buxton, S.S. Williams, S. Garven and R.W. Davies RW, Biotechnology, 5 (1987) 713.
10. M. Wesolowski-Louvel, K.D. Breunig and H. Fukuhara, In K. Wolf (ed.), Nonconventional yeasts in Biotechnology. Springer-Verlag, Berlin (1996) 139.
11. O. Ozier-Kalogeropoulos, A. Malpertuy, J. Boyer, F. Tekaia and B. Dujon, Nucleic Acids Res.,26 (1998) 5511.
12. M.M. Bianchi, C. Falcone, X.J. Chen, M. Wésolowski-Louvel, L. Frontali and H. Fukuhara H, Curr. Genet., 16 (1987) 95.
13. D.J. Walsh and P.L. Bergquist, Appl. Environ. Microbiol., 63 (1997) 3297.
14. D.J. Walsh, M.D. Gibbs and P.L. Bergquist PL, In K.-E. Eriksson and A. Cavaco-Paulo (eds.), Enzyme Applications in Fiber Processing. Am. Chem. Soc. Symp., 687 (1998a) 155.
15. D.J. Walsh, M.D. Gibbs and P.L Bergquist, Extremophiles 2 (1998b) 2.
16. G. Gellissen and C.P. Hollenberg, Gene, 1190 (1997) 87.
17. C. Tanguy-Rougeau, M. Wésolowski-Louvel and H. Fukuhara, FEBS Lett., 234 (1988) 464.
18. D.J. Saul, L.C. Williams, R.A. Reeves, M.D. Gibbs and P.L. Bergquist, Appl. Environ. Microbiol., 61 (1995) 4110.
19. K.A.G. Donald, A. Carle, M.D. Gibbs and P.L. Bergquist, Appl. Microbiol. Biotechnol., 42 (1994) 309.
20. A. Mäntylä, M. Paloheimo and P. Suominen, In G.E. Harman and C.P. Kubicek (eds.), *Trichoderma and Gliocladium,* Taylor and Francis Ltd., London and Bristol, .Vol. 2 (1998) 291.
21. M.E. Penttilä, (1998) In G.E. Harman and C.P. Kubicek (eds.), *Trichoderma and Gliocladium,* Taylor and Francis Ltd., London and Bristol, Vol. 2 (1998) 365.
22. V.S.J. Te'o, A.E. Cziferszky, P.L. Bergquist and K.M.H. Nevalainen, FEMS Microbiol. Lett., 190 (2000) 13.
23. M. Lorito, C.K. Hayes, A. Di Pietro A and G.E. Harman, Curr. Genet., 24 (1993) 329.
24. T.C. White and C. Hinde C, In Thermophiles '96 Conference, 4-9 September, Athens, Georgia, 1996. Poster abstract P2 K4.

25. B.W. Hazell, V.S.J. Te’o, J.R. Bradner, P.L. Bergquist and K.M.H. Nevalainen, Lett. Appl .Microbiol., 30 (2000) 282.
26. V.S.J. Te'o, P.L. Bergquist and H. Nevalainen, (2000) In The 5th European Conference on Fungal Genetics, 25-29 March, Arcachon, France, 2000. Poster abstract V-24.
27. D.D. Morris, M.D. Gibbs, C.W. Chin, M.H. Koh, K.K. Wong, R.W. Allison, P.J. Nelson and P.L. Bergquist, Appl. Environ. Microbiol., 64 (1998) 1759.
28. Goller, SP, Gorfer M, Mach RL, Kubicek CP, In P.C. Bridge, D.K. Arora, C.A. Reddy and R.P. Elander (eds.), Applications of PCR in Mycology, CABI Publishing, CAB International, New York (1998) 21.
29. R.J. Gouka, P.J. Punt, J.G.M. Hessing and C.A.M.J.J. van den Hondel, Appl. Environ. Microbiol., 62 (1996) 1951.
30. M. Bailey and K. Poutanen, Appl. Microbiol. Biotechnol. 30 (1989) 5.
31. P. Bergquist, V. Te’o, M. Gibbs, A. Cziferszky, F. Paula de Faria, M. Azevedo and H. Nevalainen, Extremophiles, in press, 2001.
32. F.P. Faria, V.S.J. Te'o, P.L. Bergquist, H. Nevalainen and M.O.Azevedo, In Sociedade Brasileira de Bioquímica e Biologia Molecular, XXIX Reunião, 27-30 May, Brasil, 2000. Poster abstract G-69.
33. E.M. Almeida, M.L. Polizeli, J.A. Jorge and H.F. Terenzi, FEMS Microbiol., Lett., 130 (1995) 171.
34. F.P. Faria, V.S.J. Te'o, P.L. Bergquist, M.O. Azevedo and K.M.H. Nevalainen, Submitted to Lett. Appl. Microbiol., 2001.

Biotechnology in the Pulp and Paper Industry
L. Viikari and R. Lantto (Editors)

Evaluation of the bleach-enhancing effect of xylanases on bagasse pulp

Sadhvir Bissoon[a,b], Suren Singh[a] and Lew Christov[b,c]

[a]Department of Biological Sciences, ML Sultan Technikon, Box 1334, Durban 4000, South Africa

[b]Sappi Biotechnology Laboratory, Department of Microbiology and Biochemistry, University of the Free State, PO Box 339, Bloemfontein 9300, South Africa

[c]Sappi Management Services, PO Box 3252, Springs 1560, South Africa

Until recently enzymatic bleaching of non-woody materials has received scant attention. In this study three commercial xylanases were evaluated in both ECF and TCF bleaching to enhance the bleachability of bagasse soda pulp. The enzyme pretreatment was carried out at various charges under optimized conditions. In both bleaching sequences, Xylanase P (Iogen Corp.) was the most efficient bleach-enhancing enzyme, when compared on an equal enzyme charge on pulp (as U/g) basis. This enzyme was able to improve brightness of DED bleached bagasse pulp by up to 8.2 brightness points or alternatively reduce the consumption of chlorine dioxide by 33%. Ecopulp TX-200C (Röhm Enzyme) and Cartazyme NS-10 (Clariant) were equally efficient in the ECF bleaching of bagasse pulp producing chlorine dioxide savings of 20%. In all instances, however, the biobleaching effect induced due to xylanase pretreatment was dependent on both the enzyme and chlorine dioxide charges used. At similar brightness levels, the TCF biobleaching in sequence X-OqPaP was less efficient than the ECF DED biobleaching. Mechanical pretreatment of the pulp prior to the ECF biobleaching slightly increased final brightness, indicating "activation" of pulp to facilitate enzyme and chemical bleaching. Xylanase pretreatment had an insignificant effect on the physical properties of the ECF bleached pulp.

1. INTRODUCTION

Growing concerns on environmental issues have prompted governments to develop stringent regulations against polluting processes. The use of chlorine and chlorine chemicals during the bleaching process results in the formation of chlorinated organic substances, some of which are toxic, mutagenic and non-biodegradable (1).

Xylanases hydrolyze the xylans redeposited on pulp fibers during kraft pulping. This is thought to facilitate the diffusion of lignin fragments degraded and solubilized during subsequent bleaching (2).

It has been already demonstrated that xylanases can enhance the bleaching of kraft pulps on a large scale without major capital investments (3). Xylanases can provide a cost effective way to reduce the use of chlorine-containing compounds and other bleaching chemicals.

There are reports of 8-15% savings of the total amount of chemicals across the bleach plant and a decrease in the adsorbable organic halogen levels of 12–25% in the bleach effluent (1). Increased tear strength and pulp throughput have also been reported (4).

Several commercially available xylanase preparations, most of which are active at slightly acidic or neutral pH, have been investigated in pulp bleaching. Pulpzyme HA (Novo Nordisk, Denmark), produced by *Trichoderma reesei,* was the first commercial xylanase to be used in biobleaching. It achieved a 20% kappa number decrease of oxygen-delignified birch kraft pulp (5). Cartazyme (Clariant) also improved the brightness of kraft pulps (6). The effectiveness of xylanases have been well documented for wood pulps, however, very little attention has been given to the application of xylanases to non-woody materials such as bagasse pulp. The aim of this study was to evaluate and compare the bleaching potential of three commercial xylanase preparations on bagasse soda pulp.

2. MATERIALS AND METHODS

2.1. Pulp

Unbleached bagasse soda pulp was obtained from Sappi Stanger, South Africa. The pulp was thoroughly washed with distilled water until a neutral pH of the wash waters was attained. The washed pulp had a moisture content of 84% (w/w) and it was stored in sealed plastic bags at 4°C until used.

2.2. Enzymes and enzyme assays

Three commercial enzymes, Xylanase P (Iogen Corp., Canada), Ecopulp TX-200C (Röhm Enzyme, Finland) and Cartazyme NS-10 (Clariant, UK), were evaluated in bleaching of bagasse pulp. All three xylanases belonged to family 11 of glycosyl hydrolase enzymes and were produced from genetically modified *Trichoderma reesei* (Xylanase P and Ecopulp TX-200C) or *Bacillus* strains (Cartazyme NS-10).

The xylanase activities were determined according to the method of Bailey *et al.* (7) using a 1% birchwood xylan solution. Reducing sugars (RS) were assayed by adding 1.5 ml DNS reagent to the reaction tube, boiling for 5 min, cooling, and measuring the absorbance against the reagent at 540 nm. One unit of enzyme activity was defined as the amount of enzyme that catalyzes the release of one μmole of xylose equivalents per minute under the assay conditions.

2.3. Xylanase treatment of pulp

Washed unbleached bagasse pulp was treated at a charge of 5 U xylanase/g pulp under various pH (4.0-9.0) and temperature (40-80°C) conditions for 1 h. The optimum conditions for the use of each enzyme preparation (Table 1) were determined based on the release of RS from pulp (data not shown).

Table 1. Conditions used for enzyme treatment of bagasse pulp

Enzyme	Temperature (°C)	pH	Time (h)	Consistency (%)
Xylanase P	60°C	5.0	2	10
Ecopulp TX-200C	65°C	7.0	2	10
Cartazyme NS-10	60°C	8.0	2	10

Washed unbleached bagasse pulp was treated at various xylanase charges (1, 5 and 10 U/g) under the optimum pH and temperature of each enzyme and 10% pulp consistency for 2 h. Controls were run under the same conditions however using boiled enzymes. After the incubation period, the pulp was filtered and washed. Pulp and enzyme filtrates were retained for further analyses and use.

2.4. Mechanical treatment of pulp

Prior to use of xylanase at a charge of 1 U/g, washed unbleached bagasse pulp was pretreated mechanically using a pulp homogenizer (Heidolph, Germany). Pretreatment was carried out at 26 000 rpm, room temperature and 1.5% pulp consistency for 15 min.

2.5. Analyses of pulp filtrates

The enzyme mediated release of ultraviolet (UV) absorbing and chromophoric material from pulp was monitored in filtrates by measuring the absorbance at 280 nm and 465 nm, respectively (8). The amount of RS released from pulp was determined spectrophotometrically at 540 nm according to the DNS method (9).

2.6. Chemical bleaching of bagasse pulp

Xylanase pretreated pulp underwent chemical bleaching. The conditions for the elemental chlorine-free (ECF) bleaching are summarized in Table 2. The charges of chlorine dioxide were based on a kappa factor (kf) of 0.3, 0.4, 0.5, 0.6 and 0.75. The conditions for the totally chlorine-free (TCF) bleaching are shown in Table 3.

Table 2. Conditions used for ECF bleaching of bagasse pulp

Bleaching step	Charge (%)	T (°C)	Time (h)	Consistency (%)	pH
Chlorine dioxide (D_1)	kappa no x kf	70	1	10	3.0
Alkali extraction (E)	2.5% NaOH	70	2	10	-
Chlorine dioxide (D_2)	0.5 x D_1	70	3	10	4.0

Table 3. Conditions used for TCF bleaching of bagasse pulp

Bleaching step	Charge (%)	T (°C)	Time (h)	Consistency (%)	pH
Oxygen (O)	1.2% NaOH; 400 kPa O_2	95	1	10	>10.0
Chelated peracetic acid (qPa)	0.3% Pa; 0.3% H_2O_2	75	2	10	8.5
Hydrogen peroxide (P)	3% H_2O_2; 1.8% NaOH	75	3	10	11.0

2.7. Pulp properties

The kappa number and physical properties of bagasse pulp were determined according to the Tappi Test Methods. Brightness was measured using an Elrepho photoelectric reflectance photometer (Carl Zeiss, Germany). The kappa number and brightness of unbleached bagasse pulp were 8.7 and 39.3%, respectively. All analyses were carried out in duplicate.

3. RESULTS AND DISCUSSION

3.1. Analyses of enzyme filtrates

Xylanase P was most effective in releasing RS, UV and visible light absorbing material from bagasse pulp while Ecopulp TX-200C and Cartazyme NS-10 followed in a decreasing order of efficiency (Table 4). The amount of UV absorbing material detected in the filtrates was significantly greater than that of the visible light absorbing material. Overall, the concentration of material released from pulp increased with increasing enzyme charges. Xylanase P at a charge of 10 U/g was able to release 62% and 69% more UV and visible light absorbing material, respectively, and 2.14 times more RS than the control.

Table 4. Analyses of enzyme filtrates following xylanase pretreatment of bagasse pulp

Enzyme	Charge (U/g)	$A_{280\,nm}$	A_{465nm}	RS (mg/g)
	1	4.354	0.198	0.345
Xylanase P	5	4.503	0.210	0.421
	10	4.636	0.240	0.587
	1	4.194	0.186	0.304
Ecopulp TX-200C	5	4.369	0.195	0.367
	10	4.495	0.223	0.453
	1	3.956	0.184	0.296
Cartazyme NS-10	5	4.212	0.201	0.355
	10	4.305	0.219	0.395
Control	0	2.864	0.151	0.187

3.2. Impact of xylanase pretreatment on kappa number and brightness of bagasse pulp

Xylanase pretreatment at a charge of 1 U/g appeared to have very little effect on the kappa number of bagasse pulp (Table 5). An increase in the enzyme charge to 10 U/g produced a slight decrease in kappa number with reductions of 6%, 7 % and 12% achieved with Cartazyme NS-10, Ecopulp TX-200C and Xylanase P, respectively. All three enzymes induced a direct brightening effect on bagasse pulp when used at all three charges. At 10 U/g, Xylanase P was the superior enzyme increasing the brightness by 2.2 brightness points while Cartazyme NS-10 and Ecopulp TX-200C caused brightness increases of 1.4 and 1.8 points over control, respectively.

Table 5. Impact of xylanase pretreatment on kappa number and brightness of bagasse pulp

Enzyme	Charge (U/g)	Kappa no reduction (%)	Brightness increase (points)
	1	5	0.6
Xylanase P	5	8	1.3
	10	12	2.2
	1	0	0.4
Ecopulp TX-200C	5	5	1.0
	10	7	1.8
	1	0	0.0
Cartazyme NS-10	5	0	0.6
	10	6	1.4

3.3. ECF biobleaching of bagasse pulp

The effect of xylanase pretreatments was analyzed under varying conditions of chlorine dioxide charges. Xylanase P appeared as the most efficient enzyme to enhance bleachability of bagasse pulp. The biobleaching effect induced by this enzyme was between 0.3 and 1.6 points brightness gain over control at a kf of 0.75 (Fig. 1) and between 4.9 and 8.2 points

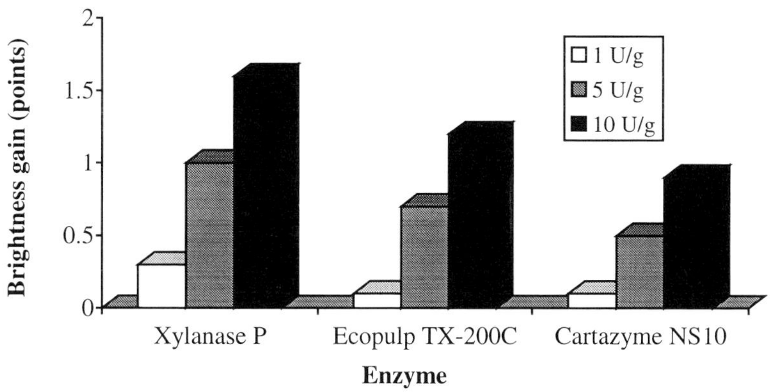

Figure 1. X-DED bleaching of bagasse pulp (kf of 0.75)

when pulp was bleached at a kf of 0.3 (Fig. 2). This clearly indicates that the biobleaching effect due to xylanase pretreatment is dependent on both the enzyme and chlorine dioxide charges used. Brightness of bleached pulp progressively increased with the increase of the enzymes charges. However, as the chlorine dioxide charges increased, the bleach-enhancing efficiency of the xylanases decreased.

Using Xylanase P, the bagasse pulp could be bleached with a kf of 0.5 to the same brightness (83.8%) as the control bleached with a kf of 0.75 without enzyme (Fig. 3).

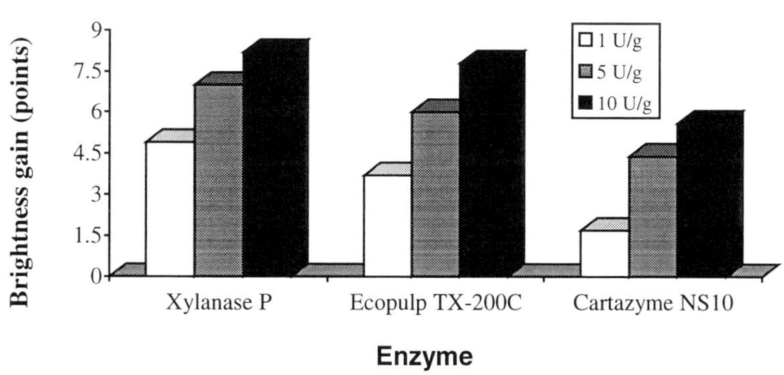

Figure 2. X-DED bleaching of bagasse pulp (kf of 0.3)

This represents a reduction in the chlorine dioxide consumption of 33.3%. The brightness gain induced by both Ecopulp TX-200C and Cartazyme NS-10, on the other hand, could be

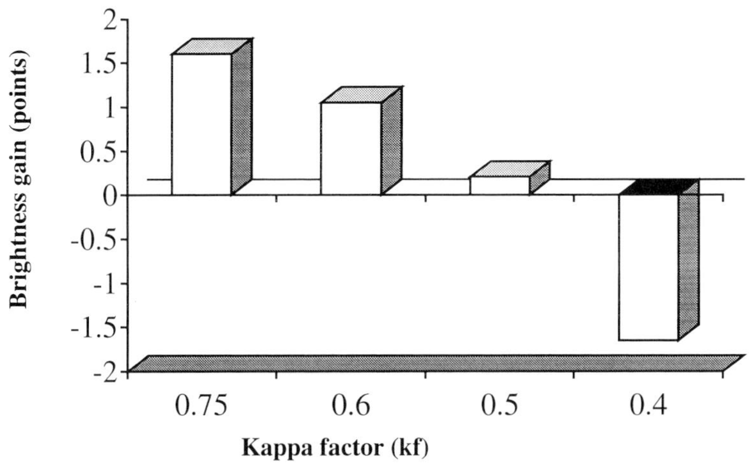

Figure 3. Reduction of chlorine dioxide consumption following X-DED bleaching using Xylanase P (10 U/g)

translated into chlorine dioxide savings of 20% (data not shown).

Mechanical pretreatment of pulp prior to xylanase (1 U/g) rendered the pulp even more susceptible to bleaching. A notable increase in brightness was observed after final bleaching of the mechanically pretreated pulp compared to the pulp bleached without pretreatment (Fig. 4). These results indicate that mechanical pretreatment did "activate" the pulp to a degree thereby enhancing the bleaching effect of the enzymes and the chemicals.

3.4. TCF biobleaching of bagasse pulp

Application of xylanases to the sequence OqPaP revealed that the TCF biobleaching was less efficient than ECF. Using Xylanase P, brightness was improved by up to 2.3 points (Fig. 5) to a final brightness of up to 71.9%. For comparison, using Xylanase P in the ECF bleaching, a brightness increase of 8.2 points was gained to produce a final brightness of

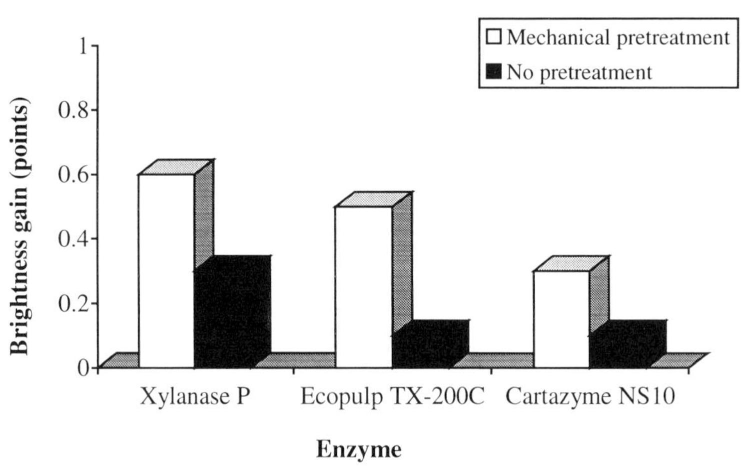

Figure 4. Impact of mechanical pretreatment on X-DED bleaching of bagasse pulp (1 U/g; kf of 0.75).

76.1% (Fig. 2). Ecopulp TX-200C and Cartazyme NS-10, when used at 10 U/g, brightened pulp by 2.1 and 2.0 points, respectively (Fig. 5).

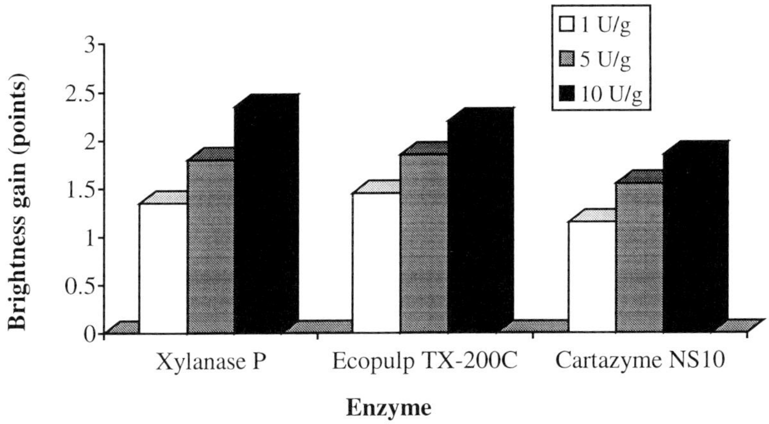

Figure 5. TCF biobleaching of bagasse pulp in sequence X-OqPaP

3.5. Physical properties

Analysis of the physical properties of bagasse pulp indicated minimal changes of the enzyme treated pulp compared to the control (Table 6). Xylanase P and Ecopulp TX-200C, however, did increase the tensile strength of the paper with a corresponding increase in the amount of energy required to rupture the test strip. The percentage stretch of the paper strips was slightly lower with the enzyme treated samples. The stretchability values obtained with Cartazyme NS-10 were lower than the other xylanases under study. All enzymes did slightly decrease the viscosity of the pulps. However, there were no significant changes in the physical properties of bagasse pulp with the incorporation of xylanases into the bleach sequence.

Table 6. Effect of xylanase pretreatment (10 U/g) on the physical properties of X-DED bleached bagasse pulp (kf of 0.75)

Enzyme	Tensile strength (N/15mm)	Stretch (%)	TEA (J/m^2)	Viscosity (mPa s)
Control	13.76	2.72	13.83	48.4
Xylanase P	14.12	2.55	14.56	46.9
Ecopulp TX-200C	13.87	2.58	14.03	47.2
Cartazyme NS-10	13.56	2.31	13.65	47.3

4. CONCLUSIONS

Comparison of enzyme efficiency in this work was made based on equal activity unit enzyme charges on pulp (U/g) rather than equal enzyme costs, therefore conclusions drawn herein do not represent a commercial evaluation of enzyme products or enzyme manufacturers. With respect to the above, Xylanase P, when compared to Ecopulp TX-200C and Cartazyme NS-10 at equal enzyme doses as U/g, appeared as the most efficient bleach-enhancing enzyme on bagasse pulp in both ECF (DED) and TCF (OqPaP) bleaching. In DED bleaching (kf of 0.75), a brightness gain of 1.6 points over control could be attained using Xylanase P. Overall, Ecopulp TX-200C and Cartazyme NS-10 followed Xylanase P in order of decreasing biobleaching efficiency. Xylanase P was able to reduce the use of chlorine dioxide by 33% as compared to 20% reduction of the chlorine dioxide consumption achieved with the other two enzyme preparations. The biobleaching effect induced due to xylanase pretreatment depended on both the enzyme and chlorine dioxide charges used. Overall, use of high kappa factors were required for a satisfactory bleaching which could be due to a low initial brightness and a relatively low brightness ceiling of bagasse pulp. Mechanical pretreatment of the pulp prior to ECF biobleaching did produce a slight increase in final brightness, indicating "activation" of pulp to enzyme and chemical bleaching. At similar brightness levels, the TCF biobleaching in sequenceX-OqPaP appeared as less efficient than the ECF biobleaching in terms of biobleaching effect. Examination of the physical properties of the paper strips revealed insignificant differences between the control and the xylanase treated samples following ECF bleaching.

ACKNOWLEDGEMENTS

We acknowledge the National Research Foundation, ML Sultan Tecknikon Research Fund and Sappi Management Services for financial support, Sappi Management Services for giving permission to publish this work and Sappi Biotechnology Laboratory for excellent technical assistance.

REFERENCES

1. P. Bajpai and P.K. Bajpai, Adv. Biochem. Eng. Biotechnol., 57 (1997) 213.
2. Kantelinen, B. Hortling, J. Sundquist, M. Linko and L. Viikari, Holzforschung, 47 (1993) 318.
3. L. Viikari, A. Kantelinen., J. Sundquist and M. Linko, FEMS Microbiol. Rev., 13 (1994) 335.
4. K. Shah, D. Cooper., R. Adolphson and K.-E.L. Eriksson, J. Pulp Pap. Sci., 26 (2000) 8.
5. B.L. Zamost, H.K. Nielsen and R.L. Starnes, J. Ind. Microbiol., 8 (1991) 71.
6. A.P. Garg AP, J.C. Roberts and A.J. McCarthy AJ, Enzyme Microb. Technol., 22 (1998) 594.
7. M.J. Bailey, P. Biely and K. Poutanen, J. Biotechnol., 23 (1992) 257.
8. K.K.Y. Wong, E. de Jong., J.N. Saddler and R.W. Allison, Appita J., 50 (1997) 415.
9. G.L. Miller, Anal. Chem., 31 (1959) 426.

Biotechnology in the Pulp and Paper Industry
L. Viikari and R. Lantto (Editors)

Boosting of LMS-bleaching with hemicellulases

T. Oksanen[a], J. Buchert[a], M. Amann[b], A. Candussio[b] and L. Viikari[a]

[a]VTT Biotechnology, P.O.Box 1500, FIN-02044 VTT, Finland, E-mail: liisa.viikari@vtt.fi

[b]Consortium für Electrochemische Industrie GmbH, Zielsttattstr 20, D-81379 München, Germany

The simultaneous or sequential combination of hemicellulases with laccase-mediator bleaching systems was shown to result in additional enhancement of pulp bleachability. The simultaneous combination of the two enzymatic steps was clearly more succesful with NHA than with HBT due to the more pronounced inactivation of enzymes by HBT. The lowest kappa number after alkaline extraction was obtained with the sequential use of the two enzyme systems. The effect of mannanase was less pronounced on this pulp, but the combination of xylanase, mannanase and LMS using NHA as mediator in a simultaneous treatment resulted in the highest brightness value and lowest kappa number. The decrease of kappa number obtained with combined treatments corresponds closely to the individual effects of the hemicellulase and laccase-mediator treatments, indicating that the mechanisms of the treatments are different and thus additional effects of treatments could be gained.

1. INTRODUCTION

Xylanase-aided bleaching of chemical pulp is one of the main biotechnical applications used today in the pulp and paper industry. The xylanase-aided bleaching is an indirect method, which does not directly degrade lignin and has thus a limited effect (1, 2). The xylanase treatment results in enhanced extractability of lignin in subsequent bleaching stages. Several alternative and obviously concurrent mechanisms have been proposed to be involved in the xylanase-aided bleaching. The enhanced leachability of lignin in fibre wall has been suggested to be due to hydrolysis of reprecipitated xylan or to removal of xylan from the lignin-carbohydrate complexes (LCC) in fibres. Removal of xylan by xylanases from softwood kraft fibers was found to uncover lignin, thus supporting the first mechanism (3). The hexenuronic acid, containing a double bond, gives rise to the consumption of bleaching chemicals and permanganate, increasing the apparent kappa number of pulp (4). The partial removal of hexenuronic acid substituted xylan by xylanase treatment consequently can also result in a lower kappa number. The effect of xylanase treatment on kappa number has, however, been reported to be relatively small (5).

In the laccase-mediator concept, the enzyme oxidized mediator acts directly on lignin and results in efficient delignification. In the initial study, the common substrate of laccases, 2,2'-azino-bis (3-ethylbenzthiazoline-6-sulfonate), ABTS, was used as mediator (6). Presently, the most extensively studied mediators are 1-hydroxybenzotrizole (HBT) and N-hydroxy-N-phenylacetamide, NHA (7, 8, 9, 10). Although HBT is very efficient, it suffers from some draw backs, such as the production cost and limited biodegradability. Also NHA results in extremely fast delignification with no significant impact on cellulose structure. The LMS system has been shown to be able to replace either the oxygen delignification or ozone stages (11, 12).

The successive combination of the two enzymatic methods; the hydrolytic xylanase and the oxidative laccase-mediator treatments, has previously been shown to increase the delignification efficiency (5, 9, 13, 14). The simultaneous application of LMS system using HBT as mediator with xylanase treatment in one single stage was found to be relatively ineffective, apparently due to the inactivation of xylanase by the mediator HBT. This inactivating effect of HBT has also been observed towards laccase (15). The HBT radicals formed can undergo chemical reactions with the aromatic amino acids of laccases, as demonstrated by quantitative amino acid analysis of the purified laccase II of *Trametes versicolor* (15). Contrary to HBT, NHA caused much less damage to the laccase. As a consequence, the most effective combination of xylanase and LMS has been with NHA as mediator. Thus, the type and specificity of the mediator used affects the compatibility of these two enzymatic steps.

In this work, the simultaneous or sequential action of two hemicellulases, *i.e.* xylanase and mannanase, with the laccase-mediator system on kraft pulps was studied. The two mediators, HBT and NHA were compared. In addition, the effects of hemicellulase-LMS treatments on the hexenuronic acid content and the fibre surface chemistry using ESCA was analyzed.

2. MATERIALS AND METHODS

Pulp and enzymes

The pulp used in this study was a commercial pine kraft pulp with kappa number 23.5, obtained from a Finnish pulp mill. The carbohydrate composition of the pulp was analysed after enzymatic hydrolysis as described previously (16). Hexenuronic acid was determined according to Tenkanen *et al.* (17).

The purified xylanase, pI 9, (18) and mannanase from *Trichoderma reesei* (19) were used for the separate and simultaneous treatments with the laccase-mediator system. The laccase was produced by *Trametes hirsuta.* The activities of xylanase, mannanase and laccase were determined according to Tenkanen (18), Stålbrand *et al.*, (19) and Niku-Paavola *et al.* (20), respectively. HBT was obtained from Aldrich and NHA was synthesized according to Oxley *et al.* (21).

Enzymatic treatments

The xylanase treatments were carried out at a pulp consistency of 4% at 45 °C for 2 h. The pH of the pulp slurry was adjusted to 5 prior to enzymatic treatments with sulphuric acid. The xylanase and mannanase dosages used were 100 nkat/g. After hydrolyses, pulps were washed with deionized water and the carbohydrates solubilized during the enzymatic treatments were analyzed by HPLC after a secondary enzymatic hydrolysis to monomers as described by Buchert *et al.* (22).

The laccase-mediator treatments were carried out under similar conditions as the xylanase treatments except that the pulp was oxygen aerated in order to ensure sufficient amount of oxygen in the reaction. The dosage of laccase in the delignifications was 30 IU/g pulp and the dosage of mediators 10 mg/g pulp. The LMS treatments were carried out either simultaneously or sequentially with hemicellulases. Following the enzymatic treatments, the pulps were alkaline extracted at 8% consistency with 2% NaOH for 1 hour at 60 °C. Before peroxide bleaching, the pulps were chelated using 0.4% EDTA at 5% consistency, 60 °C, pH

4.5 for 1 h. After each stage, the pulps were washed with deionized water. One-stage hydrogen peroxide bleaching was carried out using 3% H_2O_2, 1.75% NaOH, 0.2% DTPA and 0.5% $MgSO_4$. The kappa numbers of the pulps were determined following the SCAN-C1:77 standard and brightness was measured according to ISO standard 2470.

ESCA (Electron Microscopy for Chemical Analysis) was used to determine the coverage of lignin on the fibre surface after different treatments as described by Buchert *et al.* (3).

3. RESULTS AND DISCUSSION

The effect of mediators on the hydrolytic activities

The mediators, HBT and NHA, oxidized by laccase, clearly reduced the activity of both purified hemicellulases; xylanase and mannnase (Figs 1 A and B). Even without oxidation by laccase, the mediators inhibited these activities, although to a lesser extent than in the presence of laccase (results not shown). The inhibitory effect of mediators seemed to be even more pronounced towards xylanase, as compared with mannanase. The xylanase activity practically disappeared within half an hour in the presence of either HBT or NHA, whereas the mannanase activity was more resistant towards the mediators. The mannanase activity was decreased only by about 30% by NHA even after one hour's incubation. It should, however, be observed that the inhibitory effect of oxidized mediators may be less pronounced when lignin is present, offering sites for the radical attack. Fibres are expected to have a protective effect towards enzymes for factors like temperature or pH.

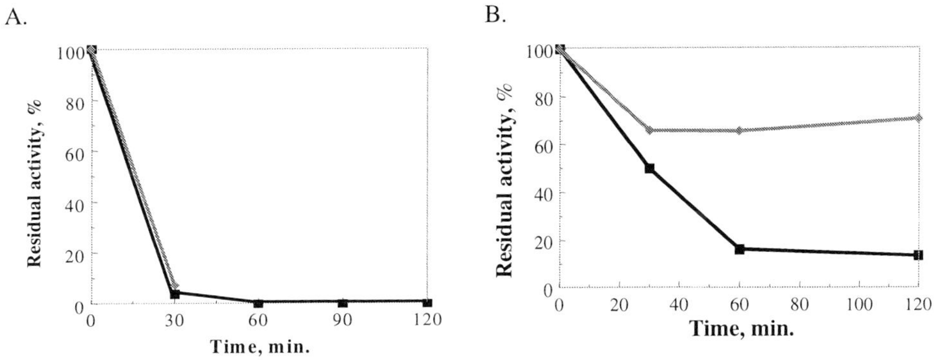

Figure 1.The effect of LMS on xylanase (A) and mannanase (B) activities using two mediators: HBT (—■—) and NHA (—◆—)

The pine kraft pulp was treated with xylanase using a dosage of 100 nkat/g alone or in combination with the laccase-mediator system (Table 1). The xylanase treatment alone solubilized 1.7% of the carbohydrates after two hours hydrolysis. Both laccase-oxidized mediators, HBT and NHA, clearly reduced the hydrolytic action of xylanase. The decrease was, however, less pronounced as compared to treatment without fibres (Fig 1). Both mediators decreased the amount of released xylose by about 50%. When compared in terms of total carbohydrates released, NHA seemed to be less inhibitory (Table 1). The hexenuronic acid content in the xylanase hydrolysates remained practically constant after combined laccase-mediator treatments. The degree of substitution was in all cases about 6 moles of hexenuronic acid per 100 moles of xylan, indicating that these mediators do not specifically remove hexenuronic acid from pulps.

Table 1. Composition of hydrolysis products after xylanase and simultaneous LMS treatments, hydrolysis time 2 hours.

	Solubilized carbohydrates, % of d.w.							
Treatment	**Total**	**Glc**	**Xyl**	**Ara**	**Man**	**Gal**	**MeGlcA**	**HexA**
XYL	1.68	0.04	1.28	0.18	0.03	0.02	0.03	0.10
XYL+LMS (HBT)	0.85	0.05	0.61	0.07	0.03	0.02	0.02	0.07
XYL+LMS (NHA)	0.94	0.09	0.63	0.10	0.03	0.02	0.02	0.05

The lignin content of the outer surface of the fibres after different treatments was analyzed by ESCA (Table 2). The lignin coverage of pine kraft pulp fibres was clearly first increased by the action of xylanase and decreased after the alkaline extraction, as observed earlier (3). When using the laccase-mediator system alone, a slight reduction was observed before the alkaline extraction stage, whereas the reduction was expectedly more pronounced after the alkaline extraction. When comparing these values with the kappa number values, it can be expected that the action of xylanases is more focused on fibre surfaces, revealing more lignin to be extracted by alkaline, whereas the LMS acts more uniformly throughout the fibre cell wall. The combination of xylanase with laccase-mediator treatments further reduced the lignin content on the fibres surfaces, especially when the treatments were carried out sequentially.

Table 2. Lignin coverage of fibre surfaces after xylanase (XYL), LMS (with HBT as mediator), sequential or simultaneous XYL-LMS treatments. XYL $\rightarrow$ LMS: sequential treatment, XYL + LMS: simultaneous treatment. Analyses by ESCA before and after alkaline extraction step.

	Lignin coverage, %				
	REF	**XYL**	**LMS**	**XYL → LMS**	**XYL + LMS**
Before extraction	13.3	14.2	12.5	11.5	13.3
After extraction	13.5	10.0	8.5	7.0	9.8

The effect of hemicellulases and LMS system on the bleachability of kraft pulp

The effects of the two mediators on the kappa number after an alkaline extraction step were compared in different combinations with the xylanase treatment. Pretreatment with xylanase alone increased the delignification from 22% to 32% (Fig. 2). The kappa numbers decreased by 2,5 and 3 units, corresponding to a delignification of 33% and 35%, by the laccase-mediator treatments alone, using HBT and NHA, respectively. NHA-laccase seemed to act slightly more efficiently than HBT-laccase also when combined with xylanase both in successive and simultaneous treatments. The use of xylanase clearly improved the efficiency of the LMS, by increasing the delignification to 39% and 44% when used separatedly and to 35% and 40% when used simultaneously, with HBT and NHA as mediators, respectively. The combination of the two enzymatic steps was thus clearly more succesful with NHA than with HBT. The lowest kappa number after alkaline extraction was, however, obtained after the successive use of the two enzyme systems, indicating some inactivation of enzymes in simultaneous treatments. The increased delignification obtained with NHA corresponded closely to the individual effects of the xylanase and laccase-mediator treatments (Fig. 2). Thus, these treatments seem to be additive.

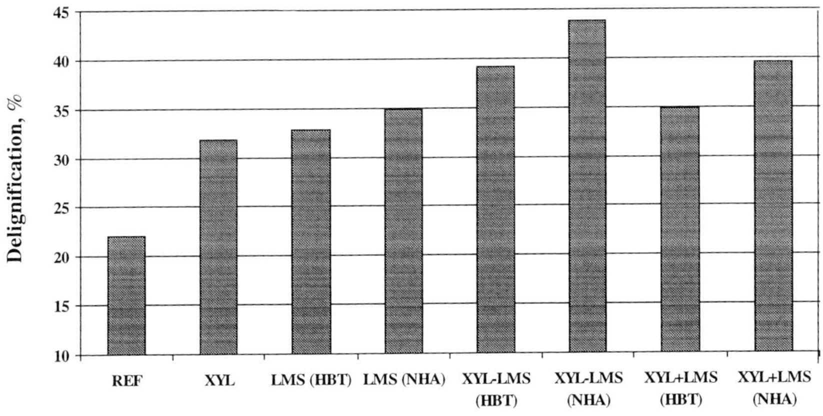

Figure 2. Sequential and simultaneous treatments of kraft pulp with xylanase and LMS, delignification after alkaline extraction of lignin. Original kappa number 23.5.

The additive effect of mannanase treatment on delignification by the hemicellulase-LMS system using only NHA as mediator was further studied. The mannanase treatment was less efficient than the xylanase treatment, decreasing the kappa number by one unit, and resulting to a modest increase in delignification, *i.e.* 25%. In the simultaneous treatment of pulp with mannanase and NHA, no further improvement in delignification was observed. However, when combining all treatments, the delignification was highest, about 42% after only alkaline extraction (Fig. 3).

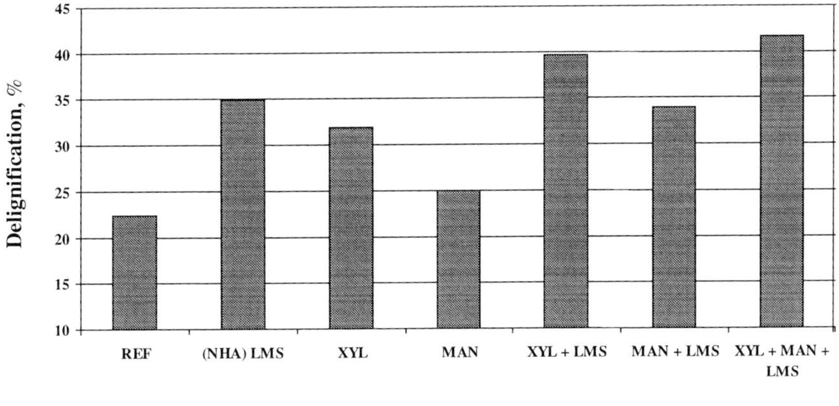

Figure 3. Simultaneous treatments of kraft pulp with xylanase, mannase and LMS, using NHA as mediator; delignification after alkaline extraction of lignin. Original kappa number 23.5.

The kappa number values followed the same trend after one stage hydrogen peroxide bleaching (Table 3). Again, the lowest kappa numbers were obtained after the successive xylanase and LMS treatments, using NHA as mediator. Even the simultaneous enzyme treatments, however, improved the efficiency of the LMS bleaching. The beneficial effects of combined treatments could also been seen in the brightness values of the pulps. The highest brightness values were observed in pulps treated sequentially with xylanase and the LMS using NHA as mediator. However, even the simultaneous treatment resulted in an almost as high brightness value using xylanase alone, and when enriched by mannanase an equally high brightness level was obtained as in the sequential treatments. In practise it would be beneficial to combine the two enzymatic treatments. These results also clearly imply that this could be most efficiently carried out by using NHA as mediator. The viscosities of the pulps treated with mediators studied here have previously been shown to be high, implying low or absent action on carbohydrates (12).

Table 3. The effect of xylanase, mannnase and LMS in different treatments on ISO brightness and kappa number after one-stage peroxide bleaching.

Treatment	Brightness ISO, %	Kappa number	Delignification, %
REF	52.1	13.5	42.5
XYL	55.4	11.4	51.5
MAN	54.1	12.7	46.0
XYL → L(HBT)	61.1	8.1	65.5
XYL → L(NHA)	62.2	6.9	70.6
L(HBT)	58.4	9.7	58.7
L(NHA	59.5	8.5	63.8
XYL + L(HBT)	58.3	9.4	60.0
XYL + L(NHA)	61.3	7.7	67.2
MAN + L(NHA)	61.6	7.8	66.8
XYL + MAN + L(NHA)	62.5	7.3	68.9

The effect of hemicellulases and LMS system on the hexenuronic acid content

As indicated by the analysis of hydrolysis products (Table 1), the hexenuronic acid (HexA) content of the pulp remained practically constant after the laccase-mediator treatments. The original content, 530 mg of hexenuronic acid per 100g of pulp, corresponding to about 6 moles HexA per 100 moles of xylan, did not decrease by the action of the two mediators tested (Table 4). Thus, the new mediator NHA was as specific towards lignin as HBT. The xylanase treatment resulted in a small decreasing effect of the hexenuronic acid content, about 15 %, corresponding to a kappa number decrease of 0.5 units.

The contribution of HexA to the kappa number of the pulps after the alkaline extraction was calculated (Table 4). The HexA content of the reference pulp contributed theoretically to 3.2 kappa units, whereas the partial hydrolysis of xylan in the pulp by the xylanase treatment decreased this value to 2.7, implying that the decrease of 0.5 unit in the kappa number by the xylanase treatment was due to the hexenuronic acid. The decrease of the kappa number after the xylanase treatments was, however, higher than this, due to the previously observed enhanced removal of lignin. The contribution by removal of lignins, attached to xylans, can neither be excluded.

Table 4. The content of hexenuronic acid in pulps after enzymatic treatments and alkaline extraction.

Treatment	HexA, mg/100g	Kappa no. after E	Contribution to kappa number by HexA	Contribution to kappa number by Lignin
REF	530	18.3	3.2	15.1 (82.5%)
XYL	450	16.0	2.7	13.3 (83.1%)
L(HBT)	570	15.8	3.4	12.4 (78.5%)
L(NHA)	520	15.3	3.1	12.2 (79.7%)

4. CONCLUSIONS

Hemicellulases and laccase-mediator systems could be succesfully combined and their effects were shown to be additive when the treatments were used sequentially. The mediator NHA was shown to be superior to the previously used mediator, HBT. The additive effect of the two enzymatic systems was slightly impaired when used simultaneously, due to the inactivating effect of the mediator on the enzymes. It can be concluded that the decrease of kappa number observed by the combined enzymatic system is based on three distinct mechanisms; enhancement of extractability of lignin by xylanases and mannanases, partial removal of hexenuronic acid by the xylan hydrolysing action of xylanase and direct degradation of residual lignin, accomplished by the laccase-mediator system.

REFERENCES

1. L. Viikari, A. Kantelinen, J. Sundquist and M. Linko. FEMS Microbiol. Revs. 13 (2-3), 335, 1994.
2. L. Viikari, M. Tenkanen and A. Suurnäkki. *In:* Biotechnology, Rehm, H-J. & Reed, G. (Eds) Vol 10: Special Processes. Weinheim-New York: Wiley-VCH, 523-546, 2001.
3. J. Buchert, G. Carlsson, L. Viikari and G. Ström. Holzforschung 50, 69-74, 1996.
4. T. Vuorinen, P. Fagerstöm, J. Buchert, M. Tenkanen and A. Teleman. J. Pulp Paper Sci. 25(5), 155, 1999.
5. L. Viikari, T. Oksanen, J. Buchert, M. Amann and A. Candussio. Proc. Int. Symp. Wood and Pulping Chemistry, Yokohama, June 7-10, Vol 1. 504-508, 1999.
6. R. Bourbonnais and M. Paice. Appl. Microbiol. Biotechnol., 36, 823, 1992.
7. H. P. Call and I. Mücke. Proc. Int. Non-Chlorine Bleaching Conf. Florida, USA, 1994.
8. H. P. Call. Proc. Int. Non-Chlorine Bleaching Conf., March 5-9, Florida, USA, Paper 4-3, 16 p., 1995.
9. J. Freudenreich, M. Amann, E. Fritz-Langhals and J. Stohrer. Proc. Int. Symp. Wood and Pulping Chemistry, June 9-12, Montreal, p. 71-76, 1997.
10. M. Amann, Proc. Int. Symp. Wood and Pulping Chemistry, June 9-12, Montreal, F4:1-5, 1997.
11. K. Poppius-Levlin, W. Wang, T. Tamminen, B. Hortling, L. Viikari and M-L. Niku-Paavola, J. Pulp Paper Sci, 25 (3), 90-94, 1999.
12. K. Poppius-Levlin, W. Wang, M. Ranua, M-L. Niku-Paavola and L. Viikari. Biol. Sci. Symp. Tappi Proc., 327-333, 1997.
13. T. Oksanen, M. Amann, A. Candussio, J. Buchert and L. Viikari. Proc. Int. Symp. Wood and Pulping Chemistry, June 9-12, Montreal, 83, 1-4, 1997.

14. P. Nelson, C. Chin, L. Viikari and M. Tenkanen. Appita J., 51(6), 451-455, 1998.
15. R. Pfaller, M. Amann and J. Freudenreich. 7th Int. Conf. on Biotechnology in the Pulp and Paper Industry, June 16-19, Vancouver, A99-A102, 1998.
16. M. Tenkanen, T. Hausalo, M. Siika-aho, J. Buchert and L. Viikari. Proc. Int. Symp. Wood and Pulping Chemistry, Helsinki, Finland, (III) p. 189-194, 1995.
17. M. Tenkanen, G. Gellerstedt, T. Vuorinen, A. Teleman, M. Perttula, J. Li and J. Buchert. J. Pulp Paper Sci. 25(9), 306, 1999.
18. M. Tenkanen, J. Puls and K. Poutanen. Enzyme Microb. Technol. 14, 566, 1992.
19. H. Stålbrand, M. Siika-aho, M. Tenkanen, and L. Viikari. J. Biotechnol. 29, 229-242, 1993.
20. M-L. Niku-Paavola, E. Karhunen, P. Salola, P. and V. Raunio. Biochem. J. 254, 877-884, 1988
21. P. W. Oxley, B. M. Adger, M. J. Sasse and M. A. Forth. Organic Synthesis, 67, 187-191, 1989.
22. J. Buchert, M. Siika-aho, J. Pere, A. Valkeajärvi and L. Viikari. Biotechnol. Techn., 7, 789-790, 1993.

Biotechnology in the Pulp and Paper Industry
L. Viikari and R. Lantto (Editors)

Mill usage and mechanistic studies of xylanase to enhance bleaching

Jeffrey S. Tolan and Corina Popovici
Iogen Corporation
300 Hunt Club Rd.
Ottawa, Ontario, Canada K1V 1C1

Xylanase enzyme-aided bleaching is the most widely used example of biotechnology in actual mill operations. About 20 bleached kraft mills in Canada, Finland, the US, and other countries use xylanase on a continuous basis. The mills use xylanase to decrease costs, decrease chlorine dioxide usage, increase pulp brightness, decrease effluent discharges, or obtain any of several other benefits. All of these benefits are obtained with low capital costs.

Xylanase is added to the pulp prior to bleaching. The enzyme does not bleach, brighten, or (for the most part) delignify the pulp. Rather, the removal of a portion of the xylan increases the efficiency of the subsequent bleaching chemicals. More specifically, xylanase treatment increases the alkaline extractability of the lignin, which decreases the amount of conventional bleaching chemicals required to bleach the pulp.

Xylanase is added to brownstock, typically just prior to the high density brownstock storage tower. The enzyme is a solution of active protein in water and is added to the brownstock decker, to the repulper, to the chute below the repulper, or to the suction head of the stock pump. The key issues in choice of location include corrosion of equipment, safety of acid use, and mixing of the enzyme into the pulp. The enzyme is usually added with acid and dilution water, to ensure control of the pH and adequate dispersion of the enzyme into the pulp, respectively. After enzyme addition, the pulp is pumped to the brownstock storage tower, where it traverses the tower while xylanase acts on the pulp. When the pulp emerges from the brownstock storage tower, it is ready to be bleached.

The treated pulp requires 10% to 20% less bleaching chemicals to reach a given brightness than the untreated pulp. Alternatively, the treated pulp can be bleached to a higher brightness than an untreated pulp. In some mills, a combination of higher brightness and lower chemical usage is obtained.

The enhanced bleachability by xylanase offers mills several options. The decreased use of chemicals results in a decrease in overall bleaching costs. This is the most common and most important benefit from using xylanase. The enzyme treatment can also extend the capacity of a ClO2 generator, in mills that are limited by ClO2 production. The increase in brightness ceiling can enable a mill to make a new grade of pulp. The decrease in ClO2 usage corresponds to a decrease in effluent discharges, including dioxin and AOX. Xylanase also decreases the TOX in pulp. Other mill-specific benefits are often observed with xylanase.

Table 1: Characteristics of mills with xylanase treatment

Property	Mills running xylanase
Chip furnish	Hardwood, softwood, eucalyptus
Digester	Batch, continuous, MCC
Additives	Surfactant, AQ
Brownstock	Conventional, O2-delignified
Bleaching	Chlorine, ECF, TCF
Bleach plant	3,4,5 stages
Extraction stages	E,Eo,Ep,Eop
Pulp	Semi-bleached, fully bleached

A primary benefit of xylanase is the low capital cost required for implementation of the technology. A second important feature is the versatility of the technology, as described in Table 1.

Figure 1 shows the 5 important areas of xylanase treatment in a bleach plant: 1) Pulp susceptibility; 2) Enzyme selection; 3) Mixing and dispersion; 4) Reaction conditions, and 5) Bleach plant control. We will summarize the mill issues involving each area and report data from mechanistic studies of xylanase in pulp susceptibility and selectivity. The paper concludes with mill studies of xylanase effects and bleach plant control.

1. PULP SUSCEPTIBILITY

By pulp susceptibility, we mean the amount of xylose released from the pulp for a given dosage and time of xylanase enzyme treatment. Tolan and Thibault **(1)** showed that the susceptibility of hardwood is higher than oxygen-delignified softwood, which is higher than softwood.

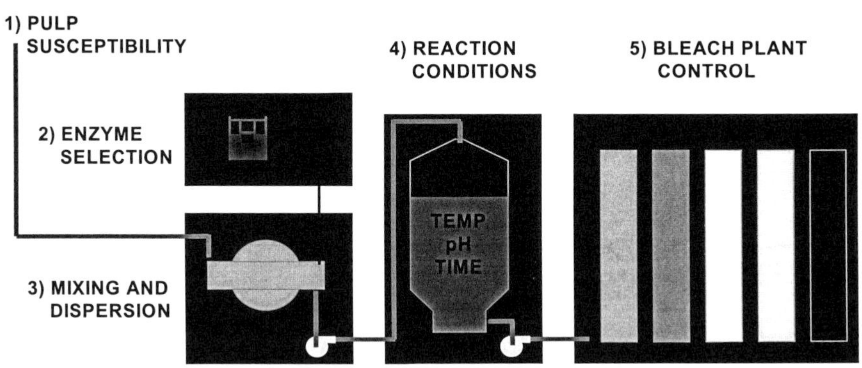

Figure 1. Xylanase implementation

A second property of the system is the selectivity, which relates the bleaching benefit to the amount of xylose released during xylanase treatment. The bleaching benefit is expressed as the decrease in the amount of bleaching chemical required to achieve a given level of brightness. A system with a high selectivity has a relatively high bleaching chemical savings per unit of xylose released.

This paper reports the results of measuring the selectivity of pulps from 28 mills. To measure selectivity, each pulp was treated with 0.1 to 5.0 xylanase units (xu) per gram of pulp of *Trichoderma* xylanase II, genetically modified for increased resistance to high temperature (**2**). Xylanase activity was measured by production of reducing sugars from oat spelt xylan; the xylanase preparation has an activity of 455 xu/milligram protein. Xylanase treatment was carried out at 10% consistency for 1 hour at 60°C, pH 7.5, after which time the amount of xylose released from the pulp was measured. The pulp was then bleached according to the mill's bleaching sequence and compared with an untreated pulp brought through the mill's bleaching sequence, to determine the percentage decrease in bleaching chemical requirements.

The results for softwood pulps are shown in Figures 2-4 and are grouped by kappa number as high (31-35), medium (27-30), and low (22-26), respectively.

In general, the low kappa number pulps exhibit the highest selectivity. The very good performance by the lowest kappa number pulps is somewhat surprising, in that one usually associates a high extent of delignification with xylan degradation, but not xylanase selectivity. There is a wide range of selectivity among all three groups of pulps.

The oxygen-delignified softwood (Figure 5) and the hardwood (Figure 6) also exhibit a wide range of selectivity. In general, the selectivity is better for softwood than for oxygen delignified softwood or hardwood. This is surprising, in that we often associate higher chemical savings with hardwood and oxygen delignified pulps than conventional softwood pulps.

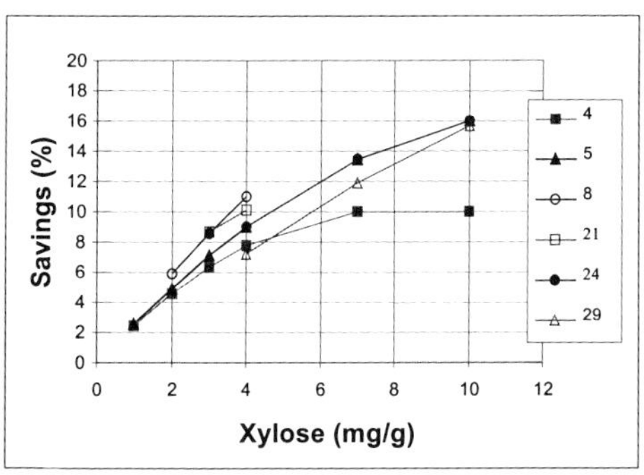

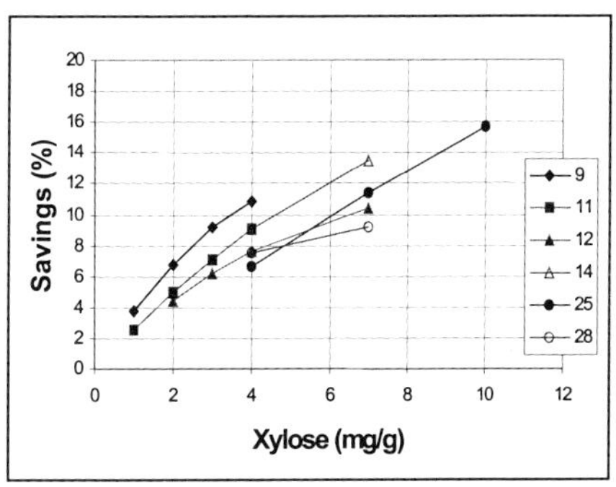

Figure 2. Selectivity of high Kappa number softwood treated with *Trichoderma* xylanase II.

Figure 3. Selectivity of medium kappa number softwood treated with *Trichoderma* xylanase II.

Note: The savings in bleaching chemicals is plotted as a function of the amount of xylose released by xylanase Treatment. For Figures 2-6, the numbers in the legend represent the number assigned to each mill whose pulp was used in the study.

kg/t, with a 5 kg/t scatter around these averages. During the early stages of xylanase treatment, no changes were made to the bleach plant control strategy. There was a slight decrease in chemical use, particularly around kappa number 16.0. This decrease in chemical use resulted from an increase in the D1 brightness, which caused the on-line sensors to automatically decrease the ClO2 charge to this stage. The ClO2 savings increased once the operators gained confidence in the xylanase treatment, and decreased their set points in the Do stage.

The second mill case study concerns a softwood mill in the southern US. Figure 8 shows the total ClO2 usage across the bleach plant, as a function of the brownstock kappa number. Xylanase treatment saves 2 kg/t of ClO2, but a very wide range of kappa numbers, from 19 to 27, are encountered and must be treated to observe this savings.

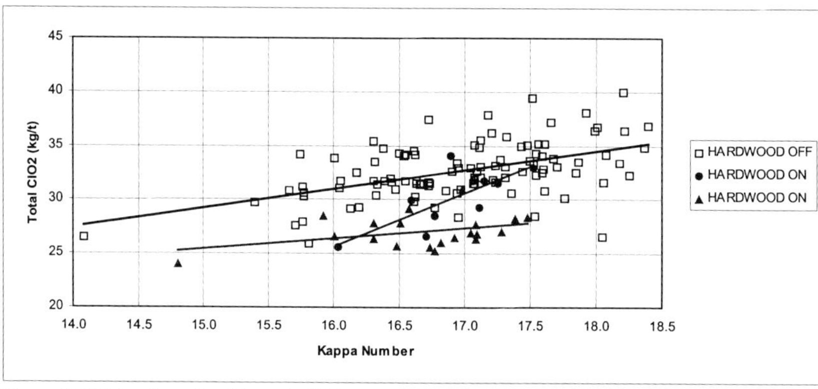

Figure 7. US hardwood mill: total ClO2 versus Kappa number. In the initial stages of the enzyme trial, ClO2 savings were moderate (circles). A higher level of savings was achieved after the Do set point was changed (triangles).

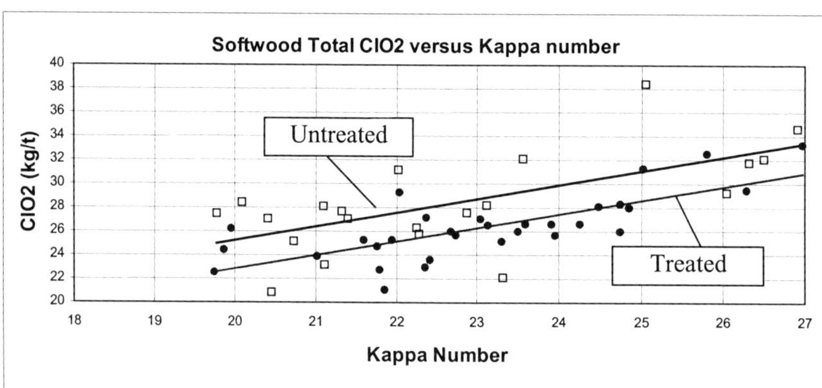

Figure 8. US southern softwood mill: total ClO2 versus Kappa number. A savings of 2.5 kg/t of ClO2 was achieved with xylanase treatment.

The third case study is a US softwood mill that tracks total active chlorine as a function of the brownstock kappa number (Figure 9). Xylanase treatment increases the bleached brightness, and this is taken into account in compensating the data in Figure 9.

The fourth case study involves a western Canada mill with oxygen delignification, that follows bleached brightness as a function of the total kappa factor across the bleach plant (Figure 10). A low dosage of xylanase of 120 ml/t increases the final brightness by 0.6 points ISO. A higher dosage of 200 ml/t increases the brightness by 1.0 point. This type of dosage optimization can be helpful.

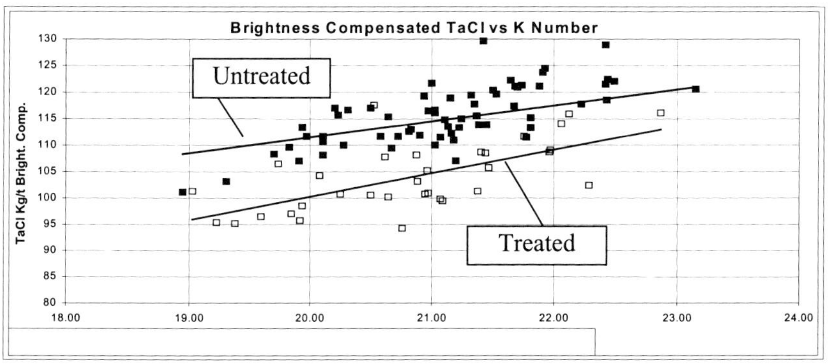

Figure 9. Softwood (500 ton/day): total active Cl versus K number. A bleaching chemical savings of about 11% was achieved with xylanase treatment.

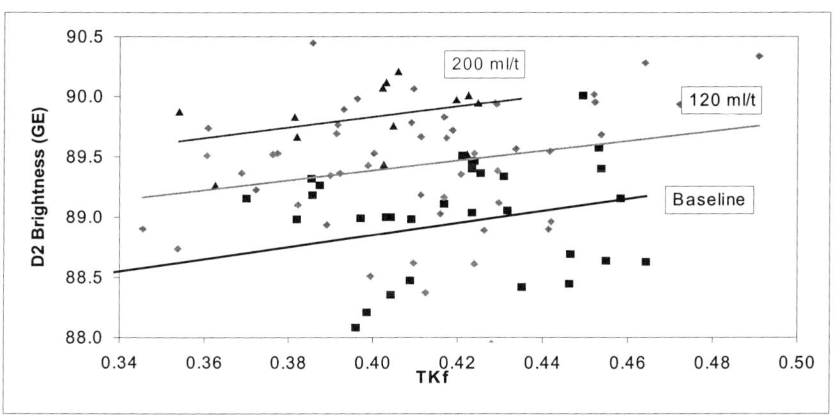

Figure 10. Northern softwood with O2 delignification (1000 ton/day). After an initial gain in brightness was achieved with 120 ml/t of xylanase, the dosage was increased to 200 ml/t and a bigger benefit was observed.

CONCLUSIONS

Xylanase aided bleaching is the most widely used example of biotechnology in actual mill operations. About 20 bleached kraft mills in Canada, Finland, the US, and other countries use xylanase on a continuous basis. The mills use xylanase to decrease costs, decrease chlorine dioxide usage, increase pulp brightness, decrease effluent discharges, or obtain any of several other benefits. All of these benefits are obtained with low capital costs.

The five important areas of xylanase treatment in a bleach plant: 1) Pulp susceptibility; 2) Enzyme selection; 3) Mixing and dispersion; 4) Reaction conditions, and 5) Bleach plant control. Increasing our understanding of selectivity and other aspects of the mechanisms of xylanase in bleaching offers the potential for improving benefits in a mill.

REFERENCES

1. TOLAN, J.S. and THIBAULT, L., Proc. 2000 TAPPI Pulping Conference, Boston, Mass.
2. SUNG, W., YAGUCHI, M., ISHIKAWA, K. US patent 5,866,408 (1999).
3. SUURNAKI, A., Li, T.-Q., BUCHERT, J., TENKANEN, M., VIIKARI, L., VURONINEN, T., ODBERG, L., *Holzforschung* 51:27-33 (1997).
4. TOLAN, J.S., TECH 2001 PAPTAC Pulp Bleaching Course Notes, section 15, October 2001.
5. TOLAN, J.S. and SPENCE, M., *Pulp and Paper Canada*, 2:36-40 (1998).

Biotechnology in the Pulp and Paper Industry
L. Viikari and R. Lantto (Editors)

Enzymatic control of dissolved and colloidal substances during mechanical pulping

J. Buchert[a], A. Mustranta[a] and B. Holmbom[b]

[a]VTT Biotechnology, PO BOX 1500, FIN-02044 VTT, Finland

[b]Åbo Akademi University, Process Chemistry Group, FIN-20500 Turku/Åbo, Finland

The presence of dissolved and colloidal substances (DCS) in the process waters during manufacture of wood-containing paper grades in an integrated paper mill may increase the chemical consumption in the wet end and affect the runnability of the process. Enzymatic modification of DCS has potential due to the specificity of the enzyme catalysis. In this paper the suitability of different enzymes for DCS control is discussed and reviewed.

1. INTRODUCTION

The presence of dissolved and colloidal substances (DCS) in the process waters during manufacture of wood-containing paper grades in an integrated paper mill may increase the chemical consumption in the wet end and affect the runnability of the process. The colloidal substances present in particles ranging in the size range from 0.1 - 1.5 μm are mainly composed of lipophilic extractives (1, 2, 3). These lipophilic extractives, commonly refered to as wood pitch, can cause deposit formation, web breaks and subsequently production downtime and need for extra cleaning. Pitch also impairs the product quality by causing dirt, holes and picking in the final sheet (4). The dissolved substances are mainly composed of carbohydrates released from the pulp during processing. In the pulping stage mainly glucomannan is dissolved in the process waters, whereas during alkaline peroxide bleaching changes in the DCS occur as dissolved glucomannan is partially precipitated to the fibres due to deacetylation and simultaneously pectin is dissolved to the waters after to demethylation (2, 5). These anionic pectin polymers consume cationic paper chemicals (6).

Enzymatic modification of DCS has potential due to the specificity of the enzyme catalysis. Lipase treatment is currently used e.g. in Japan and in China (7, 8). Pergalase treatment is also industrially used to improve the runnability, but this is method is based on the enzymatic hydrolysis of the amorphous cellulosic material from the fibres with concomitant effect on the glucomannans stabilizing the extractives (9, 10). In this paper the suitability of different enzymes for DCS control is discussed and reviewed.

2. MATERIALS AND METHODS

2.1. TMP pulp and DCS fraction

Unbleached thermomechanical pulp (TMP) produced from Norway spruce (*Picea abies*) was obtained from a Finnish paper mill. TMP was sampled after the second refiner at about 35% consistency. The DCS fraction was prepared from the pulp by diluting it to 1% consistency with distilled water whereafter the suspension was agitated for 3 hours at 60°C and 150 rpm. The suspension was centrifuged at 500 g for 30 min and the DCS fraction (supernatant) was separated (11). pH adjustment was carried out using either 1N HCl or 1N NaOH (the initial pH value of the DCS fraction was close to 5).

2.2. Enzymatic treatments

Laccase was partially purified from *Trametes hirsuta* (12). The laccase activity was measured according to Niku-Paavola *et al.* (13). Mannanase was purified from *Trichoderma reesei* culture filtrate as described previously (14). The mannanase activity was measured with locust bean gum as substrate according to Ståhlbrand *et al.* (15). Lipase from *Aspergillus* sp (Resinase A) was purchased from Novozymes (Denmark). The lipase activity was assayed by the olive oil emulsion method (16).

DCS fractions were treated with mannanase or laccase whereafter the treated DCS waters were combined to untreated TMP fibres and fines to a final consistency of 1% as described by Mustranta et al (17). In addition, a TMP suspension (1%) was treated with enzymes. Mannanase and lipase treatments were carried out at 50 °C and pH 5.0 for 2 h. Laccase treatments were carried out at room temperature with oxygen bubbling through the reaction vessel. The enzyme dosages were 1000-10000 nkat/l DCS water, or 100-1000 nkat/g TMP pulp. Reference treatments were performed under identical conditions without the addition of enzymes. Handsheets were prepared according to SCAN M 5:76 using Polymon PES-6/5/SR wire cloth in order to ensure retention of the fines in the sheets.

2.3. Analyses

The chemical composition of extractives in the TMP waters were analyzed by gas chromatography (GC) after extraction with MTBE (18). Technical sheet properties (brightness, light scattering coefficient and tensile index, wet strength) were measured according to SCAN P 3:93, SCAN C 27:76, SCAN P 38:80 and SCAN-P 20:95 standards, respectively. HPLC analysis of glucomanno-oligosaccharides was carried out according to Tenkanen et al (19). Friction of paper was measured with a PAAR RWP-apparatus. The principle of the measurement is to set a sledge covered by paper on an inclined plane that also is covered by the same paper. The angle of the plane is increased until the sledge starts to move. Static friction is the tangent of the measured angle.

3. RESULTS AND DISCUSSION

3.1. Enzymatic modification of extractives in DCS

Use of lipases for modification of lipophilic extractives

Lipases can efficiently hydrolyze triglycerides present in DCS (7, 17, 20). A commercial lipase product (Resinase) is currently on the market for this purpose and it is predominantly used for pitch control of groundwood pine pulp in Japan and also in China (7,

8, 20). Using lipase treatment it is possible to produce mechanical pulp from fresh pine wood without any seasoning. The lipase treatment allows savings in the consumption of white carbon, surface active chemicals and results in higher dynamic friction coefficient and higher brightness. The cleaning frequence and the number of stops has also been reported to decrease (7). The drawback of the Resinase is that it is unable to hydrolyze steryl esters and thus the lipase effect is restricted to about half of the lipophilic extractives (17).

The hydrolysis of triglycerides by Resinase has been reported to result in improved hydrophilicity of the fibres as measured by contact angle (17). This in turn resulted in improved bonding ability and tensile index (17). Increased amount of lipophilic extractives in the sheets is known to have a negative impact on strength properties of mechanical pulps (21). Lipase treatment has also been found to increase the friction coefficient, apparently due to decreased amounts of triglycerides in the sheets (Table 1). According to Sundberg et al (21) static friction decreases when the wood resin content increases. Lipase treatment of the extractives offers easily applicable means to render these components less harmful to the strength properties. According to the Resinase product sheet the optimal working conditions of the Resinase preparations are: temperature 50-70°C and pH 5-8. More thermophilic lipases are, however, needed to be easily incorporated into the current processes.

Table 1. Effect of lipase (Resinase) treatment of 1 % TMP pulp suspension (2h, pH 5, 60°C) on static friction of sheets.

Lipase dosage nkat/g	Fatty acids mg/g	Triglycerides mg/g	Lipophilic extractives mg/g	Static friction coeff.
0	0.1	0.2	0.5	0.71
200	0.4	0.1	0.9	0.83
500	0.2	0.1	0.6	0.85
1000	0.2	0	0.4	0.84

In addition to Resinase several other lipases have been tested for extractive control (7, 22, 23). All lipases could hydrolyze the triglycerides efficiently, whereas only partial hydrolysis of steryl esters could be obtained with certain lipases. Screening for novel steryl esterases is currently ongoing (24, 25, 26, 27). According to the results efficient hydrolysis with isolated or synthetized steryl esters in the presence of surface active agent can be obtained, whereas steryl esters present in water dispersion, such as in DCS, are only hydrolyzed to a lower extent (27, 28).

Modification of extractives with oxidative enzymes

The impact of laccase on DCS was also investigated. Model DCS water was treated with *T. hirsuta* laccase with an enzyme dosage of 1.0 nkat/ml. The final dosage thus corresponded to 100 nkat/g fibres.The laccase treatment was found to decrease the brightness, apparently due to the precipitation of polymerized lignan and lignin fragments to the fibres. The laccase treatment of the TMP pulp suspension resulted in impaired tensile strength of the sheets in both cases (Table 2). The wet strength of the sheets was, however, improved due to the laccase action and this improvement could be noticed in both pulp treatment and in DCS treatment (Table 3). The effect of laccase on the strength properties has been found to depend on the treatment conditions to a great extent and thus no clear conclusions on the impact on the strength properties can be drawn. Laccase-catalyzed modification of lignin containing fibres is currently investigated for different purposes and thus understanding of the role of

Mannanase treatment of the DCS resulted in decreased tensile indices of the sheets in both cases due to the destabilization of pitch particles to the fibres (Table 3). A decrease in the degree of bonding was also indicated by an increase in the light scattering coefficient. Thus, it seems that the hydrophilicity and also the location of the extractives present in the sheet plays a major role in the strength properties. The extractives can be rendered more hydrophilic by the lipase treatment, whereas mannanase treatment only decreases the colloidal stability of pitch particles thus facilitating their attachment to the fibres. According to Holmbom et al (2) glucomannan is sorbed to TMP fibres together with lipophilic extractives after peroxide bleaching. In this case, however, the strength properties were increased due to the coverage of the pitch particles with glucomannan.

Table 3. Effect of the mannanase treatment of DCS fraction or 1 % TMP pulp suspension on the sheet properties. Enzyme dosages: 10 nkat/ml DCS fraction, 1000 nkat/g pulp. Treatment conditions: 2 h, pH 5, 50°C.

Treatment	Enzyme	Density kg/m^3	Brightness %	Light scatt. coefficient m^2/kg	Tensile index Nm/g
DCS fraction	ref.	342	62.1	56.7	25.0
	mannanase	349	62.8	58.7	20.7
TMP pulp	ref.	309	61.3	57.7	24.3
	mannanase	325	62.3	59.7	17.3

The potential of side-group cleaving enzymes acting on O-acetyl-galactoglucomannans for DCS control has also been elucidated by Thornton et al (34) and Tenkanen et al (35). Enzymatic treatment of a TMP suspension with an isolated acetyl mannan esterase produced by *Aspergillus oryzae* resulted in about 87% deacetylation and subsequent decrease in the solubility of the dissolved galactoglucomannan. As a result half of the galactoglucomannan was adsorbed onto the TMP fibres. The yield gain was calculated to be 1% of which half was due to the precipitation of galactoglucomannan and the other half most probably due to the co-precipitation of lignin or extractives (34). Similarly to the enzymatic deacetylation, the chemical deacetylation of the solubilized galactoglucomannan during alkaline peroxide bleaching has been found to result in its adsorption onto the pulp fibres (2, 5, 36). Chemical deacetylation has certain disadvantages, such as nonspecific liberation of all esterified acids and dissolution of other wood components, for example polygalacturonic acids. Thus, in certain cases it might be beneficial to deacetylate the dissolved galactoglucomannan with a selective enzyme instead of using alkaline treatment.

Enzymatic modification of pectins

Cationic demand is known to be significantly increased after peroxide bleaching due to dissolution of anionic pectic acids (36). These pectic acids represent a major part of the anionic trash formed during mechanical pulping. By treating a DCS sample obtained from peroxide bleached TMP with a commercial pectinase preparation (Pectinex Ultra SP-L) a depolymerization of the pectic acids was obtained with subsequent decrease of the cationic demand from 431 μeq/l to 248 μeq/l (36). This due to the fact that low DP pectin oligosaccharides ar not able to react with cationic polymers. Reid and Ricard (38) have further investigated the impact of pectinase treatment on cationic demand using Canadian spruce pulp. The decrease in the cationic demand led to increased effectiveness of cationic retention aids in the treated pulps. The enzyme treatment was found to be effective when applied to either the bleached TMP or to the mixed stock with no damage to the pulp fibres.

The benefit of the pectinase treatment is that according to current knowledge pectinases are unable to act on fibre-bound pectin and thus the action of the enzyme is limited to the dissolved polygalacturonic acids (39). Subsequently the pectinase treatment does not damage the strength properties of the pulp (39). The drawback of the commercial pectinase preparation has been the low T optimae of the industrial pectinase preparations. However currently new pectinases with higher pH and T optimae are available for textile purposes (40) thus enabling their use for control of DCS.

4. CONCLUSIONS

Enzymes acting on DCS can be used to modify the properties of DCS. The chemical structure of the DCS affects, however, the choice of potential enzymes. The raw material used for the mechanical pulping as well as process conditions during pulping also effect the type of components that are dissolved. As an example, when mechanical pulp is produced from hardwoods, such as aspen, different types of components are dissolved as compared to spruce or pine pulping.

Enzymatic treatment of DSC may offer environmentally safe methods to improve pulp yield, decrease the effluent load and also to enable the closure of the water systems by decreasing the DCS present in the waters. Screening for more efficient enzymes, which would be compatible to the existing mechanical pulping process is also required. The effects of different enzymatic treatments on dissolved and colloidal subtances are summarized in Table 4.

Table 4. Potential enzymes for DCS control.

Acting on	Component	Enzyme	Chemical reaction	Technical effect
Lipophilic extractives	Triglycerides	Lipase	Hydrolysis of esters such as TGs: Liberation of fatty acids and glycerol	Strength improvement, runnability improvement
	Steryl esters	Steryl esterase	Hydrolysis of ester linkage	not determined
Hydrophilic extractives	Lignans	Laccase	Oxidation-> polymerization	Brightness decrease, wet strength improvement
Carbohydrates	Galactogluco-mannan	Mannanase	Hydrolysis	Decrease in turbidity of TMP filtrates, Destabilization of lipophilic extractives
		Acetyl mannan esterase	Liberation of acetyl groups -> precipitation of glucomannan	Yield increase
	Pectin	Polygalac-turonase (pectinase)	Decrease in DP	Decrease in cationic demand

5. ACKNOWLEDGEMENTS

This work is part of the top competence area "Enzymology of utilization of renewable materials" at VTT Biotechnology. The financial support of the National Technology Agency (TEKES) through "Enzymatic treatment of paper mill process waters" (EKI) -project is gratefully acknowledged. Riitta Alander is thanked for technical assistance.

REFERENCES

1. Allen LH (1975) Pitch in wood pulps. Pulp Pap Can 76(5): 70-77.
2. Holmbom B, Ekman R, Sjöholm R, Eckerman C, Thornton J (1991) Chemical changes in peroxide bleaching of mechanical pulps. Papier 45(10A): V16-V22.
3. Sundberg K (1995) Effects of wood polysaccharides on colloidal wood resin in papermaking. PhD thesis, Laboratory of Forest products Chemistry, Åbo Akademi University, Åbo 1995.
4. Dreisbach D, Michalopoulos D (1989) Understanding the behaviour of pitch in pulp and paper mills. Tappi J. 72(6):129-134.
5. Thornton J, Ekman R, Holmbom B and Örså F (1994a) Polysaccharides dissolved from Norway spruce in thermomechanical pulping and bleaching. J Wood Chem. Technol. 14: 159-175.
6. Thornton J, Ekman R, Holmbom B and Eckerman C (1993) Release of potential "anionic trash" in peroxide bleaching of mechanical pulp. Pap. Puu 75: 426-433.
7. Fujita Y, Awaji H, Taneda H, Matsukura M, Hata K, Shimoto H, Sharyo M, Sakaguchi H and Gibson K (1992) Recent advances in enzymatic pitch control.Tappi J. 75:117-112.
8. Chen S, Lin Y, Zhang Y, Wang X and Yang JL (2001) Enzymatic pitch control at Nanping paper mill. Tappi J. 84(4): 44-47.
9. Kantelinen A, Jokinen O, Sarkki M-L, Pettersson C, Sundberg K, Eckerman C, Ekman R and Holmbom B. (1995) Effect of enzymes on the stability of colloidal pitch. Proc. The 7th Int Symp on Wood and Pulping Chemistry ISWPC, Vol 1, pp.605-612.
10. Sundberg K, Thornton J, Holmbom B and Ekman R (1996) Effects of wood polysaccharides on the stability of colloidal wood resin.J. Pulp Paper Sci. 22(7): 226-230.
11. Ekman R, Eckerman C and Holmbom B (1990) Studies on the behaviour of extractives in mechanical pulp suspensions. Nord. Pulp Pap. Res. J 5: 96-102.
12. Poppius-Levlin K Wang W, Tamminen T, Hortling B, Viikari L, Niku-Paavola M-L (1999) Effects of laccase/HBT treatment on pulp and lignin structures. J Pulp Pap Sci 25: 90-94.
13. Niku-Paavola M-L, Karhunen E, Salola P and Raunio V (1988) Ligninolytic enzymes of the white-rot fungus *Phlebia radiata*. Biochem. J. 254: 877-844.
14. Rättö M, Siika-aho M, Buchert J, Valkeajärvi A and Viikari L (1993) Enzymatic hydrolysis of isolated galactoglucomannans from pine-wood and pine kraft pulp. Appl. Microb. Biotechnol. 40: 449-454.
15. Ståhlbrand H, Siika-aho M, Tenkanen M and Viikari L (1993) Purification and characterization of two endo-β-mannanases from *Trichoderma reesei*. J. Biotechnol. 29: 229-242.
16. Mustranta, A, Forssell P and Poutanen K (1993) Applications of immobilized lipases to transesterification and esterification reactions in nonaqueous systems. Enzyme Microb. Technol. 15:133-139.

17. Mustranta A, Buchert J, Spetz P, Holmbom B (2001) Treatment of mechanical pulp and process waters with lipase. Nordic Pulp and Paper Res. J. 16(2): 125-129.
18. Örså F and Holmbom B (1994) A convenient method for determination of wood extractives in papermaking process waters and effluents. J Pulp Paper Sci 20: J361-366.
19. Tenkanen M, Makkonen M, Perttula M, Viikari L and Teleman A (1997) Action of *Trichoderma reesei* mannanase on galactoglucomannan in pine kraft pulp. Journal of Biotechnology 57:191-204.
20. Hata K, Matsura M, Taneda H and Fujita Y(1997) Mill scale application of enzymatic pitch control during paper production. In: Advances in Biochemical Engineering/ Biotechnology, Vol 57. Ed. T. Scheper, Springer-Verlag, Berlin, Heidelberg, pp. 280-296.
21. Sundberg A, Holmbom B, Willför S, Pranovich A (2000) Weakening of paper strength by wood resin. Nord Pulp Pap Res J. 15(1):46-53.
22. Mustranta A, Fagernäs L ja Viikari L (1995) Effects of lipases on birch extractives. Tappi J. 78(2): 140-146.
23. Gao Y and Breuil C (1998) Properties and substrate specificities of an extracellular lipase purified from *Ophiostoma piceae*. World J. Microbiol. Biotechnol. 14: 421-429.
24. Buchert J, Mustranta A, Vuorentaso H, Spetz P, Ekman R and Holmbom B (2000) A novel enzymatic process for controlling pitch problems during paper manufacture. PCT/FI00/00180.
25. Kontkanen H, Mustranta A, Spetz P, Buchert J, Holmbom B and Tenkanen M (2001) Production of esterases hydrolysing steryl esters, Proc. 8th International Conference on Biotechnology in the Pulp and Paper Industry, Helsinki, Finland, June 4-8, 2001. pp. 189-190.
26. Kontkanen H, Tenkanen M, Fagerström R and Reinikainen T. Evaluation of steryl esterase activity of various commercial lipase preparations. Manuscript in preparation.
27. Tenkanen M., Kontkanen, H., Isoniemi, R., Spetz, P., Holmbom, B. and Buchert, J. (2001) Degradation of steryl esters by a lipase (Lip3) from *Candida rugosa* , Proc. 8th International Conference on Biotechnology in the Pulp and Paper Industry, Helsinki, Finland, June 4-8, 2001, pp. 191-192.
28. Buchert J, Mustranta A, Kontkanen H, Karlsson S, Tenkanen M and Holmbom B (2001) Enzymatic control of wood extractives. Proc. 11th ISWPC, Nice, June 11-14, Vol. 1, pp. 375-378.
29. Buchert J, Mustranta A, Spetz P, Ekman R, Luukko K (1999) Use of enzymes for modification of dissolved and colloidal substances in process waters of mechanical pulping. proc. Pre-symp. of the 10th ISWPC, Korea, pp. 115-119.
30. Buchert J, Mustranta A, Tamminen T, Spetz P, Holmbom B (2002) Modification of spruce lignans with *Trametes hirsuta* laccase. Holzforschung, accepted for publication.
31. Zhang X (2000) The effects of white-water dissolved and colloidal fractions on paper properties and effects of various enzyme treatments on the removal of organic components. Pulp & Paper Can 101(3): 59-62.
32. Karlsson S, Holmbom B, Spetz P, Mustranta A and Buchert J (2001) Reactivity of *Trametes* laccases with fatty and resin acids, Appl. Microbiol. Biotechnol. 55: 317- 320.
33. Pere J, Lappalainen A, Koljonen K, Tenkanen M & Mustranta A (2001) Combination of chemical and immunomicroscopical methods for surface characterization of mechnical pulp fibres. proc. 11th Int. Symp. on Wood and Pulping Chem. (ISWPC), June 11-14, 2001, Nice, France, Vol 1:259-262.

34. Thornton, J, Tenkanen M, Ekman R, Holmbom B and Viikari L (1994b). Possibility of increasing mechanical pulp yield by enzymatic treatment. Holzforschung 48:436 - 440.
35. Tenkanen M, Thornton J and Viikari L (1995). An acetylglucomannan esterase of *Aspergillus oryzae*; purification, characterization and role in the hydrolysis of O-acetyl-galactoglucomannan. J. Biotechnol. 42:197-206.
36. Thornton J, Eckerman C. and Ekman R. (1991) Effects of peroxide bleaching of spruce TMP on dissolved and colloidal organic substances. 6th Int. Symp Wood Pulping Chem. Proc. Appita 1: 571-577.
37. Thornton, JW (1994) Enzymatic degradation of polygalacturonic acids released from mechanical pulp in peroxide bleaching, Tappi J. 77(3):161-167.
38. Reid I. and Ricard M (2000) Pectinase in papermaking: solving retention problems in mechanical pulps bleached with hydrogen peroxide. Enzyme Microbial. Technol. 26: 115-123.
39. Mustranta A, Koljonen K, Holmbom B, Stenius P and Buchert J (1998) Characterization of the surface chemistry of PGW spruce pulps. Proc. 5th European Workshop Lignocellulosic and pulp, Aveiro, 30 Aug. -2. Sept. pp. 11-14.
40. Protonentis L and Lamm E Jr. (2000). Single bath pectinase biopreparation and dying of cotton. In: AATCC IC&E, Sep 17-20, 2000. Benton Convention Center Winston-Salem, NC. pp1-11

Biotechnology in the Pulp and Paper Industry
L. Viikari and R. Lantto (Editors)

Enhancement of TMP reject refining by enzymatic modification of pulp carbohydrates - A mill study

Jaakko Pere[a], Jouni Ellmén[b,*], Jukka Honkasalo[c], Pasi Taipalus[d] and Taisto Tienvieri[e]

[a]VTT Biotechnology, P.O.Box 1500, FIN-02044 VTT, Espoo, Finland
[b]Helsinki University of Technology, Laboratory of Pulping Technology, Espoo, Finland
[c]UPM-Kymmene, Rauma Paper Mill, Finland
[d]AB Enzymes, Rajamäki, Finland
[e]UPM-Kymmene, Corporate R & D, Valkeakoski, Finland

In this work, surface carbohydrates of TMP pulp were enzymatically modified prior to reject refining in mill scale and the effects on specific energy consumption and pulp quality were evaluated. The modification of pulp with CBH was performed in the reject tank at slightly acidic pH for 6-8 hours. The enzymatic treatment was followed by analysing the soluble carbohydrates liberated into white water. Minor modification of carbohydrates on TMP reject pulp brought about an increase of refiner load, which resulted in intensification of reject refining and to energy savings of about 10 to 15%. Importantly, the pulp quality was maintained and no deterioration of strength properties was detected. This enzymatic method turned out to be a reasonable altenative while searching new means for decreasing the energy consumption in mechanical pulping.

1. INTRODUCTION

The main drawback and hence a limitation to increase the use of mechanical pulps in papermaking is their high energy requirement. This concerns refining and especially thermomechanical pulping (TMP), which has a 30-40% higher specific energy consumption (SEC), as compared with grinding at the same freeness level (1). The reason for this high energy consumption in refining is still mainly unsolved, although improvements in energy efficiency has been obtained by process modifications.

From a chemical point of view, lignin is the wood component on which most attention in refining research has been paid. It is well described that reactions and softening of lignin in the phase of chip defibration are of key importance (1, 2). However, after defibration a substantial amount of energy is still needed for development of fibers suitable for papermaking, especially for production of wood containing paper grades, characterised by low freeness. In reject refining a part of the pulp is subjected to additional mechanical treatment in order to increase the flexibility of fibers and to create fines. The proportion of rejects is mill-specific, but in general rejects account for 30-50% of the total pulp in production of high grade printing papers, such as super calendered (SC) and light wet coated (LWC).

* Present address: Jaakko Pöyry Oy, Vantaa, Finland

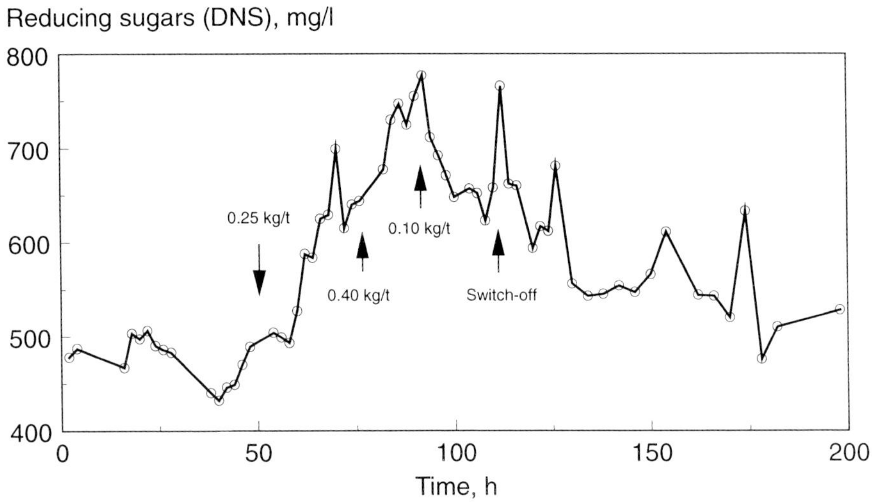

Figure 2. Concentration of soluble carbohydrates in white water samples taken from the bow screen during the first mill trial.

The white water samples taken during the first period were further analysed for cello-oligosaccharides by HPLC. It is noteworth that no soluble cello-oligosaccharides (DP $\leq$ 6) were identified in the white water prior to start-up of the enzyme period. In addition to cellobiose, which is the main hydrolysis product for cellobiohydrolases, also cello-tetraose was detected in the samples (Fig. 3). Cello-tetraose was present in the white water already before cellobiose and cello-tetraose could be detected in the samples also after cessation of enzyme addition. Obviously, dilution of hydrolysis products and/or inactivation of the enzyme in the circulating water system was rather slow. Comparison of data on cello-oligosaccharides and reducing sugars (Fig. 2) in the white water samples gave consistent results and thus, the activity of the enzyme in the process could be easily monitored by the DNS method.

In the second trial, the target was to reach a steady-state in the enzyme concentration present in the white water by decreasing stepwise the pumping rate, *i.e.* to find out the minimum effective enzyme dosage needed for reaching the improvement of refining. The enzyme addition compensates the part of enzyme which is adsorbed on pulp or further inactivated in the refiners and recirculating waters. The trial was started with the highest dose (0.4 kg/t) and thereafter the dosage was gradually decreased to 0.1 kg/t. Similar to the first trial, the accumulation of hydrolysis products in the white water was detected a few hours after the start-up of enzyme addition, whereafter the concentration of soluble products followed the changes in the enzyme dosage (results not shown). After three days´ trial with the dosage of 0.1 kg/t, the concentration of reducing sugars in the white water was stabilised at a level, which was about 30% higher than the background concentration. After termination of the enzyme addition, the concentration of soluble carbohydrates returned to the background level within two days.

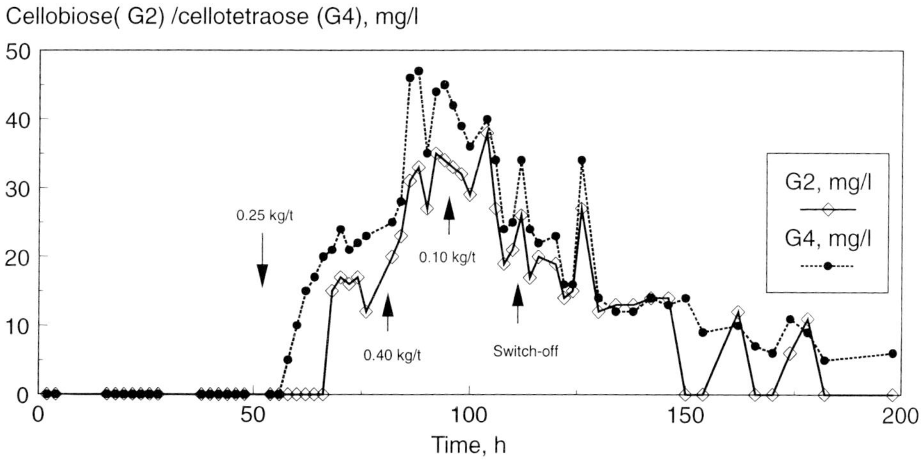

Figure 3. Concentrations of cellobiose (G2) and cello-tetraose (G4) in the white water samples taken from the bow screen during the first mill trial.

3.2 Refiner motor load and SEC

During the trials on-line data on refiner loads, SEC and freeness was obtained from the data acquisition system of the TMP line. Simultaneously with the appearance of soluble carbohydrates in the white water, an increase in motor load of the two reject refiners was detected (Fig. 4). Within few hours the refiner loads (1st and 2nd refiners) reached a level, which was 10-15% higher than before enzyme addition.

While the refiners were normally run for a constant freeness, the increased refiner load enhanced fiber development and consequently, a decrease in CSF was observed (Fig. 5). The reduction in CSF was compensated by opening of plate gap (Fig. 6). Thereafter, the CSF of the pulps started to increase back to the set values. The plate positions could be kept open till the cessation of enzyme addition, but thereafter they were tightened in order to maintain the target freeness (Fig. 6).

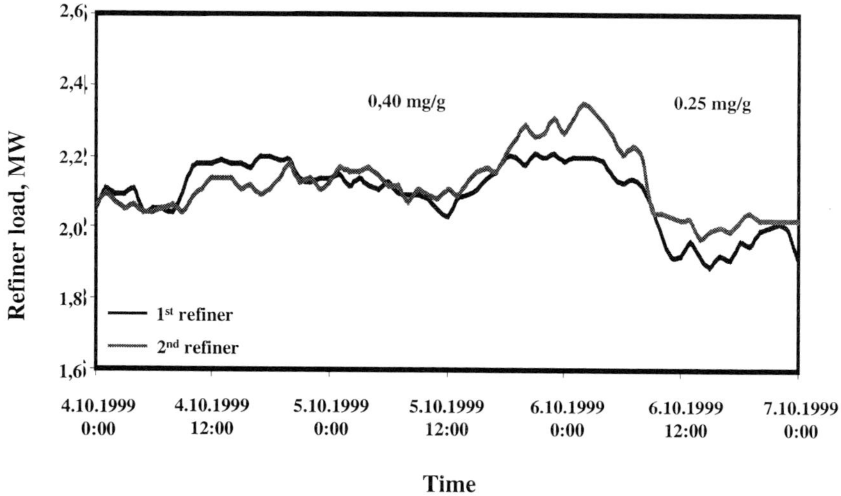

Figure 4. Data on motor loads of the refiners at the beginning of the second trial.

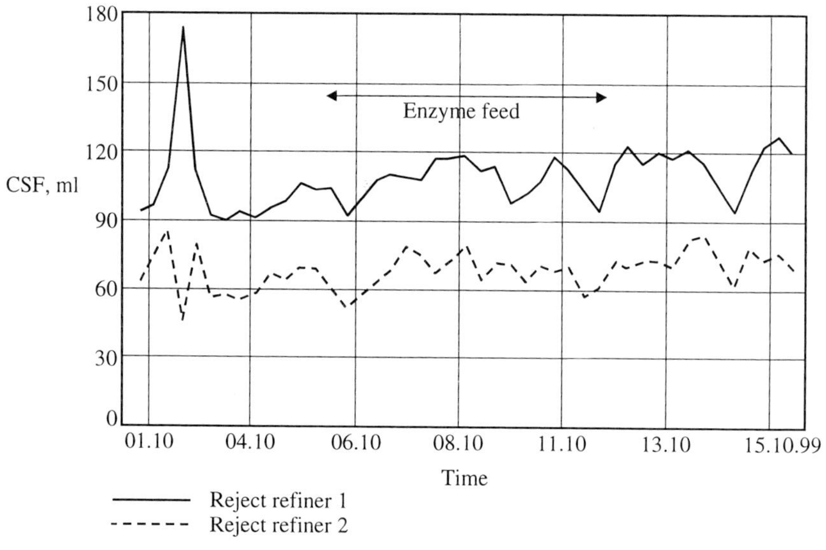

Figure 5. Freeness (CSF) of the pulp after the reject refiners during the second trial..

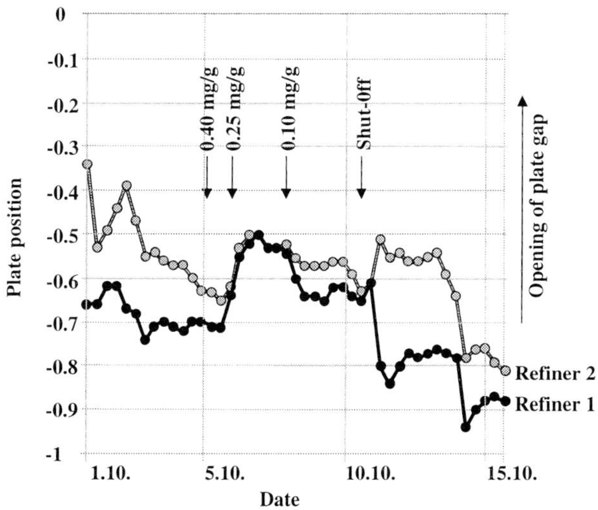

Figure 6. Plate postion of the refiners during the second trial. Changes in the dosage of the enzyme are shown in the figure.

As a consequence of the plate gap opening, the specific energy consumption (SEC, MWh/t) of the reject refiners decreased by 10-15%, as compared with the periods without enzyme addition. This is shown in Fig. 7, where the SEC of the reject refiners is shown for the second experimental period. First, SEC started to raise due to increased motor load, but decreased clearly after release of the refiner plates. Thereafter, the SEC was retained at a reduced level even with the lowest enzyme dosage used (0.1 kg/t). According to these results, 0.1 kg/t was a sufficient make-up dose of enzyme to generate and maintain the energy saving boosting of refining. After termination of the enzyme feed the SEC reached the original level, indicating a gradual dilution of the residual enzyme activity out of the system.

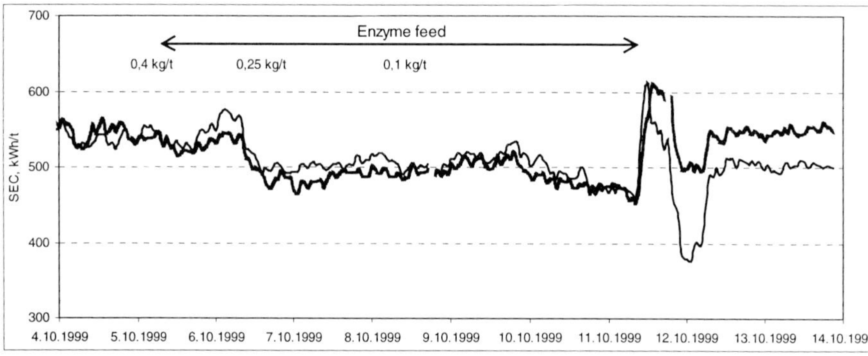

Figure 7. The specific energy consumption (SEC) of the reject refiners during the second trial.

3.3 Confirming refining tests

Because under normal mill operation on-line data collection of *e.g.* CSF and production is sensitive to disturbances, additional refining trials were performed, where data on motor load, production and CSF were collected off-line with varying refining intensities (plate gaps). These trials were carried out prior to, during and after the experimental period (enzyme addition) and the pulps were analysed in laboratory for CSF and handsheet properties.

As seen in Fig. 8 for SEC, scattering of the experimental points was rather large. This concerned especially the reference points, which deviated extensively from each other during the long observation period (about three weeks). However, average energy savings of 10 to 15% in the specific energy consumption (SEC) could be verified between the reference and enzyme periods.

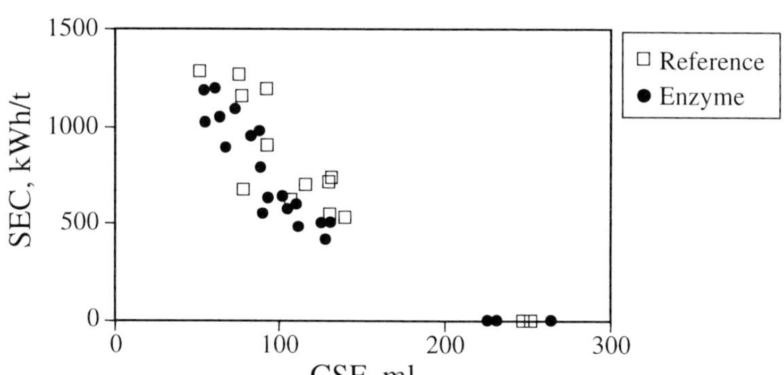

Figure 8. Energy consumption (SEC) and CSF during the enzyme and reference periods as obtained in the additional refining tests.

The quality of accept pulp was maintained good during the trials, and differences between the experimental and reference pulps were rather small. The tensile index of the handsheets is shown as a function of SEC in Fig. 9. As can be seen, no improvements in strength properties of laboratory sheets were detected unlike previously in the laboratory refinings, in which the unrefined reject was characterised by a high freeness, 600 ml CSF (4). Obviously, the main reason for the results in this work was the low CSF of the unrefined reject pulp (200-250 ml CSF), which reduced the modifying power of the enzyme. Obviously, the fines originally present in the unrefined reject adsorbed a part of the enzyme, diminishing its effects on the long fibres. The optical properties were in the same range with the reference samples, as well (results not shown). Additionally, no negative effects on the normal operation of the TMP plant or the paper-machine were observed during the trials.

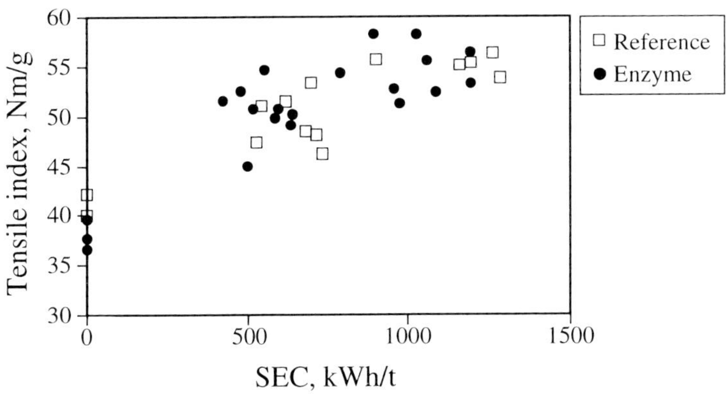

Figure 9. Tensile index of the handsheets as a function of SEC.

The mechanism of the enzymatic treatment is still unsolved. The increase in refiner loads, which was the primary effect of the enzyme, was in the same magnitude as earlier detected in laboratory with a wing defibrator (4). Interestingly, the same phenomenon was observed in two totally different refiners. The wing defibrator was operated batchwise at a high consistency (23%) and with a long residence (refining) time (up to 24 min), as compared with the continuously fed disk refiner. In both cases an increase of 10-15% in motor load in the beginning of the runs was observed for the enzymatically treated pulp as compared with the reference periods. It can be speculated that changes in the interfibre friction or modifications in the properties of flocs might be behind the observed effects. In conclusion, the observations might indicate a more universal phenomenon related to fibre development in reject refining, which needs to be studied in more detail.

4.CONCLUSIONS

Biomechanical pulping with the CBH rich preparation was verified to be a viable approach and applicable in reject refining of TMP pulp. The enzymatic pretreatment of pulp enhanced reject refining resulting in energy savings of 10-15%. These savings were not obtained without deteriorating the pulp quality. The enzymatic treatment was easy to combine

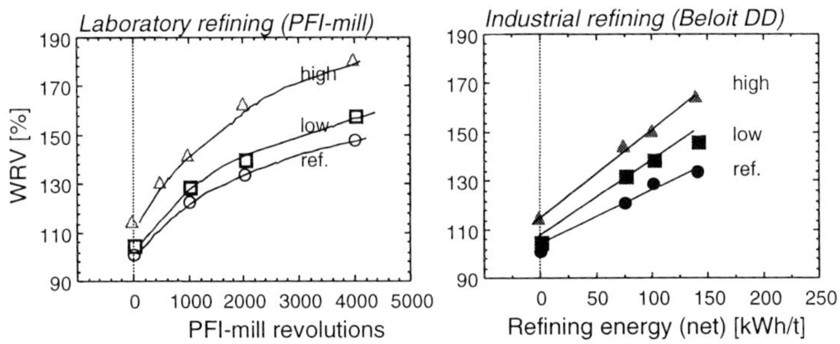

Figure 3. WRV of enzyme-treated pulps vs. energy input in PFI-mill refining (left diagram) and industrial refining (right diagram).

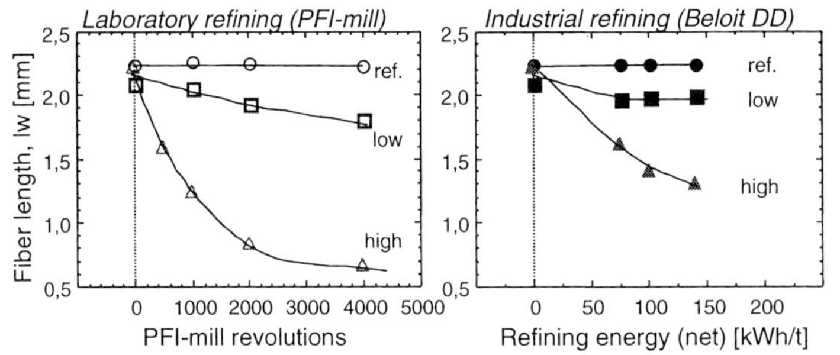

Figure 4. Fiber length of enzyme-treated pulps vs. energy input in PFI-mill refining (left diagram) and industrial refining (right diagram).

The energy reduction depended both on target WRV and the enzyme charge. During the industrial refining the need for energy was reduced with 65 kWh/t at WRV 140% (-40%) and with 45 kWh/t at WRV 120% (-70%) for an enzyme charge of 1 unit(CMC)/g. The WRV was, as will be shown later, a good indicator for sheet bonding.

Refining also reduced fiber length for the enzyme-treated pulps, but not for the reference pulp, *Figure 4*. The fiber length reduction was more severe for the high enzyme charge than for the low enzyme charge. Furthermore, the industrial refining did not reduce fiber length as much as the PFI-mill.

This is an observation that goes against most practical experience. It suggests that the fiber shortening obtained as a result of enzymatic treatment does not follow the same mechanism as the fiber shortening obtained in normal industrial refining. The laboratory refining treats the fibers in a more homogenous manner than the industrial refiner and will cause the fibers to break in all weak sites. In the industrial refiner some fibers will pass through the refiner without being treated.

The enzymatic treatment did not significantly affect fiber curl in itself, only indirectly because the new fiber segments created at fiber breakage had lower fiber curl than the original fiber.

3.2. Handsheet properties

Handsheet properties were studied both for the laboratory and for the industrially refined pulps. An important difference between laboratory and industrial refining is that the laboratory refining in the PFI-mill remove fiber curl and fiber kinks and generate a much straighter fiber than the industrial refining. The industrial refining can remove some fiber curl but are usually not as effective as the laboratory refining. The difference in tensile strength between the laboratory and the industrially refined reference pulps can, when compared at the same WRV, be explained by the differences in fiber curl (Mohlin, Miller 1992).

For the laboratory refined pulps, large differences were observed between the reference pulp and the enzyme treated pulps, *Figure 5*. The differences were much less for the industrially refined pulps. The low enzyme charge did not affect the relationship between WRV and tensile index or tensile stiffness index for the industrially refined pulps.

When the fiber shortening occurs in sites where a kink already exists, the effect on tensile strength and tensile stiffness will only be evident after laboratory refining as straightening occurs. For the industrially refined pulps the load-bearing ability of the fiber is already reduced by the kinks, and will not be impaired further by fiber breakage as a result of enzyme treatments.

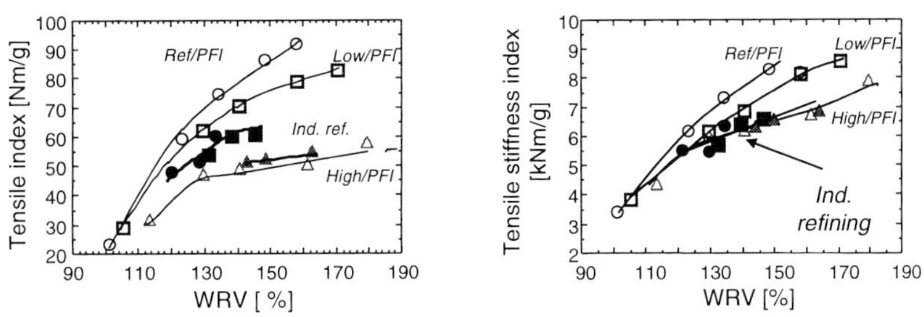

Figure 5. Tensile index and tensile stiffness index vs. water retention values for handsheets made from laboratory refined pulps (open symbols) (PFI) and industrially refined pulps (filled symbols). Legend: reference ●, low enzyme charge ■ and high enzyme charge ▲.

Tear strength is a property that is strongly influenced by fiber strength and fiber length, but not so much by fiber kink and curl. The reduction in tear strength due to the enzyme treatment was found to be similar for both types of refining, *Figure 6*. However, tear reduction was somewhat less for the industrially refined pulps, as the fiber length reduction also was less.

3.3. Industrial-like papermaking - drainage and sheet uniformity

Drainage and dryness after wet pressing are important aspects in papermaking process economy. Cellulase treatments have been used to improve drainage rates, but in those cases the enzyme treated pulps were not exposed to refining after the treatment. In this study no large effects on drainage rate could be observed.

Manual observations of the drainage on the wire gave the impression that the drainage was somewhat slower for the enzyme treated pulps. Dryness after wet pressing on the other hand was somewhat improved when compared at the same WRV, *Figure 7*. The enzyme treated fibers had about one percent-unit higher dryness than the untreated fibers. However, enzyme charge seemed to have little effect on the dryness.

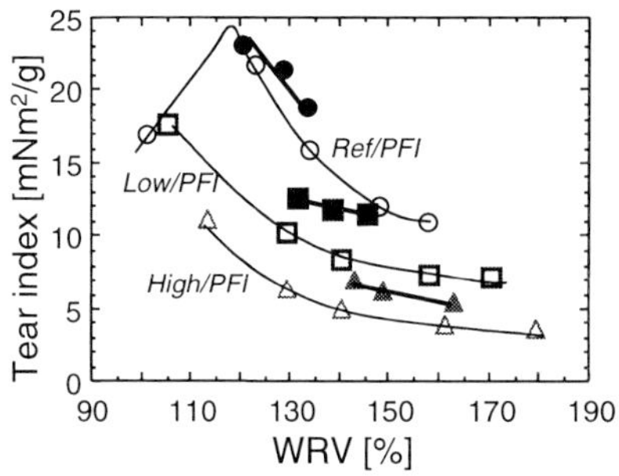

Figure 6. Tear index vs. WRV for handsheets made from laboratory and industrially refined pulps. Symbols see Fig 5.

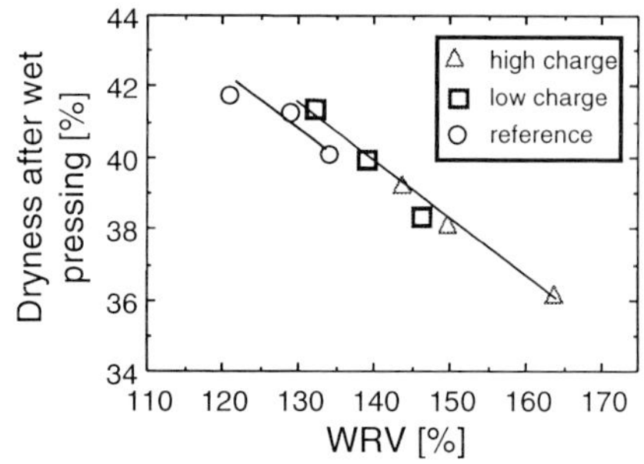

Figure 7. Dryness after wet pres-sing versus WRV.

Figure 8. β-radiogram for the untreated pulp to the left and high enzyme-treated pulp to the right, both refined to 100 kWh/t.

Sheet uniformity, formation, was evaluated by determining the coefficient of variation for basis weight, called formation number (Norman 1986). The method employed presents the formation number representing the total basis weight variation in the sheet and the numbers representing different size classes. These can be used to separate the effect of fiber flocculation from other effects of formation. The ratio of formation numbers for the two wavelength regions, 3 - 30 mm and 0,3- 3 mm respectively is used and is called the floc formation ratio (Mohlin 2001).

The fiber length reduction obtained by the enzyme treatment before refining had a large positive effect on sheet uniformity. Figure 8 shows the sheet basis weight variation for untreated pulp and for the high enzyme charge pulp, as revealed by the β-radiograms, while the relationship between the formation number and fiber length is shown in Figure 9.

Fiber length is very important for fiber flocculation, and Kerekes (1992) has derived an expression to describe the effect of consistency and fiber dimensions on fiber flocculation, the crowding number. According to his equation the crowding number is proportional to consistency and (fiber length)^2 and inversely proportional to fiber coarseness. In this trial fiber coarseness was not changed, so a linear correlation was observed between floc formation ratio and forming consistency times (fiber length)^2, see right diagram in figure 9. The sheet uniformity for the high enzyme charge is on the same level as would be observed for a 100 % hardwood sheet.

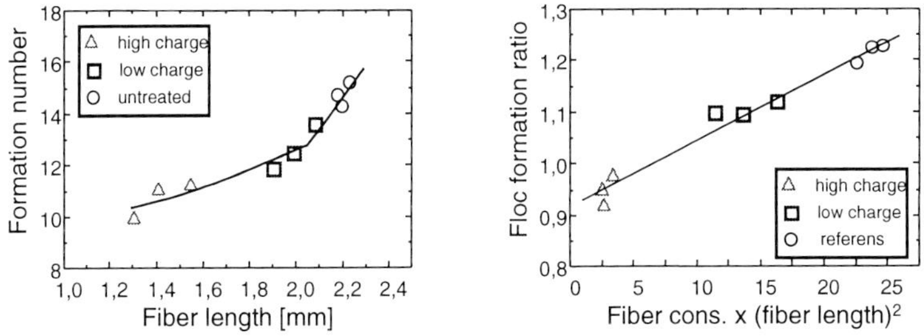

Figure 9. Formation number decreased with decreasing fiber length. Floc formation ratio was linearly related to fiber consistency x (fiber length $)^2$ as predicted by Kerekes crowding number.

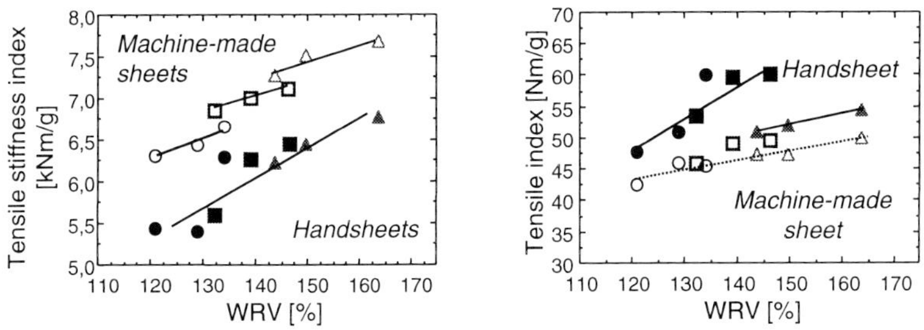

Figure 10. Tensile stiffness index and tensile index for the industrially refined pulps as a function of WRV. Machine-made sheets has open symbols and geometric means for MD and CD strength are used. Handsheets have filled symbols. Legend: reference● , low enzyme charge ■ and high enzyme charge▲ .

3.4. Properties of machine-made sheet

The properties of the machine-made-sheets further emphasized that the negative effects of the fiber weakening by the cellulase treatment was overvalued in laboratory evaluations. The shorter fibers decreased fiber flocculation, and this had a positive effect on strength properties. The effect was most pronounced for the high enzyme charge pulp. In the following figures strength of both machine-made sheets and handsheets illustrates how the sheet forming conditions changed the relation between reference pulp and enzyme treated pulps.

Tensile stiffness index for the machine made sheets showed an improved tensile stiffness index at similar WRV-levels for the enzyme-treated pulps, *Figure 10.*The higher overall level for the machine-made sheets is due to more restrained drying conditions for the machine-dried sheets. For tensile index all pulps showed the same relationship with WRV for the machine-made sheets. The general lower level for the machine-made sheets was due to the differences in forming consistency between the handsheets and the machine-made sheets.

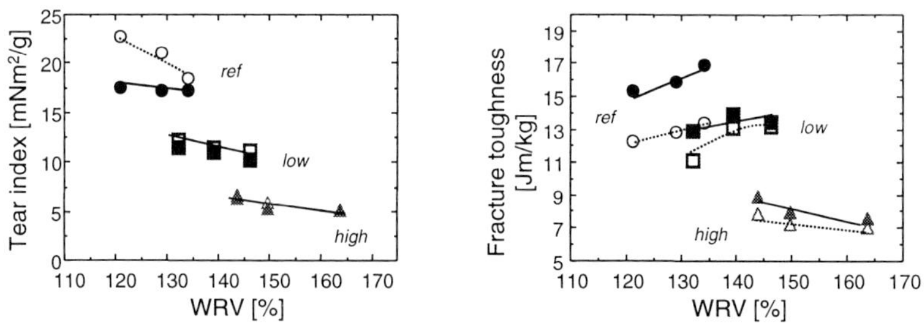

Figure 11.Tear index (geom. mean) and fracture toughness index (geom. mean) as a function of WRV for machine-made sheets (open symbols) and handsheets (filled symbols).

For tear index, which is very sensitive to the reduction in fiber length, very similar results were obtained for the machine-made sheet and for the handsheets, Figure 11. Another measure for runnability that is judged to be more relevant than tear index is fracture toughness. For this measure a difference between machine-made sheets and handsheets was observed for the untreated pulps but for the enzyme treated pulps similar results were obtained for the two sheet making methods. The results indicate that the low enzyme charge would have been acceptable also from a runnability point of view.

4. CONCLUSIONS

Enzyme treatment of the fibers had several positive effects on papermaking properties. The most important was the improved formation due to the fiber shortening. Another positive effect, was the lower need for energy in refining, as the cellulase treatment decreased the energy needed to the target WRV during refining. For enzyme used here, the reduction was 65 to 45 kWh/t for an enzyme charge of 1 unit(CMC)/g.

The enzyme treatment is suitable for sheets where formation, tensile stiffness and tensile strength are important, but should not be used where runnability and toughness are critical. Further trials are needed to evaluate different enzymes to find optimal combinations of enzyme preparations. It can also be expected that the intensity of refining can have some effect and that different fiber types will need different types of treatments. The benefits of an enzyme treatment would be decreased need for energy both in refining and in drying. The possibility for improved formation can in many cases be an interesting factor.

REFERENCES

Boucai, E. (1971). Pulp Paper Mag Can **72**(10): 73-80.

Gurnagul, N., D. H. Page, et al. (1992). Nord. Pulp Paper Res. J **7**(3): 152- 154.

Kantelinen, A., O. Jokinen, et al. (1997). Biological Sciences Symposium, Tappi.

Karlsson, H., P.-I. Fransson, et al. (1999). 6th Int. Conf. on New Available Technologies, Stockholm, SPCI, p 367

Kerekes, R.J., Schell, C.J:, (1992). J Pulp Paper Science **18**(1): J32-J38

Mohlin U.-B., Alfredsson, C., (1990)). Nord. Pulp Paper Res. J **5**(4): 172- 179
Mohlin U.-B., Miller, J., (1992) 4th Int Conf New Available Techn. and Current Trends, SPCI, Bologna, vol 2, 274-283
Mohlin U.-B., (2001). Nord. Pulp Paper Res. J, accepted for publication
Norman, B. (1986). in Paper: Structure and Properties.Ed. J. A. Bristow and P. Kolseth. New York, Marcel Dekker Inc: 123.
Pere, J., M. Siika-aho, et al. (1995). Tappi Journal **78**(6): 71-78.
Pere, J., M. Siika-Aho, et al. (2000). Tappi Journal **82**(11): 1-8.
Suurnäkki, A., M. Siika-aho, et al. (1997). 1997 Biological Sciences Symposium, Tappi.
Wong, K. K. Y. and S. D. Mansfield (1999). Appita Journal **52**(6): 409-414.

Biotechnology in the Pulp and Paper Industry
L. Viikari and R. Lantto (Editors)

The Effects of Recombinant *Cellulomonas fimi* β-1,4-glycanases on Softwood Kraft Pulp Fibre and Paper Properties

Shawn D. Mansfield[1], Neil R. Gilkes[2], R. Antony J. Warren[2] and Douglas G. Kilburn[2]

[1] Department of Wood Science and [2] Department of Microbiology and Immunology University of British Columbia, Vancouver, British Columbia, Canada

ABSTRACT

Recombinant *Cellulomonas fimi* β-1,4-endoglucanases (Cel5A and Cel6A) and cellobiohydrolases (Cel6B and Cel48A) were assessed for their capacity to selectively modify the physical and optical properties of handsheets derived from a fully bleached, never-dried softwood pulp. The isolated binding domain (Cel6A CBD) and catalytic domain (Cel6A CD) were also evaluated. Treatment with endoglucanases, particularly Cel6A CD, caused substantial damage to the intrinsic fibre strength of the kraft furnish, and consequently compromised the physical handsheet properties. In contrast, treatment with Cel48A produced beneficial modifications, including improvements in handsheet tensile strength following mechanical refining and various degrees of wet pressing. Cel6B and Cel6A CBD had very limited effects on the fibre characteristics, and therefore did not alter the quality of the handsheets produced. An examination of carbohydrate solubilization and changes in the degree of polymerisation of the polysaccharides indicated that the capacity for Cel48A to beneficially modify pulp and paper characteristics seems to be related to its capacity to selectively degrade cellulose-hemicellulose linkages, and release and modify xylan moieties. Furthermore, this enzyme preparation demonstrated limited carbohydrate dissolution, indicating that the observed modifications were attained with very little yield loss. These results suggest that Cel48A, and enzymes with similar glycanase activity should be considered and further evaluated for their potential to treat pulp fibres to improve paper properties.

1. INTRODUCTION

The process of mechanically beating or refining pulp imparts desirable alterations to fibres morphology and enhances the papermaking properties. These changes in fibre morphology (*i.e.* cutting and fibrillation) and the disruption of bonds within the secondary wall all contribute to the improvements in paper strength and optical parameters. However, this process requires substantial energy input, particularly in mills manufacturing paper containing mechanical pulps.

The potential of hydrolytic enzymes to decrease refining energy requirements and enhance fibrillation of pulps has been recognized for many years. A process using a cellulase complex from the white-rot fungus *Trametes suaveolens* to treat chemical pulp fibres was patented in 1968 [1]. Later, a mixture of xylanases and cellulases was shown to reduce the refining energy requirement for secondary fibre processing [2-4]. However, the use cellulase mixtures never gained acceptance because reductions in refining energy were always accompanied by lowered strength and yield due to the aggressive attack on fibres produced by the synergistic action of endoglucanases and cellobiohydrolases.

Subsequently, various researchers evaluated individual monocomponent cellulases in attempts to modify pulp fibres specifically and circumvent the reductions in strength or yield [5-7]. Studies on isolated *Trichoderma reesei* cellulase monocomponents suggest that endoglucanases are more destructive than the predominantly exo-acting cellobiohydrolases. For example, no major differences in strength resulted from treatment with Tr Cel7A (previously called cellobiohydrolase I), while Tr Cel5A (previously endoglucanase II) caused severe fibre damage that could not be restored by beating [5]. Further research confirmed that endoglucanase treatment lowered both fibre and paper strength, but also showed that the extent of degradation varied with the type of endoglucanases employed [7]. These data provide a rationale to screen other, related enzymes for useful fibre modifying properties.

To date, the search for appropriate "fibre modifying" enzymes has focused on the components of fungal cellulase systems, while bacterial cellulases remain relatively unexplored. The extent to which bacterial cellulase systems resemble those of fungi is still unclear. However, some differences between fungal and bacterial systems are evident. For example, cellulolytic fungi such as *Trichoderma* and *Humicola* spp. use the concerted action of cellobiohydrolases belonging to glycosyl hydrolase families 6 and 7, together with a range of endoglucanases, to solubilise cellulose. Although many cellulolytic bacteria produce a family 6 cellobiohydrolase, they appear to lack family 7 enzymes. However, several well-characterized bacterial systems contain cellulases from family 48, not represented in fungi, which appear to be functional equivalents of the family 7 enzymes [8]. Given such differences, bacterial cellulase systems may prove to be promising sources of enzymes with novel fibre-modifying properties.

This study examines the effects of cellulases from the bacterium *Cellulomonas fimi* on the properties of kraft pulp. The enzymes studied include Cel48A and Cel6B (cellobiohydrolases belonging to families 48 and 6, respectively) and Cel6A and Cel5A (endoglucanases belonging families 5 and 6). In the case of Cel6A, the effects of its isolated catalytic and cellulose-binding domains were also examined independently. The properties of handsheets prepared from enzyme-treated pulp, with and without refining, and after application of various levels of wet pressing pressure, were examined. The observed changes in handsheet properties were related to changes in the degree of polymerisation and amounts of polysaccharide dissolution in an attempt to elucidate mechanisms of pulp fibre modification at the molecular level.

2. MATERIALS AND METHODS

2.1. Pulp

Never-dried, fully bleached kraft pulp containing approximately 85% Douglas-fir was obtained from Weyerhaeuser Inc. (Tacoma, Washington). The carbohydrate composition of the pulp was 0.6% arabinose, 0.8% galactose, 82.4% glucose, 7.0% xylose, and 7.2% mannose residues, as determined by HPLC following secondary acid hydrolysis [9].

2.2. Enzymes

C. fimi genes encoding Cel6A (previously called CenA), Cel5A (previously CenD), Cel6B (previously CbhA) and Cel48A (previously CbhB) were expressed in *Escherichia coli*, and the corresponding enzymes purified as previously described [10-13]. The Cel6A catalytic domain (Cel6A CD) and cellulose-binding domain (Cel6A CBD) were isolated by digestion with *C. fimi* protease and size-exclusion chromatography [10].

2.3. Enzymatic treatment of pulps

Pulp slurry in 5 mM potassium phosphate buffer was adjusted to pH 7 with sulphuric acid and autoclaved for 5 min. Enzyme (1.0 mg/g dry weight of fibre for Cel6A, Cel5A, Cel6A CD and

Cel6A CBD; 0.5 mg/g for Cel6B and Cel48A) were added in the potassium phosphate to bring the preheated (37°C) pulp sample to the target consistency of 3% (w/v). The slurries were incubated for 2 h at 37°C with continuous stirring. Following enzymatic treatment, the filtrates were collected for analysis and the pulps were washed 3 times with water to remove residual protein, and then autoclaved for 15 minutes to inactivate any residual enzyme. Control pulps were prepared under the same conditions, without enzyme addition.

2.4. Papermaking and testing

Handsheets were prepared and tested according to standard Tappi Test Methods, with the white water recirculated to ensure that both the primary and secondary fines were included in the resultant handsheet. Handsheets of unrefined and refined furnishes were subject to wet pressing at 40, 90 and 140 psi during sheet making. A PFI laboratory refiner (Tappi Test Method T 248 cm-85) was used to prepare the refined furnish to 2000 revolutions. Pulp freeness was measured at 20°C according to Tappi Test Method T227 om-94.

2.5. Fibre analysis

Fibre coarseness was determined by passing more than 20,000 fibres, in exact aliquots of approximately 5 mg of pulp through a Kajaani FS-200 (Kajaani, Finland) fibre analyzer.

2.6. Analysis of soluble carbohydrates

Filtrates were analysed by high performance anion-exchange chromatography (HPAEC) after secondary acid hydrolysis on a CarboPac PA-1 column using a Dionex DX-500 system equipped with a pulsed amperometric detector (Dionex, Sunnyvale, CA, USA).

2.7. Determination of degree of polymerisation

Molecular mass distributions were determined by gel permeation chromatography of tricarbanyl derivatised samples. Pulp samples were carbanylated [14] and the derivatised material was recovered by evaporation. Samples were redissolved in tetrahydrofuran to approximately 0.2 mg/mL and filtered through a 0.45 μm pore Teflon membrane. Samples were analysed on a Waters 625 liquid chromatography system (Millipore Corp., Milford, MA) equipped with four TSK-gel columns (Varian, Sunnyvale, CA). The columns (G1000 HXL, G3000 HXL, G4000 HXL and G6000 HXL) were connected in series and had nominal molecular weight cut-offs of 1×10^3, 6×10^4, 4×10^5, and 4×10^7, respectively. Samples were eluted with tetrahydrofuran at a flow rate of 1 mL/min and tricarbanylates quantified by absorption at 254 nm using a Waters 486 UV spectrophotometer. The column series was calibrated using polystyrene standards [15]. The degree of polymerisation (DP) was calculated by dividing the molecular weight of the tricarbanylated derivative by the molecular weight of tricarbanylated anhydroglucose ($DP = M_w/519$).

3. RESULTS AND DISCUSSION

The primary function of refining is to modify the structure of the fibre in order to enhance their papermaking potential. Subsequently, wet pressing removes water and consolidates the wet web prior to drying. Wet pressing results in several physical and topographical changes in the fibre network, which affect the properties of paper and the recycling potential of the fibre. Refining and wet pressing both significantly influence sheet formation and wet web strength and hence machine performance. The investigations described below are concerned with the effects of individual bacterial cellulases (or their isolated functional domains) on handsheets produced from unrefined and refined fibres using varying degrees of wet pressing during sheet formation.

3.1. Effect of enzymatic treatments on properties of unrefined handsheets

The most significant effect of wet pressing is sheet densification [16], and it is clear that higher pressing pressure resulted in increased sheet density in all samples (Fig 1A). Pretreatment with endoglucanase monocomponents (Cel5A, Cel6A or Cel6A CD) enhanced densification at both low and medium pressing pressures, but had no significant effect at higher pressure. Similar effects following endoglucanase pretreatment have been reported previously [5-7]. Pretreatment with cellobiohydrolases (Cel6B and Cel48A) had little or no effect on sheet density (Fig.1A), in agreement with previous studies using fungal cellobiohydrolases [5, 7]. Cel6A CBD was similarly ineffective.

Air resistance increased with sheet density in all samples, in agreement with previous data showing that wet pressing results in substantial intra- and interfibre pore closure [17]. It was apparent that at a given sheet density, the air resistance of the resultant handsheets was enhanced slightly by selective enzyme treatment, namely Cel5A, Cel6A CD and Cel48A (Fig. 1B).

The degree of interfibre bonding within a handsheet can be inferred from measurement of light scattering: a decrease in the scattering coefficient indicates greater fibre-fibre contact within the sheet, as light is scattered at air-fibre interfaces [18]. As expected, scattering coefficients were reduced in all samples at high sheet densities due to enhanced bonding of densely packed fibres (Fig. 1C). However, treatment with the endoglucanases (Cel5A, Cel6A, Cel6A CD) and the cellobiohydrolase Cel48A resulted in significantly larger reductions at higher sheet densities, relative to control handsheets.

The observed changes in densities and scattering coefficients suggested that some enzyme treatments enhanced intrinsic fibre flexibility, making the fibres more collapsible during papermaking. To investigate these effects further, various handsheet strength parameters were determined and plotted against apparent sheet density. In general, all enzyme-treated samples showed improved tensile strength at a given density relative to the control at the lowest pressing pressure (Fig. 2A). However, at the mid-range pressing pressure (90 psi), only the Cel6A, Cel6A CD, Cel5A and Cel48A preparations showed improvements, while at the high pressing pressure (120 psi) only the Cel6A CD demonstrated higher values than the corresponding control. The greatest improvements observed (Cel6A, Cel6A CD and Cel5A), corresponded to those enzymes shown to produce the largest increases in densification. In contrast, tear strength (Fig. 2B), a parameter directly influenced by intrinsic fibre strength and enhanced inter-fibre bonding, was adversely affected by endoglucanase treatment, while treatment with cellobiohydrolase Cel48A or the Cel6A CBD gave marginal improvements. Wet zero-span breaking strength determinations (Fig. 2C) showed that the observed tear strength reductions in endoglucanse-treated samples were the direct result of significant losses in fibre strength. No losses in fibre strength were seen in other samples.

3.2. Effect of enzymatic treatment on properties of refined handsheets

Having established that treatment with *C. fimi* cellulases caused significant modifications to handsheet properties; the response to refining was investigated. Samples were subjected to 2000 revolutions in a laboratory PFI refiner, and then wet pressed at different levels prior to drying.

As with unrefined fibre, treatment with endoglucanases (Cel6A, Cel6A CD and Cel5A) produced significant increases in sheet density relative to the control, while the effects of Cel6A CBD and the cellobiohydrolases Cel6B were minimal (Fig. 3A). Generally, the degree to which an enzyme affected density was correlated with its ability to increase air resistance (Fig. 3B) and decrease the scattering coefficient (Fig. 3C), strongly supporting the theory of enzyme-induced fibre collapsibility and flexibility previously proposed [19].

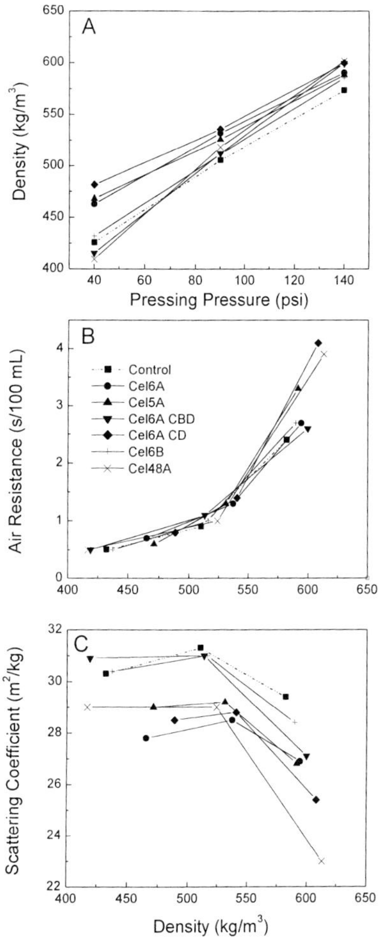

Figure 1. Properties of handsheets prepared from cellulase-treated kraft pulps without refining

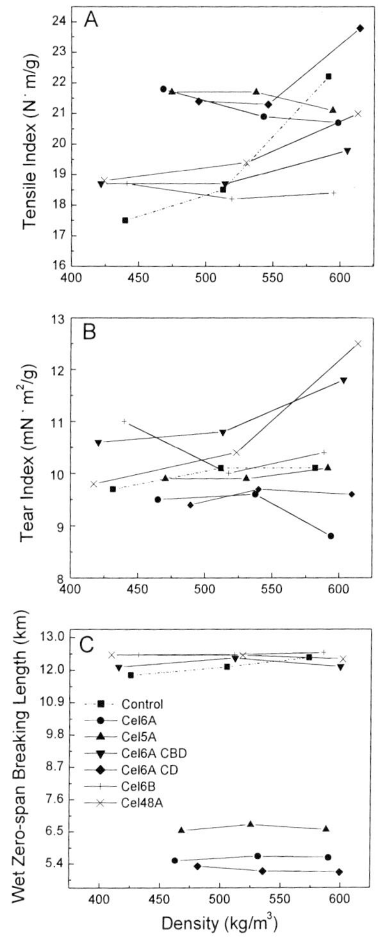

Figure 2. Properties of handsheets prepared from cellulase-treated kraft pulps without refining

Following refining, the extent of wet pressing did not significantly alter the strength properties of either the enzyme-treated control fibres (Fig. 4). However, it was clear that the tensile strength of the sheets was significantly compromised by the application of endoglucanase preparations with the greatest damage resulting from treatment with Cel6A CD, followed by Cel6A and Cel5A (Fig. 4A). In contrast, treatment with the cellobiohydrolase Cel48A resulted in improved tensile strength, while Cel6B treatment was ineffective. As with unrefined fibre, the tear strength (Fig. 4B) and intrinsic fibre strength (Fig. 4C) of the refined furnish was substantially degraded by endoglucanase treatments, while the cellobiohydrolase-treated and Cel6A CBD-treated samples retained similar properties to those of the control.

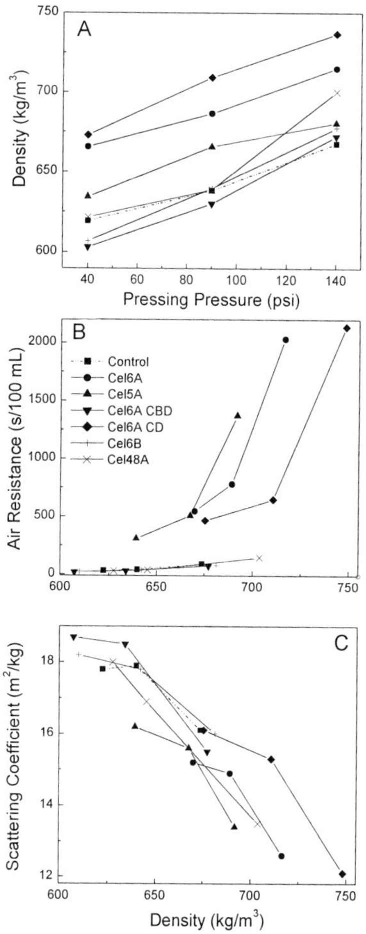

Figure 3. Properties of handsheets prepared from cellulase-treated kraft pulps after 2000 revolutions PFI refining.

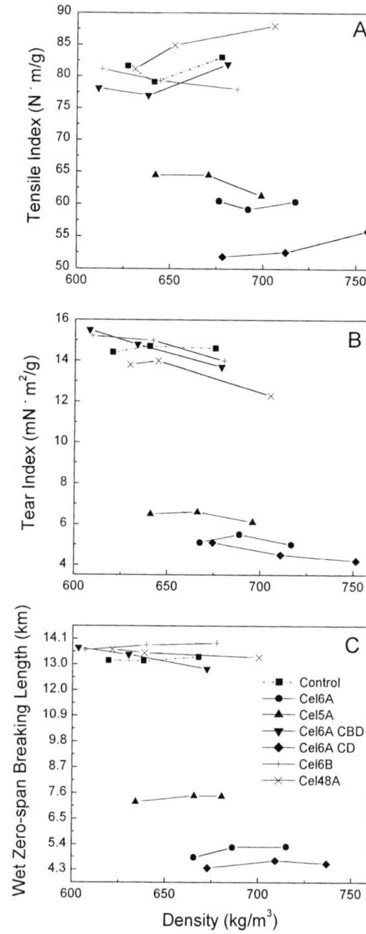

Figure 4. Properties of handsheets prepared from cellulase-treated kraft pulps after 2000 revolutions PFI refining.

3.3. Fibre modifications

The influence of fibre morphology on the physical and optical properties of paper has been thoroughly investigated. In general, finer fibres are more flexible and collapsible and produce denser sheets during papermaking. It has previously been shown that reductions in the coarseness of Douglas-fir kraft pulp by treatment with fungal cellulase enhanced subsequent densification [9, 20]. Similarly, in this investigation, *C. fimi* cellulases also caused small reductions in the coarseness of the unrefined furnish (Table 1). PFI refining alone slightly reduced control pulp coarseness, but the combination of refining and endoglucanase treatment resulted in significant (up to 20% with Cel6A CD) reduction in fibre coarseness (Table 1) and fibre length distribution (data not shown). In contrast, no discernable changes were detected with cellobiohydrolase treatments. These data indicate that the endoglucanase component(s) of complete cellulase complexes may be responsible for the reductions in fibre coarseness previously observed with cellulase complexes.

Table 1. Fibre coarseness and freeness of kraft pulp treated with *C. fimi* cellulases

Enzyme	Unrefined coarseness (mg/m)	Refined coarseness (mg/m)	Unrefined freeness (mL)	Refined freeness (mL)
Control	0.310 ± 0.02	0.293 ± 0.06	738	457
Cel6A	0.294 ± 0.05	0.242 ± 0.04	734	135
Cel5A	0.282 ± 0.01	0.253 ± 0.01	747	176
Cel6A CBD	0.286 ± 0.06	0.298 ± 0.05	717	464
Cel6A CD	0.286 ± 0.07	0.229 ± 0.07	755	128
Cel6B	0.295 ± 0.04	0.289 ± 0.04	744	450
Cel48A	0.300 ± 0.09	0.303 ± 0.07	752	498

3.4. Pulp Freeness

Pulp freeness or drainability, as measured by Canadian Standard Freeness (CSF) is an empirical measure of the ease with which water drains from a pulp suspension. This parameter is directly influenced by both the extent of refining and the amount of recycled fibre found within pulp suspensions, as the presence of fines and/or highly fibrillated fibers decreases pulp freeness.

Comparative studies have been initiated to determine what effects individual cellulase monocomponents had on freeness [6]. These results indicated that endoglucanases were more effective than cellobiohydrolases. Furthermore, cellobiohydrolases have been shown to act synergistically with endoglucanases to increase freeness, and their combined actions resulted in increased sugar liberated and yield loss [21]. The employment of *C. fimi* cellulase components did not reveal any discernible differences in the effects that endoglucanases and cellobiohydrolases have on the freeness of the unrefined furnish. Pulp freeness following refining indicated that only the endoglucanases (Cel5A and Cel6A) and the catalytic domain (Cel6A CD) could significantly alter this process parameter. Pretreatment with Cel6B and Cel48A, as well as the purified cellulose binding domain (Cel6A CBD) had practically no effect on the development of pulp freeness (Table 1), corroborating with previous studies [21, 22] examining the effects of *Trichoderma reesei* monocomponent cellulases on pulp properties.

3.5. Carbohydrate solubilisation and modification

Analysis of the carbohydrates solubilised by *C. fimi* cellulases showed that all enzymes, with the exception of Cel48A, were highly selective for glucan (Table 2). Trace quantities of arabinose and xylose liberated by Cel6A and Cel5A indicates either low level cross-reactivity for arabinoxylan or the solubilisation of cellulose molecules associated with soluble hemicellulose, while the absence of mannose is consistent lack of Cel6A and Cel5A cross-reactivity on glucomannan [23].

Release of glucose by Cel5A was about twice that for Cel6A (Table 2), but Cel5A was less effective than Cel6A in reducing fibre strength (Figs. 2 and 4) and degree of polymerisation (Fig. 5A). Similar results obtained by Kleman-Leyer *et al.*, [24] using cotton fibres led to the suggestion that Cel6A preferentially attacks kinks in fibres, resulting in short fibre fragments of low DP, while Cel5A degrades all regions of the fibre surface equally so that only small shifts in DP are observed, even after extensive degradation.

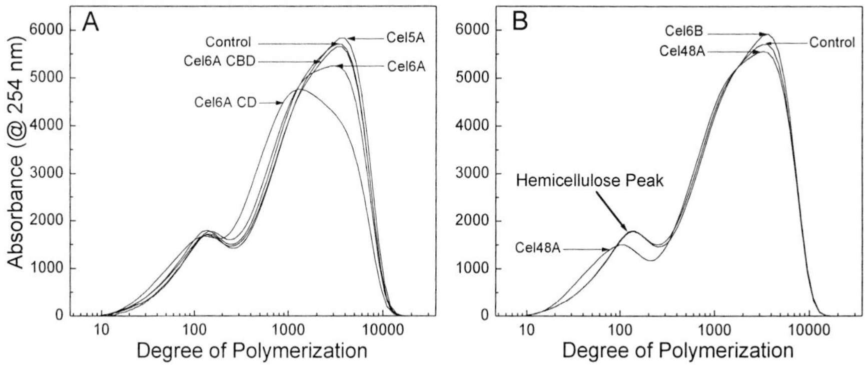

Figure 5. Degree of polymerisation of cellulase-treated kraft pulps.

The cellobiohydrolases Cel6B and Cel48A released substantially less glucose than the endoglucanases (Table 2), but Cel6B released almost two times as much glucose as Cel48A, in agreement with a previous report [25], which indicated that Cel6B was more aggressive in hydrolyzing both crystalline and amorphous cellulose substrates. An analysis of the molecular size distribution of the pulp cellulose indicated that the cellobiohydrolases had only a small effect on the degree of polymerisation of the cellulose (Fig. 5B) providing additional evidence that they are predominantly exo-acting enzymes, and confirming similar conclusions based on the analysis of relatively "pure" cellulosic substrates [25] and CMC hydrolysis experiments using viscometric measurements [12, 13].

Interestingly, the Cel48A treatments indicated that a significant proportion of the xylan component had been hydrolyzed (Table 2). Concomitantly, the DP of the xylan seems to have been substantially altered by Cel48A, as indicated by the reduction in the major xylan peak and the generation of the shoulder of lower molecular weight pentose molecules (Fig. 5B). This result was surprising because Cel48A does not exhibit any activity on xylan substrates [13].

Table 2. Carbohydrates solubilised by enzymatic treatment of kraft pulp with *C. fimi* cellulases

Enzyme	Carbohydrates Solubilised (% dry weight of pulp)*					
	Arabinose	Galactose	Glucose	Xylose	Mannose	Total Solubilisation
Cel6A	0.02	0	0.68	0.08	0	0.68
Cel5A	0	0	1.12	0.03	0	1.15
Cel6A CBD	0	0	0.05	0	0	0.05
Cel6A CD	0	0	0.38	0.01	0	0.39
Cel6B	0	0	0.29	0	0	0.29
Cel48A	0.08	0	0.18	1.39	0.04	1.69

* Values represent treatment minus control

Regarding the handsheet properties, Cel6B had a very limited effect on any of the pulp/paper properties, while Cel48A improved a number of handsheet parameters, including tensile strength. The ability to improve sheet strength was also maintained after refining, while all the other monocomponent preparations resulted in deleterious effects to both fibre and sheet properties following the mechanical processing of the fibres. The beneficial modifications observed with Cel48A treatments may be related to its ability to modify the xylan component, or the xylan-cellulose linkages, which could effectively enhance the fibre flexibility and consequently improve the resultant handsheet properties.

4. CONCLUSIONS

The effects of purified recombinant *Cellulomonas fimi* β-1,4-glucanases on the physical and optical properties of handsheets derived from a fully bleached, never-dried softwood were determined. It was apparent that endoglucanases cause substantial damage to the intrinsic fibre strength of the kraft furnish, and consequently the physical handsheet properties. The catalytic domain, Cel6A CD, proved to be even more detrimental to the fibre than its intact protein analog, while solubilising noticeably less polysaccharide. In contrast, the cellulose binding domain, Cel6A CBD, had a very limited effect on the fibre characteristics, and therefore did not compromise the quality of the handsheets produced. Both of the cellobiohydrolases (Cel6B and Cel48A) demonstrated limited carbohydrate dissolution, little or no change in cellulose DP, and had different effects on pulp properties. Cel48A demonstrated beneficial paper properties, including improvements in handsheet tensile strength following mechanical refining and wet pressing. This latter result seems to be related to the release and modification of the xylan moieties. These results suggest that Cel48A should be further considered for treating pulp fibres to improve paper properties, and future investigations should be directed at elucidating the mechanism(s) of Cel48A induced fibre modification.

5. ACKNOWLEDGEMENTS

The authors would like to thank Dr. Scott Stephens of Weyerhaeuser Inc., for his valuable input into this manuscript. We would also like to acknowledge Weyerhaeuser Technical Center (Tacoma WA) for its help with papermaking and testing.

REFERENCES

1. W. D. Yerkes, Process for the digestion of cellulosic materials by enzymatic action of *Trametes suaveolens*, US Patent No. 3,406,089 (1968).
2. J. Comtat, F. Mora P. and Noé, Process of treatment of pulps by an enzymatic solution favoring fibrillation and thus treated pulps, French Patent No. 2,557,894 (1984).
3. P. Noé, J. Chevalier, F. Mora and J. Comtat, J. Wood Chem. Technol., 6 (1986) 167.
4. B. R. Moran, Pulp and Paper 70, (1996) 119.
5. J. Pere, M. Siika-aho, J. Buchert and L. Viikari, Tappi J. 78 (1995) 71.
6. G. Stork, H. Pereira, T. Wood, E. Düsterhöft, A. Toft and J. Puls, Tappi J. 78 (1995) 79.
7. R.P. Kibblewhite and T.A. Clark, Proc. 50th Appita Ann. Gen. Conf. Vol. 1. (1996) 65.
8. N.R Gilkes, E. Kwan, D.G. Kilburn, R.C. Miller and R.A.J. Warren, J. Biotechnol. 57 (1997) 83.
9. S.D. Mansfield, K.K.Y. Wong, E. de Jong and J.N. Saddler, Tappi J. 79 (1996) 125.
10. N.R. Gilkes, R.A.J. Warren, R.C. Miller and D.G. Kilburn, J. Biol. Chem. 263 (1988) 10401.
11. A. Meinke, N.R. Gilkes, D.G. Kilburn, R.C. Miller and R.A.J. Warren, J. Bacteriol. 175 (1993) 1910.
12. A. Meinke, N.R. Gilkes, E. Kwan, D.G. Kilburn, R.A.J. Warren and R.C. Miller, Mol. Microbiol. 12 (1994) 413.

13. H. Shen, N.R. Gilkes, D.G. Kilburn, R.C. Miller and R.A.J. Warren, Biochem. J. 311 (1995) 67.
14. L.R. Schroeder and F.C. Haigh, Tappi 62 (1979) 103.
15. L. Valtasaari and K. Saarela, Paperi ja Puu - Papper och Trä 57 (1975) 5.
16. T.W. Bither and J.F. Waterhouse, Tappi J. 75 (1992) 201.
17. T.C. Maloney, T.-Q. Li, U. Weise and H. Paulapuro, Proc. 50th Appita Ann. Gen. Conf. Vol. 2 (1996) 647.
18. A.M. Scallan and J. Borch, In The fundamental properties of paper related to its uses. Trans. Symp. Cambridge (F. Bolam Ed.) Vol. 1. (1973) 152.
19. S.D. Mansfield and A.R. Dickson, Appita J. 54 (2001) 239.
20. M.S. Lumme, S.D. Mansfield and J.N. Saddler, Wood Fibre Sci. 31 (1998) 385.
21. T. Oksanen, J. Pere, J. Buchert and L. Viikari, Cellulose 4 (1997) 329.
22. Y. Kamaya, J. Ferment. Bioengin. 82 (1996) 549.
23. P. Tomme, E. Kwan, N.R. Gilkes, D.G. Kilburn and R.A.J. Warren, J. Bacteriol. 178 (1996) 4216.
24. K. Kleman-Leyer, N.R. Gilkes, R.C. Miller and T.K. Kirk, Biochem. J. 302 (1994) 463.
25. H. Stålbrand, S.D. Mansfield, J.N. Saddler, D.G. Kilburn, R.A.J. Warren and N.R. Gilkes, Appl. Environ. Microbiol. 64 (1998) 2374.

Biotechnology in the Pulp and Paper Industry
L. Viikari and R. Lantto (Editors)

Enzyme Treatments for Improved Retention in Newsprint Stocks

Ian D. Reid and Michelle Ricard

Pulp and Paper Research Institute of Canada, 570, boul. St-Jean, Pointe-Claire, QC, Canada H9R 3J9

1. INTRODUCTION

Modern papermaking benefits from the use of retention aids to keep fines and filler particles in the paper sheet and to speed the drainage of water. Polymers of various structures, either uncharged, cationic, or less frequently, anionic, are commonly used as retention aids[1]. Mechanical pulps are often not washed before they reach the papermachine. Dissolved and colloidal substances in the pulp suspension can lower runnability and product quality[2]. As the papermachine whitewater recirculates, these dissolved and colloidal contaminants accumulate. Some dissolved and colloidal substances interfere with the action of retention aids. Dissolved anionic polymers, known as deleterious substances or anionic trash, can combine with and reduce the effectiveness of cationic retention aids[3,4]. The current trend toward increasing closure of the whitewater circuit and reduced consumption of fresh water is expected to exacerbate these problems[5,6].

Alkaline peroxide bleaching, required for production of high brightness grades from mechanical pulps, causes dissolution of acidic polysaccharides which are troublesome interfering substances[7]. Prominent among these acidic polysaccharides are pectins, or polygalacturonic acids. The enzyme pectinase can depolymerize polymers of galacturonic acid, and consequently lower the cationic demand of filtrates from peroxide bleaching of TMP[8,9] and increase the effectiveness of cationic retention aids with these pulps[10].

For newsprint manufacture, however, mechanical pulps are usually brightened with hydrosulfite at moderate pH levels. The cationic demand in unbleached TMP is low, and is mainly due to uronic acids in hemicelluloses bound to the fibres and fines[11].

In addition to polysaccharides, mechanical pulp suspensions contain dissolved lignans and lignin-carbohydrate complexes[12]. These materials may contribute to cationic demand, affect the stability of dispersed pitch, or otherwise influence the papermaking properties of the pulp. Treatment of mechanical pulp filtrates with the oxidative enzyme laccase decreased the concentration of dissolved lignans, sterols and steryl esters, as well as suspended lipids[13]. The effects of the laccase treatment on papermaking or paper properties were not reported.

In this study we examined the ability of commercially available polysaccharide-degrading enzymes and laccases to improve retention in newsprint furnishes, alone and in conjunction with chemical retention aids.

2. MATERIALS AND METHODS

2.1. Pulp and whitewater

Headbox stock, a blend of hydrosulfite-brightened thermomechanical pulp (TMP; 68-80%) and deinked pulp (DIP; 20-32%) was sampled from the headbox of a paper machine producing newsprint at an eastern Canadian mill. The TMP furnish was 60% spruce and 40%

Table I. Characteristics of pulps and whitewaters

Sample	Consistency %	Cationic Demand meq/L	pH
Headbox stock	1.0-1.15	0.63-0.72	4.5-4.8
TMP	3.6-4.0	0.48-0.54	4.5-5.3
DIP	3.4-3.9	0.43	5.0-5.5
White water	0.48-0.73	0.21-0.67	4.5-4.9

fir. Deinked pulp was prepared from 60% recycled newspaper and 40% recycled magazine. Pulps and whitewater were brought back from the mill on the day of sampling and stored at 4°C until use.

To separate whitewater fines and soluble fractions, whitewater (650 g) was centrifuged (4000 rpm, 15 min) and the supernatant containing the water-soluble fraction (ca. 510 g) was immediately decanted. The pellet containing the fines was resuspended in 10 mM 2,2'-dimethylsuccinic acid (DMS; ca. 510 g) adjusted to the same pH as that of the whitewater. The fines and the water-soluble fractions were covered and stored overnight at 4°C.

2.2. Enzymes

Hydrolases: Table II lists the commercial enzyme preparations we used. The enzymes were gifts from Buckman (Buckman Laboratories of Canada, Ltd., Vaudreuil, Québec), Novozymes North America, Inc., (Franklinton, North Carolina) and Iogen (Iogen Corporation, Ottawa, Ontario). Novozym 863 was stored in 1-mL lots at –20°C and thawed just prior to use. Other enzymes were stored as recommended by the distributors at 4°C.

Polysaccharide hydrolyzing activities including cellulase, xylanase, mannanase, pectinase, amylase and arabinogalactanase were determined by the rate of release of reducing ends from carboxymethyl cellulose, birchwood xylan, locust bean gum, polygalacturonic acid, soluble starch and larch arabinogalactan, respectively. One unit of enzyme releases 1 μmol of reducing end per minute at 50°C, pH 5.3. Reducing sugars were measured with copper neocuproine as previously described[14].

Oxidases: Two Buckman laccases, 635 D and 636D, and one Novozymes laccase, NS 51022 from *Trametes villosa,* were used to treat newsprint headbox stock. Laccase activity was measured using 2,2'-azinobis(3-ethylbenzthiazoline-5-sulfonate) according to Bourbonnais *et al.*[15]

Table II. Polysaccharide hydrolyzing activities in enzyme preparations (U/mL). Dominant activities shown in **bold**.

	Cellulase	Xylanase	Mannanase	Pectinase	Amylase	Galactanase
Novozymes						
Novozym 863	53	6	44	**2634**	180	5.3
Gamanase	4	10	**585**	2	93	4
Iogen						
BRC	**1016**	84	19	18	5	3.2
Buckman						
Buzyme 2524–1[a]	60	**1150**	0.7	2.4	2.6	105
Buzyme 2524–2[b]	**798**	**2150**	17	1.2	28	1.2

[a]Received from Buckman on Nov. 19, 1998
[b]Lot CBN-1023-70A, received from Buckman on Feb. 2, 2000

2.3. Enzymatic Treatment of Pulp and Whitewater

Enzymatic treatment and subsequent analyses were carried out without adjusting the pH of the pulp samples or whitewater. Beakers containing pulp, whitewater, soluble fraction of whitewater or whitewater fines resuspended in DMS buffer were heated in water baths until contents reached 50, 60 or 70°C. Control samples were removed before adding enzymes. Sometimes, a second control sample was prepared by adding enzymes that were denatured by boiling for 15 min in a capped test tube. Diluted enzyme preparation was added to final concentrations of 0 to 5000 U of the dominant activity per L of pulp or pulp fractions (Table II). The pulp was mixed, and samples were removed at desired time intervals (0-4h) and tested immediately for first pass retention or preparation of handsheets.

2.4. Analytical Methods

First Pass Retention (FPR) was measured at 50°C with a Dynamic Drainage Jar (DDJ) as described[14]. Retention aid (Bufloc 594 or PEO) dosages providing FPR values of approximately 50% with the control pulps were chosen. Bufloc 594 was a gift from Buckman Laboratories of Canada Ltd. Bufloc 594 solution was prepared daily at 0.5% (w/v) in distilled H_2O, 1-2 h before use. Screens with mesh sizes of 70 and 45 and a stirrer speed of 1000 rpm were used in the dynamic drainage jar. Stock consistencies were determined according to PAPTAC Standard Method D.16[16]. Fibre length distribution was determined with a Kajaani FS-200 optical analyzer using TAPPI Standard Method T 271[17].

Handsheet Strength Tests: Standard handsheets were prepared from headbox stock before and after treatment with Iogen BRC at 245 U cellulase/L for 2 h at 50°C, according to PAPTAC Standard Method C.4[16], and tested according to PAPTAC Standard Method D.12[16].

In a second experiment, designed to minimize contact of Iogen BRC with the fibres, whitewater was treated separately from the TMP and DIP. A mixture of 23.8% DIP and 76.2% TMP was diluted and disintegrated as for standard handsheets. This DIP: TMP mixture was then diluted to a consistency of 0.3375% and stored at 4°C. Whitewater consistency was 0.73%. Whitewater samples were heated to 50°C and treated with Iogen BRC (100 U cellulase/L). After 30 min, the control or treated whitewater was mixed with the DIP, TMP and deionized water in the deckle of the handsheet maker. To obtain a standard handsheet weight of 1.2 g ± 0.03, 155 mL of whitewater and 135 mL of DIP:TMP mixture were added to the handsheet maker deckle. Handsheet formation took 6 min; blotting and couching of the handsheet took another 3 min. To prevent continued action of the cellulase on the fibres of the handsheet, it was immediately dried at 130°C on a handsheet speed dryer from Testing Machines International (Montreal, Canada).

The statistical significance of treatment effects was tested by analysis of variance with the computer program Prism 3.0 (GraphPad Software, San Diego, CA). The null hypothesis was rejected when its probability was less than 5%. Individual treatment means were compared by Dunnett's or Bonferroni's multiple comparison tests.

3. RESULTS AND DISCUSSION

3.1. Effect of Iogen BRC

Treating newsprint stock with the crude cellulase Iogen BRC (245 U/L, 50°C) increased its FPR. The effect increased with enzyme dose up to a cellulase activity of 100 U/L and then remained constant at higher doses (Fig. 1). Treatment at 50°C gave the best response;

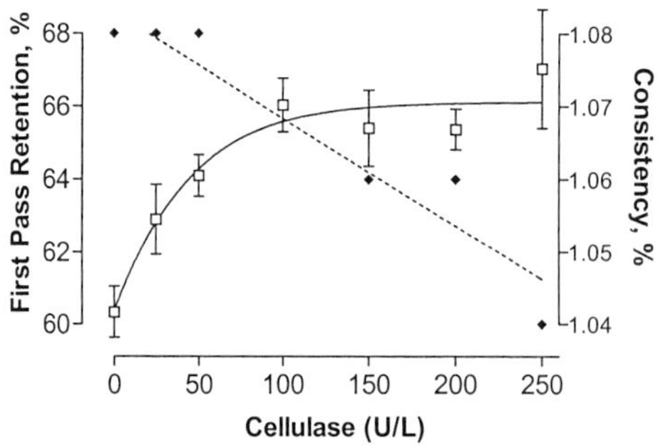

Figure 1. Effect of Iogen BRC dose on consistency(◆) and FPR(□). Headbox stock was treated with enzyme for 1 h at 50°C. FPR was measured with addition of Bufloc 594 at 0.5 kg/t, a 45 mesh screen, and stirring at 1000 rpm. Data are means of 3 replicates ± SE.

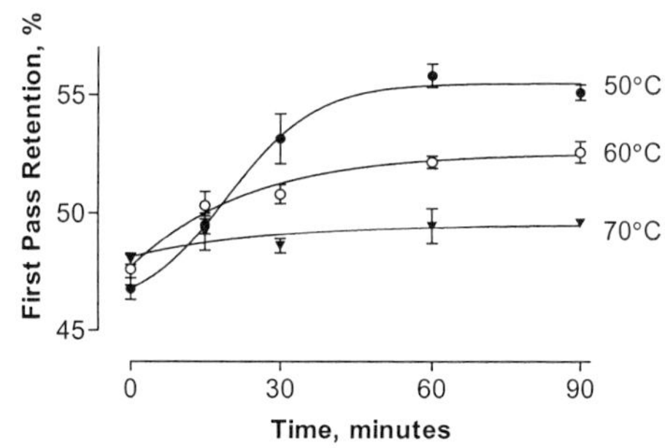

Figure 2. Effect of temperature on Iogen BRC kinetics. Headbox stock was treated with 100 cellulase U/L. FPR was measured with addition of Bufloc 594 at 0.5 kg/t, a 45 mesh screen, and stirring at 1000 rpm. Data are means of 3 replicates ± standard errors.

treatments at 60 and 70°C gave progressively smaller increases in retention. At 50°C, a treatment time of one hour was sufficient for maximum effect (Fig. 2).

When Iogen BRC treatment was combined with either cationic (Bufloc 594) or nonionic (Floc 999) retention aids, the effects on FPR were additive (Table III). This independence of effects suggests that the enzyme treatment enhances retention by a different mechanism than the chemical retention aids.

3.1.1. Separate treatment of TMP, Deinked pulp, and Whitewater

Iogen BRC treatment of headbox stock damaged the strength properties of the pulp (see below). We wanted to know if we could avoid loss of pulp strength by treating only one of the TMP or the Deinked pulp (DIP) streams. When we treated the two types of pulp, diluted to 1% consistency with whitewater, separately and mixed them in 75% TMP:25% DIP ratio just before measuring retention, we found that the effects on FPR were additive, both with and without retention aid. This suggested that treatment of the whitewater, and not of either type of pulp fibre, was the important factor for retention. Treatment of the whitewater confirmed this. When the whitewater alone was treated with the enzyme, and then used to dilute the pulps just before measuring retention, the FPR was significantly increased. Treating the pulps as well as the whitewater gave a small, non-significant, further increase in retention (Table IV).

Enzyme treatment of the whitewater significantly decreased the consistency of the TMP-DIP-whitewater mixture. Enzyme treatment of the pulps as well as the whitewater gave a

Table III. Effects of Iogen BRC treatment of newsprint stock on First Pass Retention in combination with cationic and nonionic retention aids.

Retention aid	Control	Treated
None	52.1 ± 0.8	56.6 ± 1.4
Bufloc 594 (2 kg/t)	63.1 ± 1.5	68.3 ± 0.3
Oxirez (150 g/t)/Floc 999 (75 g/t)	56.5 ± 0.6	66.6 ± 0.9

Headbox stock was treated with 245 cellulase U/L for 1 h at 50°C. FPR was measured with a 70 mesh screen. Data are means of 3 replicates ± SE.

Table IV. Effect of Iogen BRC Treatment of Whitewater and Thickstocks

Treatment			Consistency	First Pass Retention
TMP	DIP	WW	%	%
-	-	-	1.008 ± 0.005	51.4 ± 0.3
-	-	+	0.985 ± 0.003	53.9 ± 0.2
+	+	+	0.963 ± 0.003	54.5 ± 0.3

Values are means of 6 replicates ± standard error. Iogen BRC was applied at 245 cellulase U/L for 1 h at 50°C. FPR was measured with no retention aid, and a 70 mesh screen

further decrease in consistency that was statistically significant (Table IV). Enzyme treatment of either pulp type significantly increased their freeness and their fines content. Particle size analysis showed that the enzyme treatment lowered the content of very small fines (<0.1 mm) and increased the number of particles in the 0.1 – 0.4 mm size range.

To further characterize the target of the enzyme action, the whitewater was fractionated by centrifugation into a fines fraction (pellet) and a soluble fraction (supernatant). The whitewater fines and solubles were separately treated with Iogen BRC, and then recombined with pulp for retention testing. To estimate the effect of complete removal of soluble interfering substances, distilled water was used in place of the whitewater supernatant. Enzyme treatment of the whitewater supernatant alone did not increase FPR, but treatment of the fines alone did (Table V). This indicates that the enzyme is acting on particle-bound substrate(s), not on dissolved substances. Replacing the whitewater supernatant with distilled water increased FPR, especially in combination with treatment of the pellet and addition of retention aid (Bufloc 594). This shows that soluble interfering substances were present in the whitewater, but they were not affected by this enzyme.

3.1.2. Effects on strength

Standard handsheets were prepared from the headbox stocks before and after treatment with Iogen BRC, and their physical properties were measured (Table VI Stock treatment). The results clearly show that the pulp treated with Iogen BRC suffered major losses in burst, tear, and tensile strength.

We tried to avoid the damage to pulp strength from cellulase treatment by applying the enzyme to the whitewater in the absence of the fibres, and by lowering the enzyme concentration. Cellulase molecules generally include cellulose binding domains and adsorb strongly to cellulose surfaces. When Iogen BRC was added to whitewater, the cellulase activity was indeed adsorbed on the fines and disappeared from the solution phase. We hoped that the time between mixing the enzyme-bearing fines with the fibres in the TMP and deinked pulps and inactivating the enzyme by drying the formed sheet would be too short for

Table V. Effects on FPR of Treating Whitewater Particulate and Soluble Fractions with Iogen BRC

	Without Bufloc		With Bufloc 594 (2 kg/t)	
	Control Pellet	Treated Pellet	Control Pellet	Treated Pellet
Control Supernatant	51.4 ± 0.2	52.2 ± 0.5	62.1 ± 0.9	64.3 ± 0.4
Treated Supernatant	50.4 ± 0.3	53.1 ± 0.5	61.9 ± 0.4	65.0 ± 0.5
Water	49.3 ± 0.6	53.4 ± 0.7	65.6 ± 0.7	70.2 ± 0.5

Values are means of 3-6 replicates ± standard errors. Iogen BRC was applied at 245 cellulase U/L for 1 h at 50°C. FPR was measured with a 70 mesh screen.

Table VI. Effect of treatment of newsprint stock with Iogen BRC cellulase on physical properties of handsheets

Test	Units	Stock treatment[a] Control	Stock treatment[a] Iogen BRC	Stock treatment[a] Difference	Whitewater treatment[b] Control	Whitewater treatment[b] Iogen BRC	Whitewater treatment[b] Difference
CSF	mL	31.1	42.3	**36%**	nd	nd	
Grammage (O.D.)	g/m^2	62.3	60.3	**-3%**	64.3	66.3	**3%**
Caliper SS	μm	154	153	**0%**	226	231	**2%**
Apparent Density	g/cm^3	0.40	0.39	**-3%**	0.29	0.29	**0%**
Burst Index	kPm^2/g	2.27	1.43	**-37%**	1.97	1.67	**-15%**
Tear Index	mNm^2/g	6.53	4.73	**-28%**	6.47	6.02	**-7%**
Breaking Length	km	3.81	2.65	**-30%**	3.43	3.03	**-12%**
Stretch	%	1.99	1.41	**-29%**	1.92	1.73	**-10%**
Tensile Index	N.m/g	37.33	26.00	**-30%**	33.59	29.68	**-12%**
TEA Index	mJ/g	503.78	246.03	**-51%**	412.85	327.55	**-21%**
Elastic Modulus	km	393	356	**-9%**	325	316	**-3%**
Zero-Span Breaking Length	km	9.43	8.25	**-12%**	8.65	8.16	**-6%**

[a]Complete headbox stock treated with 245 cellulase U/L for 2 h at 50°C
[b]Whitewater treated with 100 cellulase U/L for 0.5 h at 50°C, then mixed with TMP and DIP just before forming handsheets

significant enzyme transfer and attack on the fibres. To test this idea, we had standard handsheets made from a mixture of Iogen BRC-treated whitewater, TMP, and deinked pulp immediately after mixing, and dried them quickly on a sheet dryer. The pulps were mixed with the same amount of whitewater as they would be in a papermachine headbox, and then diluted with pure water to the lower consistency required by the handsheet former. Unfortunately the drainage time in the handsheet former was several minutes, so that the contact time between enzyme and fibres was much longer than it would be on a papermachine. Under these conditions the enzyme treatment caused a significant decrease in the handsheet strength properties (Table VI Whitewater treatment), although the effect of the whitewater treatment was much smaller than the effect of treating the whole headbox stock. Possibly the contact time between enzyme and fibres on a papermachine would be short enough to make strength loss negligible. The relative changes in burst (-15%) and tensile (-12%) strength were larger than the relative changes in zero-span tensile strength (-6%), suggesting that differences in inter-fibre bonding contribute to the observed effect of the enzyme.

3.2. Xylanase: Buzyme 2524

In our initial enzyme screening, Buzyme 2524 showed some ability to increase FPR, so we tested it in more detail. Xylanase is the major activity in this preparation; cellulase is also present at about 10% of the xylanase activity. Treatment of newsprint stock with Buzyme 2524 increased retention, both with and without added retention aid. The effect increased with enzyme dose over a wide range (Fig. 3).

The enzyme treatment led to a dose-dependent drop in the consistency of the pulp. Buzyme 2524 was active at temperatures of 50 - 70°C. At the optimal temperature of 60°C, the reaction was rapid during the first 30 minutes, and continued more slowly thereafter (Fig. 4).

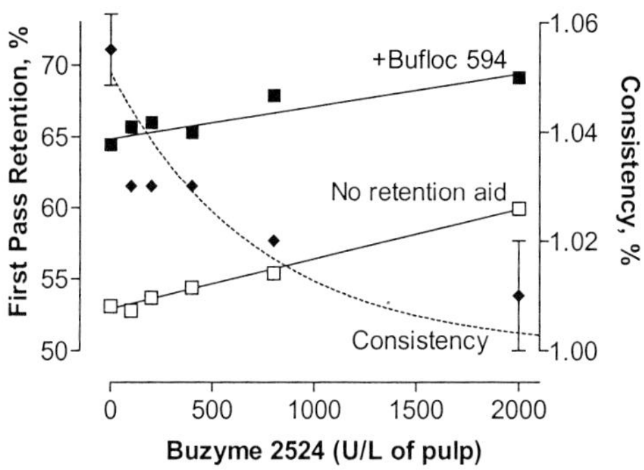

Figure 3. Effect of xylanase Buzyme 2524 dose on consistency (◆) and retention (■,□). Mill A headbox stock was treated with enzyme for 1 h at 50°C. FPR was measured with a 70 mesh screen. Bufloc 594 was added at 2 kg/t when indicated.

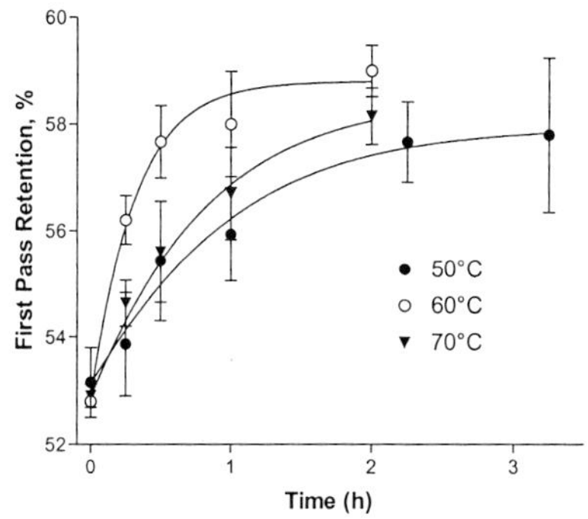

Figure 4. Temperature effects on kinetics of retention improvement by Buzyme 2524. Mill A headbox stock was treated with 1000 xylanase U/L. FPR was measured with no retention aid and a 70 mesh screen. Data are means of 3 replicates ± standard errors.

When the whitewater, TMP and DIP components of the newsprint stock were treated separately with Buzyme 2524, the whitewater showed the largest effect. In conjunction with Bufloc 594, treatment of the whitewater was almost as good as treatment of all three components (Table VII). When the particulate (pellet) and soluble fractions of the whitewater were treated separately, treatment of the particulates was most effective (Table VIII). Buzyme 2524 treatment of the particulate fraction did not change the particle size distribution. This is the same pattern as we observed with Iogen BRC.

Table VII. Effectiveness of Buzyme 2524 Treatment of Whitewater on First Pass Retention

Treatment	First Pass Retention, %	
	without Bufloc	with Bufloc 594 (2 kg/t)
Control	57.0 ± 0.9	68.7 ± 0.8
Treated Whitewater	63.9 ± 0.5	72.6 ± 0.4
Treated Whitewater & Pulps	66.2 ± 1.7	73.2 ± 0.2

Values are means of 3-9 replicates ± standard error. Whitewater was treated with 1000 xylanase U/L for 80 min at 60°C. FPR was measured with a 70 mesh screen.

Table VIII. Effect of separate treatment of whitewater pellet and supernatant fractions with Buzyme 2524[a] on First Pass Retention.

	Pellet	
Supernatant	Untreated	Treated
Untreated	56.9 ± 2.9	74.8 ± 6.6
Treated	58.6 ± 3.5	77.0 ± 3.8

Data are means of 3 replicates ± standard errors. FPR was measured with addition of Bufloc 594 at 2 kg/t and a 45 mesh screen.

[a]Batch 2, 1000 xylanase U/L, 60 min at 60°C

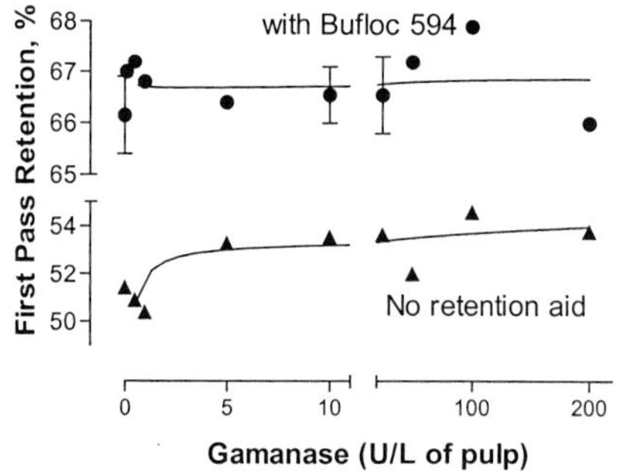

Figure 5. Effect of mannanase dose on retention of newsprint stock. Headbox stock was treated with enzyme for 1 h at 50°C. FPR was measured with Bufloc 594 at 0 or 2 kg/t, and a 70 mesh screen. Data are means of 1-3 replicates ± standard errors.

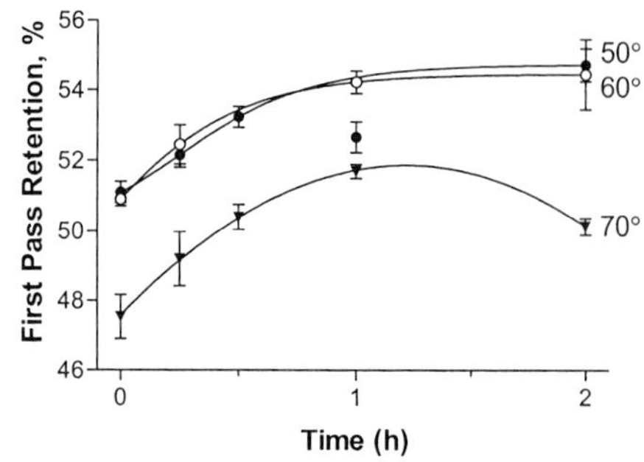

Figure 6. Effect of temperature on mannanase kinetics. Headbox stock was treated with 5 mannanase U/L of Gamanase. FPR was measured with no retention aid, and a 70 mesh screen. Data are means of 3 replicates ± standard errors.

3.3. Mannanase: Gamanase

Novo's Gamanase enzyme, in which D-mannanase is the main activity, gave a small but measurable increase in FPR at remarkably low doses, 5 mannanase U/L. There may also have been an enhancement of the effect of Bufloc 594 at even lower enzyme charges, 0.5 – 1 U/L (Fig. 5). Treatment with Gamanase was effective at 50, 60 and 70°C (Fig. 6). The 70°C treatment was tested in a separate experiment from the other temperatures, and the starting FPR of the pulp was lower. However, the increase in FPR was as large at 70°C as at the lower temperatures. At 5 U/L of mannanase, a one hour treatment was sufficient to give maximum increase in FPR without retention aids. When the whitewater, TMP, and deinked pulp were treated separately with Gamanase, the whitewater treatment gave the largest effect on retention (Table IX). Separate treatments of the whitewater fines and the whitewater solubles both gave significant increases in FPR (Table X). The effect of treatment of the supernatant (+2.0) was larger than the effect of treatment of the pellet (+1.6). In this respect, Gamanase behaves differently than Iogen BRC and Buzyme 2524; treatment of the whitewater supernatant with those enzymes did not increase retention.

Table IX. Effect of treating whitewater, TMP, and deinked pulp with Gamanase

Treatment			First Pass Retention, %	
WW	TMP	DIP	without Bufloc	with Bufloc[a]
+	+	+	42.6 ± 0.5	46.9 ± 0.7
+	-	-	40.7 ± 0.5	46.2 ± 0.5
-	+	-	39.5 ± 0.3	44.3 ± 0.4
-	-	+	38.7 ± 0.4	38.9 ± 0.4
-	-	-	38.9 ± 0.4	43.8 ± 0.2

Values are means of 3-6 replicates ± standard error. Whitewater and pulps were treated with 5 mannanase U/L for 1h at 60°C. FPR was measured with a 45 mesh screen.

[a]Bufloc 594, 1 kg/t

Table X. Effect of separate treatment of whitewater pellet and supernatant fractions with Gamanase[a] on First Pass Retention.

	Pellet	
Supernatant	Untreated	Treated
Untreated	51.5 ± 0.8	52.8 ± 0.8
Treated	53.3 ± 0.5	55.1 ± 0.3

FPR was measured with Bufloc 594 at 1 kg/t, and a 45 mesh screen. Data are means of 3-6 replicates ± standard error.
[a]5 mannanase U/L, 60 min at 60°C

3.4. Enzyme combinations

Most of our tests have been done with newsprint stock from one mill. To check whether pulp from other mills would react in the same way to enzyme treatment, we have tested newsprint stock from another mill. This stock contains only hydrosulfite-brightened TMP, no DIP.

It reacted similarly to the pulp from the first mill to treatment with the three enzymes Iogen BRC, Buzyme 2524, and Gamanase. All the enzymes produced an increase in FPR, and Iogen BRC gave the largest effect (Table XI).

We have also tested combinations of the enzyme preparations. Each enzyme was applied at sufficient concentration to give a near maximum effect. The effects of the enzymes were additive. Using Buzyme 2524 in combination with Iogen BRC gave a larger increase in FPR than BRC alone, and adding Gamanase to the mix gave a further increase in FPR (Table XII). If a common enzyme in all 3 preparations was responsible for the effect, or if the enzymes were all acting on the same substrate, we would not expect such additivity. Therefore our results suggest that there are at least three distinct enzyme substrates; in other words, there are at least three materials in the newsprint stock that lower retention. In the case of Iogen BRC and Buzyme 2524 treatment, the component whose degradation increases retention is concentrated in the whitewater fines. The substrate for Gamanase is also localized in the whitewater.

Our results also indicate that a mixture of enzymes is required for maximum increase in newsprint retention. Even though Iogen BRC is an unpurified mixture of *Trichoderma* enzymes, it apparently lacks important activities that are present in Buzyme 2524 and Gamanase. We don't know yet whether these additional activities are the xylanase and mannanase major components of these two preparations, or some minor components.

The finding that the enzymatic treatment of the fine particles has a larger effect on retention than treatment of the soluble fraction is unexpected. Possibly the enzyme treatment is

Table XI. Relative effectiveness of crude cellulase, xylanase, and mannanase in increasing FPR of newsprint

	First Pass Retention, %	
Enzyme	without Bufloc	with Bufloc 594, 1 kg/t
Control	44.1 ± 0.5	48.8 ± 0.6
Iogen BRC (245 U cellulase/L)	46.7 ± 0.2	54.3 ± 0.6
Buzyme 2524 (1000 U xylanase/L)	45.6 ± 0.6	51.5 ± 0.6
Gamanase (5 U mannanase/L)	44.9 ± 0.3	51.7 ± 0.2

Values are means of 3 replicates ± standard error. Headbox stock was treated with enzymes for 30 min at 50°C. FPR was measured with a 70 mesh screen.

Table XII. Additive effects of enzyme combinations on FPR of newsprint

Enzyme	First Pass Retention, % without Bufloc	First Pass Retention, % with Bufloc[d]
Control	44.8 ± 0.2	51.7 ± 0.6
Iogen BRC[a]	47.6 ± 0.7	55.4 ± 0.2
Iogen BRC[a] + Buzyme 2524[b]	48.0 ± 0.2	56.0 ± 0.9
Iogen BRC[a] + Buzyme 2524[b] + Gamanase[c]	48.5 ± 0.3	57.0 ± 0.6

Values are means of 2-3 replicates ± standard errors. Headbox stock was treated with enzymes for 30 min at 50°C. FPR was measured with a 70 mesh screen.
[a]245 U cellulase/L of pulp
[b]1000 U xylanase/L of pulp
[c]5 U mannanase/L of pulp
[d]Bufloc 594, 1 kg/t

changing the surface properties of these fine particles in some way that enhances their adhesion to fibres.

3.5. Laccases

Treatment of the headbox stock with laccase from *Trametes villosa* did not change the FPR of the pulp without any retention aid, and only slightly increased FPR with Bufloc 594 (Table XIII).

3.6. Costs of enzyme treatment

The single most effective enzyme preparation is Iogen BRC, and the cost of treating whitewater with this enzyme at its optimum dose of 100 U/L would be $195 per tonne of newsprint. Adding Buzyme 2524 or Gamanase would further increase the cost. This cost is much too high; the price of enzyme needs to come down by two orders of magnitude to be cost-effective.

4. Conclusions

Three commercial enzyme preparations, Iogen BRC, Buzyme 2524, and Gamanase, increase retention in newsprint stocks by novel mechanisms. The improved retention is independent of chemical retention aids, and additional to their effects. This is different from the effect of pectinase in peroxide-bleached mechanical pulps, where the enzyme acts to increase effectiveness of cationic retention aids[14].

The major target of the enzymes' action is the fines, not the long fibres or dissolved materials. Enzyme treatment does not cause a net removal of the fines, but apparently alters their surface properties to help them adhere to fibres. It is possible that some fine particles are

Table XIII. Effect of laccase treatment of headbox stock on FPR with or without retention aid

Retention aid	Control	Laccase
None	50.0 ± 0.1	49.8 ± 0.3
Bufloc 594, 0.5 kg/t	58.1 ± 0.5	59.0 ± 0.8

Headbox stock was treated with *T. villosa* laccase, 10 U/g of pulp o.d. for 2h at 50°C. FPR was measured with a 70 mesh screen. Data are means of 3 replicates ± standard errors.

degraded by the enzymes and that new fines are generated by action of the enzymes on the fibre surfaces, as previously suggested for cellulase treatment of mechanical pulps[18]; the drop in stock consistency during enzyme treatment supports this. The specific enzymes responsible for the retention improvement and the chemical changes they produce in the fines have not yet been identified. Iogen BRC, the most effective preparation, is rich in cellulase. However, another cellulase preparation, Buzyme 2523, had no effect on retention. Similarly, Buzyme 2524 contains strong xylanase activity, but other xylanase preparations, Iogen BRXH, Buzyme 2511, and Buzyme 2513, were less effective. The major activity in Gamanase is β-1,4-mannanase. Possibly certain types of cellulase and xylanase are effective and others are not. All of these enzyme preparations may also contain other activities that we didn't assay for. Each of the three enzyme preparations can add to the effects of the others, suggesting that at least three enzymes are needed for maximum effect, and that there are at least three distinct target substrates on the fines, and at least one dissolved substrate.

The enzymes have adequate stability at the temperatures of 50-70°C prevalent in stock preparation systems, but require treatment times of up to an hour to produce their effects. The enzyme treatment can be applied to the whitewater to reduce damage to fibre strength.

Treatment with the oxidative enzyme laccase does not increase retention in newsprint stock; in fact, it may increase cationic demand.

There are at least three obstacles to applying enzyme treatments in newsprint mills:

1. The enzyme cost is too high.
2. Cellulase may damage the strength of the fibres, unless the contact between treated whitewater and fibres can be made brief enough.
3. The papermachine whitewater recirculates continuously, and there is not enough time for enzymes to act on the whitewater fines.

Because the enzymes in Iogen BRC, Buzyme 2524, and Gamanase increase retention by an unknown and apparently novel mechanism, further study of their mechanism might uncover a cost-effective treatment. In the best case, individual enzymes that modify the fines to increase their retention, but that do not harm fibre strength, could be identified, cloned, and produced cheaply.

5. ACKNOWLEDGEMENTS

We thank Vincent Dubreuil and Daniel Montplaisir of Kruger, Inc. for their help in providing mill samples, Paul Scudamore and Jurgen Superga of Paprican's Analytical Services group for their willingness to adapt testing procedures and schedules to the needs of enzyme treatments, Gunar Laivins for advice on nonionic retention aids, Neal Franks of Novozymes for supplying enzymes, Michael Anstey, Gerald Dykstra, and Chip Hill of Buckman Laboratories for discussions, enzymes and retention aids, and Buckman Laboratories of Canada Ltd. for financial support.

6. REFERENCES

1. Horn, D. and Linhart, F. Retention aids. In: *Paper Chemistry* (Roberts, J. C., ed.). Blackie Academic & Professional, London, 1996, 64-82
2. Zhang, X., Beatson, R. P., Cai, Y. J. and Saddler, J. N. Accumulation of specific dissolved and colloidal substances during white water recycling affects paper properties. *J. Pulp Pap. Sci.* 1999, **25**(6), 206-210
3. Linhart, F., Auhorn, W. J., Degen, H. J. and Lorz, R. "Anionic trash": controlling detrimental substances. *Tappi J.* 1987, **70**(10), 79-85

4. Webb, L. Papermaking's problem substances interfere with machine runnability. *Pulp & Paper* 1997, **71**(11), 69-74
5. Allen, L., Polverari, M., Levesque, B. and Francis, W. Effects of system closure on retention- and drainage-aid performance in TMP newsprint manufacture. *Tappi J.* 1999, **82**(4), 188-195
6. Polverari, M., Allen, L. and Sitholé, B. Effects of system closure on retention and drainage aid performance in TMP newsprint manufacture - Part II. *Tappi J.* 2001, **84**(3), 56
7. Holmbom, B., Ekman, R., Sjöholm, R., Eckerman, C. and Thornton, J. Chemical changes in peroxide bleaching of mechanical pulps. *Das Papier* 1991, **45**(10A), V16-V22
8. Thornton, J. W. Enzymatic degradation of polygalacturonic acids released from mechanical pulp during peroxide bleaching. *Tappi J.* 1994, **77**(3), 161-167
9. Thornton, J. W., Eckerman, C. S., Ekman, R. O. and Holmbom, B. R. Treatment of alkaline bleached mechanical pulp with pectinase. U.S. Patent 5,487,812, 1996
10. Reid, I. D. and Ricard, M. Pectinase in papermaking: solving retention problems in mechanical pulps bleached with hydrogen peroxide. *Enz. Microb. Technol.* 2000, **26**(2-4), 115-123
11. Sundberg, A., Pranovich, A. and Holmbom, B. Distribution of anionic groups in TMP suspensions. *J. Wood Chem. Technol.* 2000, **20**(1), 71-92
12. Alvarado-Jacobs, F. Extractives in process waters from newsprint papermaking (TMP). Doctoral thesis, Royal Institute of Technology, Stockholm, Sweden, 1995
13. Zhang, X., Stebbing, D. W., Saddler, J. N., Beatson, R. P. and Kruus, K. Enzyme treatments of the dissolved and colloidal substances present in mill white water and the effects on the resulting paper properties. *J. Wood Chem. Technol.* 2000, **20**(3), 321-335
14. Reid, I. D. and Ricard, M. Decreasing cationic demand and increasing fines and filler retention of peroxide-bleached thermomechanical pulp with pectinase enzyme. Paprican 1998, 20
15. Bourbonnais, R., Paice, M. G., Reid, I. D., Lanthier, P. and Yaguchi, M. Lignin oxidation by laccase isozymes from *Trametes versicolor* and role of the mediator 2,2'-azinobis(3-ethylbenzthiazoline-5-sulfonate) in kraft lignin depolymerization. *Appl. Environ. Microbiol.* 1995, **61**, 1876-1880
16. Standard Methods Committee. *Standard Testing Methods* Pulp and Paper Technical Association of Canada, Montreal, 2000
17. *TAPPI Test Methods* TAPPI Press, Atlanta, 1998, 1500
18. Jackson, L. S., Heitmann, J. A. and Joyce, T. W. Enzymatic modifications of secondary fiber. *Tappi J.* 1993, **76**(3), 147-154

Potential of enzymatic deinking

L. Magnin[a], P. Delpech[b] and R. Lantto[c]

[a]Centre Technique du Papier, BP 251, 38 044 Grenoble Cedex 9, France

[b]Papeterie de Moulin Vieux, 503, avenue du Dauphiné, 38530 Pontcharra, France

[c]VTT Biotechnology, P.O. Box 1500, FIN-02044 VTT, Finland

Standard alkaline flotation deinking and neutral enzyme-aided flotation deinking were compared at pilot scale using a typical woodfree paper composition (30 % coated sheet fed offset printing, 35 % HP laser printing, and 35 % OCE copy). Speck contamination of the pulps was 88 % lower after the enzyme treatment than the alkaline treatment. Based on this result, full-scale enzymatic deinking experiment was carried out at Moulin Vieux mill, Pontcharra, France. The mill produces woodfree deinked pulp from 100 % printed coated woodfree papers. Efficiency of the enzymatic deinking was compared to that of the standard deinking. 4 - 6 additional brightness points were obtained with the enzymatic deinking compared to the mill chemistry deinking. Residual ink content in the enzyme treated deinked pulp was lower. The mechanical properties were maintained and enzymes did not cause runnability problems on the paper machine. However, COD load was observed to be 20 - 40 % higher than in the standard process. For woodcontaining recovered paper, the results obtained at pilot scale showed that good ink removal and a lower speck contamination were obtained by enzymatic treatment in neutral conditions but, in spite of a reduced amount of ink, brightness was lower. The lower brightness was recovered by a post-bleaching stage.

1. INTRODUCTION

Use of recovered paper as raw material in papermaking has increased during the last ten years. The recovery rate for wastepaper approached 50 % during the year 2000. In Europe composition of paper contains an average of about 40 % recovered paper [1]. Recovered paper is mainly used to manufacture paperboard case material followed by production of newsprint and many other paper grades. Inexpensive recovered paper can be converted to products of good quality by deinking. However, some types of recovered paper and especially the printing inks are difficult to deink. Poor ink detachment causes smeared pulp or low brightness, also toner printed papers lead often to speckled pulp. This is the most significant technical obstacle preventing the worldwide use of recover paper as raw material in papermaking.

Use of enzymes is one way to solve the deinking problem [2, 3, 4]. In principal, there are three different targets for the action of enzymes. Hemicellulases and cellulases [5, 6, 7] can attack the components on fibre surface. These enzymes alter the fibre surface by modification of chemical bonds in the vicinity of ink particles, thereby freeing ink for removal by washing or flotation. Secondly, starch based coating can be hydrolysed by amylolytic enzymes [8]. Finally vegetable oil based ink binders can be degraded by lipases [9]. The use of cellulases and hemicellulases is most widely reported in literature [10, 11, 12, 13].

In this study, the data obtained from the laboratory scale deinking experiments with several individual enzymes and paper grades were scaled up to pilot scale. Finally the enzymatic treatment of woodfree recovered paper was tested with varied raw material and process conditions at full scale.

2. MATERIALS AND METHODS

2.1. Pilot scale deinking of woodfree recovered paper

Woodfree paper raw material

As a typical mixture of woodfree paper raw material, the following paper grades were selected: 30 % coated sheet fed offset printing, 35 % HP laser printing and 35 % OCE copy. This paper mixture was chosen to represent a typical woodfree paper composition showing deinking difficulties.

Enzymes

Two enzyme mixtures shown in Table 1 were selected based on the results obtained at laboratory scale deinking experiments carried out with individual enzymes and paper grades (manuscript in preparation).

Table 1. Enzymes and dosages used in the pilot scale deinking of woodfree paper.

Enzyme product*	Main enzyme activity	Enzyme dosage, nkat** /g O.D. pulp	
		Mixture I	Mixture II
Ecopulp TX-200C	Thermoxylanase	100	-
Ecopulp M	Mannanase	100	100
Ecostone A 200	Amylase	100	100
CBH I ***	Cellulase	-	0.25 mg/g

*Enzymes were obtained from AB Enzymes Finland Oy.

**Enzyme activity unit nkat (nanokatal) is defined as the amount of enzyme activity that converts 1 nmol per second of substrate in the assay conditions.

*** Research enzyme preparation containing cellobiohydrolase I. About 90% of the protein present in the preparation was CBH I.

Operating conditions

The deinking experiments were carried out at the pilot plant of Centre Technique du Papier, CTP. A single deinking loop was performed in the pilot plant as shown in Figure 1. The pulper was run at medium consistency of 12 – 13 %.

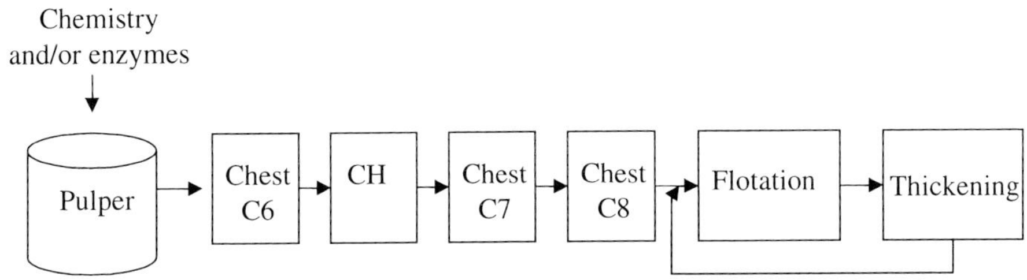

Figure 1. The pilot deinking loop used in the experiments. After pulping the pulp was sent to the retention chest C6. The pulp was screened (Lamort CH screen, 20/100 slots) before transfer to the chest C8 prior to flotation and thickening.

Neutral deinking, conventional alkaline deinking and two separate enzymatic deinkings (mixtures I and II) were performed as described below.

Neutral deinking

Pulping was carried out for 15 minutes at 45 °C. No retention time after pulping was applied. Soap of fatty acid (0.6 % SERFAX MT 90, Stephenson Group, UK) was added at the inlet of the flotation.

Conventional alkaline deinking

Pulping was carried out for 15 minutes at 45 °C. No retention time after pulping was applied. The total chemistry (1 % H_2O_2, 1 % NaOH, 2.5 % silicate, 0.6 % soap) was added to the pulper. pH was adjusted to 8.5 at the inlet of the flotation.

Enzymatic deinking

Pulping was carried out for 15 minute at 50°C. Retention time after pulping was 1 hour in C6 chest. Water pH was adjusted to 7. Enzyme mixtures (Table 1) were added to the pulper and 0.6 % soap was added at the inlet of the flotation.

Deinking efficiency

The deinking efficiency was assessed along the deinking loop by measuring the following optical properties: Brightness of the pulp (reflectance of light at 457 nm), ERIC on pads of entire and hyperwashed pulp, specks from thin handsheets. Flotation yield was also measured.

The ERIC measurement is an estimation of effective residual ink content. The measurement is based on determination of reflectance in infrared light. The ERIC measures the amount and fragmentation level of ink in the pulp. Ink specks are particles with a diameter of 50 μm or above. Specks do not affect pulp brightness.

Release of dissolved and colloidal substances

Dissolved material, COD and cationic demand (using polyethyleneimine for neutralisation) were determined from centrifuged filtrates.

Mechanical properties

Wet zero span breaking length, breaking length (ISO 1924-2 norm), burst index (NF Q03-053 norm) and tear index (NF EN 21974 norm) were measured from the handsheets.

2.2. Mill scale deinking of woodfree recovered paper

Based on the results obtained at pilot scale with the woodfree paper mixture, enzymatic deinking was performed at Moulin Vieux deinking mill, France. Production capacity of woodfree deinked pulp at Moulin Vieux mill is 16 000 ton/year.

Operating conditions

The deinking line at Moulin Vieux mill (Figure 2) contains a deinking stage by flotation followed by a thickening stage, a kneading step, a bleaching stage with hydrogen peroxide (0.8 % sodium hydroxide, 0.7 % silicate, 1 % hydrogen peroxide) and a post deinking stage by washing. Each trial lasted for 24 hours. Raw material was composed of 100 % AFNOR 7 quality papers (woodfree coated printed papers). The mill normally uses this raw material.

The operating conditions are summarised in Table 2. All enzymes except lipase (Novozymes A/S, Denmark) were obtained from AB Enzymes Oy, Finland. The enzyme mixture contained thermoxylanase (Ecopulp TX 200C), mannanase (Econase MP1000), amylase (Ecostone A200) and lipase (Resinase A2X). Two dosages of the enzyme mixture were tested. The lower dosage was chosen to avoid degradation of the fibres.

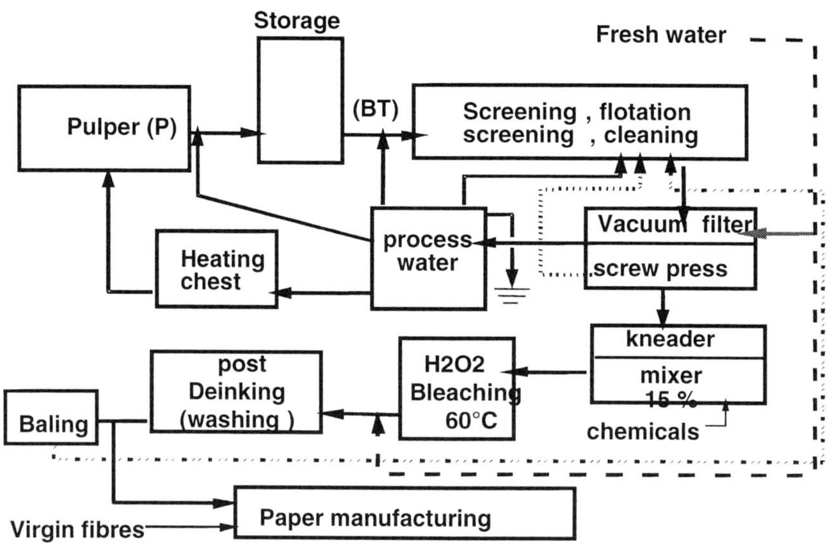

Figure 2. Deinking line of Moulin Vieux mill. BT, blow tank.

Table 2. Operating conditions of the deinking trials at Moulin Vieux mill.

Operation conditions	Standard deinking, low production rate	Enz. deinking, low production rate	Standard deinking, normal production rate	Enz. deinking, normal production rate
Production, ton / hour	1.17	1.37	1.63	1.72
Pulping chemistry				
silicate, %	0.7	0	0.6	0
H_2O_2, %	0.46	0	0.45	0
Surfactant*, %	0.05	0.05	0.05	0.05
Enzyme mixture				
Thermoxylanase		125 nkat/g		100 nkat/g
Mannanase		125 nkat/g		50 nkat/g
Amylase		125 nkat/g		100 nkat/g
Lipase		125 nkat/g		100 nkat/g

* Liptol S100, SEPPIC, France.

Deinking efficiency Deinking efficiency was monitored along the deinking line by measuring optical parameters (brightness, ERIC and specks), effluent properties (COD and dissolved material) and mechanical properties. Runnability of the deinking line and paper machine was also observed.

2.3. Pilot scale deinking of woodcontaining recovered paper

Woodcontaining paper raw material

The following woodcontaining paper grades were selected: 50 % old news offset coldset printed 25 % SC paper offset heatset printed 25 % LWC paper offset heatset printed. This paper mixture was chosen to represent a typical woodcontaining composition with various types of papers difficult to deink.

Enzymes

Two enzyme mixtures (Table 3) were selected based on the results obtained at laboratory scale deinking experiments carried out with individual enzymes and various paper grades (manuscript in preparation).

Table 3. Enzymes and dosages used in the pilot scale deinking of woodcontaining paper.

Enzyme product	Main enzyme activity	Enzyme dosage, nkat/g O.D. pulp	
		Mixture III	Mixture IV
Ecopulp TX-200C	Thermoxylanase	100	100
Resinase A2X	Lipase	100	100
Ecostone A 200	Amylase	-	100
CBH II*	Cellulase	-	0.25 mg/g

**Research enzyme preparation containing cellobiohydrolase II. About 90% of the protein present in the preparation was CBH II.

A single deinking loop as described in Figure 1 was used for conventional alkaline deinking, neutral deinking and enzymatic deinking. The same analyses were carried out as in the pilot scale deinking trials of woodfree paper.

3. RESULTS AND DISCUSSION

3.1. Pilot scale deinking of woodfree recovered paper

Conventional alkaline, neutral and neutral enzymatic flotation deinking were compared with each other. Results of the analyses are shown in Table 4. Enzyme treatments reduced the number and area of specks in the final deinked pulp. Particularly with the enzyme mixture I, speck contamination of the deinked pulp was 88 % lower than that observed with alkaline treatment. Enzyme mixture II was also efficient but at lower extent. This result is significant for mixed office waste containing a lot of toner printing papers inducing specks, which are detrimental to the visual aspect of the pulp. Enzymatic deinking with the mixture I, standard deinking and surfactant deinking gave comparable flotation yields (87 % in suspended solids). Enzymatic deinking with mixture II gave a slightly lower yield of 84 %.

Brightness of the pulps were identical after enzymatic deinking with mixture I and alkaline conventional deinking. Deinking with mixture II gave slightly lower brightness values. In all four deinking experiments, ink was easily detached and removed. Nevertheless, ink was removed slightly more efficiently after the enzymatic treatments compared with the conventional deinking (ERIC values of 14 ppm for enzymatic deinking compared to 34 ppm for conventional alkaline deinking).

Cationic demand was efficiently reduced when enzymes were used. However, COD release appeared to be higher with the enzymatic treatments. This was most obviously due to the hydrolytic property of the enzymes. The enzymes released soluble sugars from the pulp to the process water thus increasing COD.

Decrease of about 15 % in burst index and breaking length was observed in the enzymatically deinked pulps compared with the alkaline deinked pulp. However, no variation in tear index or wet zero span breaking length was observed.

Table 4. Results of the analyses carried out to the woodfree pulps deinked at pilot scale.

Analysis	Conventional	Surfactant	Enz. mixture I	Enz. mixture II
Brightness 0 % UV, %	84	82	84	82
ERIC, ppm	34	21	14	18
Number of specks, /m²	32654	15182	3891	15411
Area of specks, mm²/m²	854	956	98	374
Yield, %	88	87	88	84
Cationic demand, μ eq PEI/g paper	64	5	4	4
COD, mgO_2/g paper	53	27	60	60
Wet zero span breaking length, km	9.79	10.1	9.79	10.25
Breaking length, km	4.68	5.05	3.99	4.06
Burst index, kPa.m²/g	2.85	3.15	2.29	2.41
Tear index, mN.m²/g	9.72	9.92	9.50	10.72

3.2. Mill scale deinking of woodfree recovered paper

Thermoxylanase, mannanase and amylases were chosen to the enzyme mixture to be tested at full scale, because they turned out to be efficient at the pilot scale deinking of woodfree paper. Lipase was added to hydrolyse vegetable oil in the ink. Since the mill is integrated, production is not maintained on a constant level. This explains why two trials were run at the normal production rate and the other two at lower production rate (Table 2). Since the production rate affects the deinking efficiency, the standard deinking was compared with the enzymatic deinking in both cases at the identical production capacity.

The quality of the enzyme treated deinked pulp was compared with the quality of the deinked pulp obtained by the standard chemistry of the mill. Enzymatic deinking was efficient on printed woodfree coated papers. As illustrated in Figures 3a and 3b, enzymes improved the total output of deinking. The enzymatic procedure resulted in 4-6 units higher brightness than the standard procedure. Residual ink content was also lower in the enzyme treated pulp compared with the pulp deinked using the standard chemistry. However, the content of specks was similar probably because, the recovered paper used contained fewer toner printed papers than the recovered paper mixture used at pilot scale.

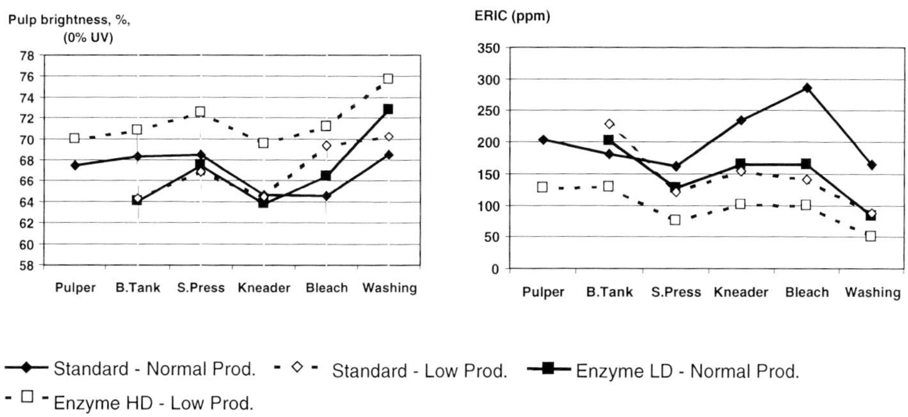

Figure 3. Development of brightness (a) and residual ink content (b) of the pulps during the deinking process. LD low dosage, HD high dosage.

The effect of the enzymatic treatment on the mechanical properties is shown in Figures 4a and 4b. Decrease in the mechanical properties was observed directly after the enzymatic treatment in the blow tank but this negative effect was compensated later in the process. Particularly with the lower enzyme dosage the mechanical properties were kept in the acceptable level.

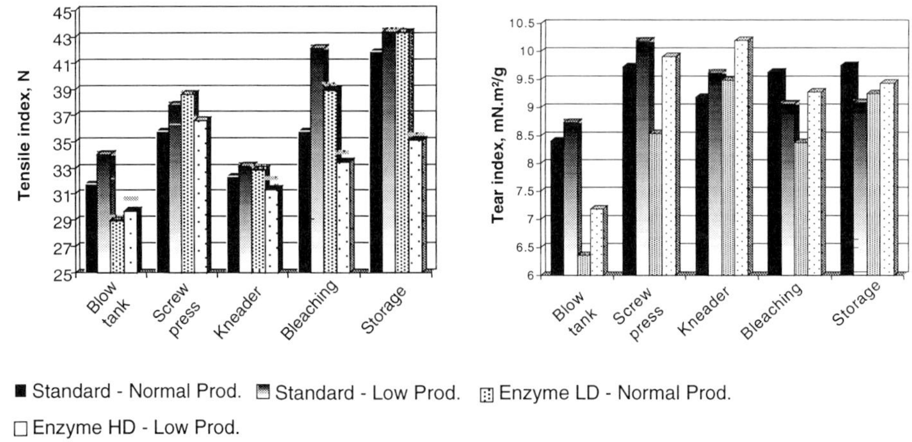

Figure 4 . Development of tensile index (a) and tear index (b) of the pulps during the deinking process.

Higher COD level was observed during the enzymatic deinking trial compared to the standard mill deinking. At normal production rate, COD of 56 g O_2/g was detected with enzymes whereas the standard mill chemistry gave 39 g O_2/g. At lower production rate, the COD levels were 64 O_2/g with enzymes and 53 g O_2/g with the standard chemistry. On the paper machine, no modification of the runnability was observed. To obtain enough data about possible consequences of enzymatic deinking on the mill performance, longer trial run than the ones reported here should be carried out.

3.3. Pilot scale deinking of woodcontaining recovered paper

In addition to woodfree recovered paper woodcontaining recovered paper was tested for enzyme-aided deinking. Conventional alkaline, neutral and neutral enzymatic deinking of woodcontaining paper raw material were compared. The enzymatic deinking reduced significantly the number and area of specks. As shown in Table 5, speck contamination was 45 % lower with enzyme mixture IV compared to the alkaline deinking. Ink detachment and ink removal were slightly better with enzymatic deinking compared to the alkaline deinking. ERIC values were lower in enzymatically deinked pulps. Less COD and anionic colloidal substances were released to the process waters during the enzymatic process. No modification of the mechanical properties was observed due to action of enzymes. Brightness values of the enzymatic deinked pulp were 3 - 4 units lower compared with the alkaline deinking. This is obviously due to lack of bleaching agent, hydrogen peroxide, in the pulper.

Table 5. Results of analyses carried out to woodcontaining deinked pulps after alkaline, surfactant (neutral) and enzymatic deinking.

Analysis	Conventional	Surfactant	Enz. mixture III	Enz. mixture IV
Brightness, 0 % UV, %	58	55	54	55
ERIC, ppm	232	192	157	170
Number of specks, /m²	291672	289154	220985	168343
Area of specks, mm²/m²	3207	3486	2601	1747
Yield, %	91	86	83	85
Cationic demand, μ eq PEI/g paper	18	4	4	5
COD, mgO_2/g paper	36	11	13	24
Wet zero span breaking length, km	6.24	6.65	6.89	6.73
Break length	3.49	3.98	4.25	3.73
Burst index, kPa.m²/g	2.16	2.16	2.32	2.08
Tear index, mN.m²/g	6.99	7.04	7.40	6.99

The lower brightness level could be compensated for by a subsequent bleaching stage as described in Table 6. Using an identical amount of bleaching agent (1 % hydrosulphite) after the enzymatic deinking procedure, brightness was 4 units higher than in the standard deinking having 1 % H_2O_2 in the pulper. Enzymatic deinking of woodcontaining paper appears thus to be feasible only in combination with a post-bleaching stage.

Table 6. Effect of the post-bleaching after the enzymatic deinking on brightness and COD.

Process	Bleaching agent	COD, mgO_2/g paper	Brightness, %
Conventional	1 % peroxide in the pulper	36	54
Enzyme mixture III	1 % hydrosulphite	20	58
Enzyme mixture IV	1 % hydrosulphite	32	58

4. CONCLUSIONS

Enzymatic deinking of woodfree recovered papers showed significant positive effects. Particularly positive result was the reduction of speck contamination in woodfree paper mixture containing toner-printing papers. This is important since specks are very detrimental to the visual aspect of the pulp. The industrial trial with 100 % woodfree recovered papers with few toner-printing papers was successful. Overall efficiency of deinking was improved (higher brightness and lower residual ink content in the enzymatic deinked pulp). The mechanical properties were maintained at the correct level. The main drawback observed was the high COD level in the process waters.

The results obtained with woodcontaining recovered paper at pilot scale showed that good ink removal and lower speck contamination could be obtained by enzymatic deinking in neutral conditions but the brightness too low because of lack of bleaching agent in the process. Enzymes could be feasible when used in combination with a post-bleaching stage. However, in the case of a deinking line producing newsprint grade, a post-bleaching could hardly be envisaged because of increasing production costs. However, in the case of a deinking line producing supercalendered paper, enzyme-aided deinking process is possible as a post-bleaching stage is included in the process.

ACKNOWLEDGEMENT

This work was conducted in the framework of a European Research Project (FAIR CT 96 1492). The authors wish to acknowledge the co-participants of the project: Institut für Papierfabrikation IfP (Germany), TNO Institute of Industrial Technology (The Netherlands), Röhm Enzyme Finland (Finland, present name AB Enzymes Oy) and Sachsen Papier Eilenburg (Germany, present name Stora Enso).

REFERENCES

1. H. Grossman, Proc. 5th Advanced training course on deinking technology, Grenoble, France, March (2001).
2. T. Welt and R.J. Dinus, Progress in paper Recycling 4 (1995) 36.
3. P. Bajpai and P. Bajpai, Tappi J. 81 (1998) 111.
4. J.-C. Pommier, J.-L. Fuentes and G. Goma, Tappi J. 72 (1989) 187.
5. D.Y. Prasad, J.A. Heitmann and T.W. Joyce, Progress in Paper recycling, May (1992) 21.
6. D.Y. Prasad, J.A. Heitmann and T.W. Joyce, Nordic Pulp and Paper Res. J. 2 (1993) 284.
7. O.U. Heise, J.P. Unwin, J.H. Klungness, W.G. Fineran Jr, M. Sykes and S. Abubakr, Tappi J. 79 (1996) 207.
8. H.K. Zollner and L.R. Schroeder, Tappi J. 81 (1998) 166.
9. A.L. Morkbak and W. Zimmermann, Progress in Paper Recycling, February (1998) 14.
10. T.J. Kim, S.S.K. Ow and T.J. Eom, Tappi Proc. Pulping Conference, Orlando, November 3-7 (1991) 1023.
11. J.A. Heitmann, T.W. Joyce and D.Y. Prasad, Proc. 5th Int Conf. on Biotechnology in the Pulp and Paper Industry, Kyoto, May 27-30 (1992).
12. W. Rushing, T.W. Joyce and J.A. Heitmann, Proc. 7th Int.Symp. on Wood and Pulping Chemistry, Beijing, China, May 25-28 (1993) 233.
13. A.L. Morback and W. Zimmermann, Prog. Pap. Recycling 7 (1998) 33.

Index of authors